Concrete Structures:
Materials, Maintenance and Repair

Concrete Design and Construction Series

SERIES EDITORS

PROFESSOR F. K. KONG
Nanyang Technological University

EMERITUS PROFESSOR R. H. EVANS CBE
University of Leeds

OTHER TITLES IN THE SERIES

Concrete Radiation Shielding: Nuclear Physics,
Concrete Properties, Design and Construction *by M. F.
Kaplan*

Quality in Precast Concrete: Design — Production —
Supervision *by John G. Richardson*

Reinforced and Prestressed Masonry *edited by Arnold W.
Hendry*

Concrete Structures: Materials, Maintenance and Repair

Denison Campbell-Allen

and

Harold Roper

Longman
Scientific &
Technical

Copublished in the United States with
John Wiley & Sons, Inc., New York

Longman Scientific & Technical,
Longman Group UK Limited,
Longman House, Burnt Mill, Harlow,
Essex CM20 2JE, England
and Associated Companies throughout the world.

Copublished in the United States with
John Wiley & Sons, Inc., 605 Third Avenue, New York, NY 10158

First published in 1991

British Library Cataloguing in Publication Data
Campbell-Allen, Denison
    Concrete structures: materials, maintenance and repair.
    1. Concrete structures. Maintenance & repair
    I. Title II. Roper, H. III. Series
    624.1834

    ISBN 0-582-05067-7

Library of Congress Cataloging-in-Publication Data
Campbell-Allen, D.
    Concrete structures: materials, maintenance, and repair / Denison
Campbell-Allen and H. Roper.
        p.   cm. — (Concrete design and construction series)
    Includes bibliographical references and index.
    ISBN 0-470-21727-8
    1. Concrete construction. 2. Concrete construction — Maintenance
and repair. 3. Concrete. I. Roper, H. (Harold) II. Title.
III. Series.
    TA681.C38   1991
    624.1'834 — dc20   91-9052
                                                            CIP

Set in 10/12 pt Times
Produced by Longman Singapore Publishers (Pte) Ltd.
Printed in Singapore

Contents

Preface

This volume is directed towards those people — generally engineers — who already have some familiarity with concrete and concrete construction, whether as practitioners or students. In selecting the topics to cover, we have drawn on our experience in meeting — and in some cases solving — problems and difficulties that have arisen over many years. The book has its origin in our collaboration in preparing a chapter of the Handbook of Structural Concrete (eds Kong, Evans, Cohen and Roll), Pitman, 1983. We acknowledge sincerely the encouragement of Professor Kew Kong for us to develop this much expanded and up-dated coverage.

We have aimed to produce a readable text which provides background and, we hope, understanding of present-day problems of achieving sound concrete construction and of remedying the defects in unsound concrete. We have left the detailed recipes on what to do in specific circumstances to others. The various chapters can, we think, be read independently. Chapter 5, which deals with the behaviour of steel in concrete, may seem at first glance to be unnecessarily detailed. We are sure, however, that the problems of reinforcement corrosion are the most widespread and least understood in the industry. The difficulties become even greater when prestressing steel is affected since, in that situation, the consequences of a tendon failure may be catastrophic. For most engineers literature explaining these problems and their solution is not readily available and we believe that our chapter will help to correct this situation.

Our narrative is founded on the extensive published literature that is now available. Where specific topics are mentioned and referenced, the reference list generally indicates the most convenient source rather than necessarily the original source of the material. Further papers and volumes on related topics, which are not specifically referenced but which are likely to be readily available in libraries, are included in the 'Bibliography' associated with each chapter. Standard codes and specifications which get a mention in the text are not included in the reference list but are tabulated separately.

We acknowledge the great assistance of Mr Ron Brew in preparing the diagrams.

DCA

HR

The University of Sydney,

1990

References, Bibliography and Standards

References are listed chapter by chapter and therefore some references may appear more than once.

Frequent reference is made to American Concrete Institute publications and particularly to the material contained in the five volumes of the *ACI Manual of Concrete Practice*. Many ACI reports originally appear in the Journals and Proceedings of ACI, but if they have found a place in the *Manual of Concrete Practice* we have referenced them only in the ACI standard form, such as *ACI 201.2R-77* (Guide to Durable Concrete). The first number (201) denotes the ACI committee responsible, the decimal indicates that the committee is responsible for more than one report, the letter 'R' denotes a report rather than a standard and the last figure is the year of latest major revision.

A bibliography of related papers and monographs, which are not specifically referenced, is listed chapter by chapter, at the end of the book.

Standard codes and specifications referred to anywhere in the text are listed for the whole book by country of origin.

Acknowledgements

for figs 3.20 adapted from fig. 5 and table 1 (Winchester, 1983), 8.25 from fig. 4 (Wilson, 1980) and 8.26 from fig. 1 (Hankins et al.); Chapman and Hall for the bottom of figs 2.5, 10-2.5.3 and figs 6.4, 6.5 & 6.15 (modified by the author from figs 1 & 2) from fig. 2.7 & 2.1 and tables 5.4 & 5.5 from tables 5 & 6 from original illustrations and tables (Neville, 1981); Institution of Civil Engineers for the bottom of figs 1-6.3 from figs 2, 4 & 3 (Pugsley (1968, 1971), 8.5 & 8.6 (modified by the author from figs 1, 3 & 5 (Fookes, 1985); Institute of Structural Engineers and Thomas Telford for figs 8.4 & 8.5 from fig. 1 8.2, Fluckiger et al. (1983)) and 7.2 (from tables (Gilbert, Murphy & Daniels, 1985), the author, A.L. Johnston for fig. 1 and 3 from figs 6 (Johnson & Parsons, 1986); the author, C.L. Marsh, ... for the figures from figs 3 (Marks), Cement & Concrete, ... 1973); ... the tables, 4 from table (Marsh et al.) ... table 3 ... (Morgan, 1964) and Temple Press for table 6 (Morgan, 1964); for figs 5.12 & 5.13 from figs 6 & 8 (Hamada, 1968); Concrete Society (London)

We are grateful to the following for permission to reproduce copyright material:

American Concrete Institute for figs 2.1 & 2.2 from figs 5 & 8 (Kirkness & Guarre, 1986), 3.2–3.4 from figs 1–3 (Bloem, 1968), 3.16–3.19 from figs 2–5 (Roy & Idorn, 1982), 3.21 & 3.24 from figs 1.7 & 1.1 (Zoldners, 1971), 4.2 from figs 1 & 2 (Clifton & Kaetzel, 1988), 4.3 from fig. 2 (Kobayashi et al., 1981), 4.12–4.15 from figs A2–A5 (Fentress, 1973), 4.16 from fig. 2.1 (ACI Committee 210, 1987), 4.26 from fig. 3.14 (Feld, 1964), 4.30 from fig. 16 (Griffiths et al., 1987), 5.10 & 5.11 from figs 1 & 2 (Gjorv & Vennesland, 1976), 5.15a–c & 5.19 from figs 6, 8–10 (Verbeck, 1975), 5.17 and table 3.1 from figs 2 & 6 (Powers et al., 1954), 7.1 & 7.2 from figs 3.3c & 4.2.6 (ACI Committee 223, 1970), 7.3 from fig. 10 (Pfeifer & Perenchio, 1982), 7.9 & 7.11 from figs 6, 16 & 17 (Burge, 1986), 7.10 adapted from fig. 5 (Ramakrishnan et al., 1981), 8.8 from fig. 11 (Fintel & Ghosh, 1988), 8.9 from fig. 3 (Hugenberg, 1987), 8.17 from fig. 8 (Houghton, Borge & Paxton, 1978), 8.18 and table 8.4 from figs 3 & 1 (Holland et al., 1986), 8.21–8.24 from figs 1, 3–5 (Arioglu et al., 1983), 8.29 & 8.30 from figs 2, 6 & 7 (Waring, 1986) and tables 3.7 from fig. 1 (Powers, 1959), 3.12 from table 6 (Whiting, Litvin & Goodwin, 1978), 3.13 from table 3.6 (ACI Committee 523, 1986), table 4.3 from table 1 (Dunstan & Joyce, 1987) and 7.2 from table 102c (ACI Committee 506, 1985); British Cement Association for figs 5.29 & 5.30 from figures, pp. 287–306 (Gooch, 1974) and 5.31 from a figure, pp. 152–157 (McGuinn & Griffiths, 1977); Building Research Establishment and the author, H.L. Malhotra for figs 8.19 & 8.20 from figs 3 & 5 (Malhotra, 1982); Bureau of Reclamation (Denver, U.S.A.) for figs 7.7 & 7.8 adapted from figs 176 & 180 (Concrete Manual, 1975); Canadian Ministry of Transport and Communications and the author, D.G. Manning for fig. 8.16 from fig. 5 and table 8.3 from table 1 (Manning & Bye, 1984); Cement Association of Japan for figs 3.10–3.12, 3.22 & 3.23 from figs 26, 29, 30, 32 & 33 (Verbeck & Helmuth, 1968); Concrete Association Research and Development Laboratories (Cement Research Division) for figs 5.12 & 5.13 from figs 6 & 8 (Hamada, 1968); Concrete Society (London) for figs 3.26 adapted from fig. 2 and table 1 (Concrete Society, 1982), 8.25 from fig. 1 (Wilson, 1986) and 8.26 from fig. 3 (Houde *et al.*, 1986); Deutscher Ausschuss für Stahbeton for figs 5.20–5.23 from figs 3, 4, 29 & 47 (Schiessl, 1976); the author, L. Grill for figs 8.4a & b from figs 2.3.2 & 2.3.2(d) (Grill, 1985); Il Cemento for table 3.6 from a table, pp. 3–14 (Collepardi, Marcialis & Solinas, 1973); International Association for Bridge and Structural Engineering and the respective authors for figs 8.1–8.3 from figs 2, 10 & 11 (Pakvor & Djurdjevic, 1989) and 8.5 from figs 5 & 6 (Grill, 1985); Institute of Structural Engineers and Thomas Telford Ltd for tables 4.1 & 4.2 from tables 1 & 2 (Turner *et al.*, 1982) and 6.1 from table 1 (Bilger, Murphy & Reidinger, 1986); the author, B.L. Johinke for figs 6.1–6.4 from figs 2–6 (Johinke & Tickner, 1985); the author, M. Marosszeky for fig. 4.30 from fig. 16 (Griffiths, Marosszeky & Sade, 1987) and table 4.4 from table 3 (Marosszeky, Chew, Gamble & Sade, 1987); the author, D.R. Morgan for tables 3.5 from a table (Morgan, 1973) and 7.3 from table 2 (Morgan, 1988); National Institute of Standards and Technology for fig. 3.8 from figs 16 & 12 (L'Hermite, 1960); Northampton County Council (Highways and Transportation Dept.) for fig. 6.5 from a figure, p. 20 (Andrews, 1986); Noyes Publications for fig. 3.9 from fig. 4.26 (Ghosh & Malhotra, 1978); Pergamon Press PLC for figs 4.8 & 4.9 from figs 1 & 5 (after Planas *et al.*, 1984) and 4.10 from fig. 6 (Elices *et al.*, 1986); Portland Cement Association for figs 5.14a–c from figs 2–4 (Gouda & Monfore, 1965), 5.16 from fig. 12 (Monfore, 1968) and 5.18 from figs 3 & 4 (Ost & Monfore, 1966); the editor, *Progress in Concrete Technology* for table 7.1 from table 6 (Dikeou, 1980); Réunion Internationale des Laboratories d'Essais et de Recherches sur les Matériaux et les Constructions for fig. 4.1 from fig. 1 (Wright & Frohnsdorff, 1985); Standards Association of Australia for figs 3.13 & 3.14 and table 3.8 from figs 4.7.3(A & B) & section 4.7.3.1(c) (AS 1481–1978); Thomas Telford Ltd, for figs 8.10, 8.11 & table 8.1 from figs 1, 4 & table 1 (Cole & Horswill, 1988) and 8.12, 8.13 & table 8.2 from figs 2, 14 & table 5 (Sims & Evans, 1988); Transportation Research Board for table 3.9 from a table (Tremper & Spellman, 1963); John Wiley & Sons Inc. for figs 3.15 from fig. 10.2 (Powers, 1968), 4.24 & 4.25 from figs 13.31 & 13.26 (Park & Paulay, 1975) and 5.6 & 5.7 from figures (Pourbaix, 1971). We are indebted to Concrete Repairs Ltd, Cathite House, 748 Fulham Road, London for supplying the front cover photograph.

Whilst every effort has been made to trace the owners of copyright material, in a few cases this has proved impossible and we take this opportunity to offer our apologies to any copyright holders whose rights we may have unwittingly infringed.

1 Serviceable and Durable Concrete

Concrete Design and Construction

The two disciplines essential for structural engineering are those of Applied Mechanics and Properties of Materials. For conventional reinforced concrete design and construction, the emphasis in texts,[1.1–1.3] codes* and in practice has been on structural mechanics. Appropriate reference is made as needed to the mechanical properties of the materials being used. Mechanical properties of concrete, such as strength, stiffness and creep, are related to the composition of the concrete by general expressions, often simplified. It is only in the consideration of creep that it has become customary to take account of many more of the features of concrete so as to provide more exact expressions into the mechanical equations.[1.4,1.5] For reinforcing material, the mechanical properties are commonly described in terms of two or three values, such as elastic modulus, proof stress and ultimate strength, as laid down in specifications. This collection of values and expressions provides a perfectly adequate description of the mechanical behaviour of the composite material which is structural concrete, whether plain, reinforced or prestressed.

The emphasis on mechanics, which has always existed, has become even heavier since computers have enabled analysts to employ more elaborate expressions in the description of behaviour, even though the more complicated analysis is not necessarily based on more reliable information. As a consequence of this careful attention to mechanical behaviour, it is only occasionally that major structural failure occurs. When it does, it may well be due to a gross oversight by the designer, or failure of communication within the design team, a major construction error or the incidence of exceptional and little-understood loadings such as blast or earthquake. Minor structural defects are often the result, not of inadequate structural calculations, but of a failure by the designer to pay proper attention to details. There is always a risk that what appears to be a minor structural defect, such as insufficient bearing of a roof beam on an edge beam, may lead to a major collapse.

* Codes and standard specifications are listed at the end of the volume (p. 349).

The aspect of design which has received far less attention in the past, partly no doubt because it is less amenable to mathematical analysis, is the response of concrete structures to actions other than loading. In the CEB-FIP Recommendations of 1970,[1.6] in which it was stated in the Preface by Levi and Rusch that 'Collaboration with RILEM has also emphasized the notion of durability, which is regarded as equal in importance to strength', almost the only reference to actions other than loading is in the paragraph under 'General Principles' stating: 'Finally, a structure may be subjected to other actions which may lead to failure, such as fire and corrosion. The treatment of these problems lies essentially in the detailing of the structure'. It is only very recently that national codes of practice have paid any serious attention to aspects of durability. For example, ACI 318−89 contains a new chapter on 'Durability requirements' which stresses the importance of special exposure requirements and discusses the requirements for concrete to resist freezing and thawing exposures and sulphate exposures and to avoid corrosion of reinforcement. Even then many codes proceed by introducing more stringent (and therefore expensive) design requirements rather than by referring to the need for a full understanding by the designer and builder of the properties of the materials involved.

It is indeed fortunate for the widespread and successful use of reinforced concrete that conventionally constructed concrete provides an environment in which steel is protected from corrosion, a situation which has probably arisen more by good luck than by good design. Fifty years ago, it was accepted wisdom that reinforced concrete was successful because concrete and steel had about the same coefficient of thermal expansion. There was no general mention of the essential alkalinity providing protection of the steel against corrosion nor much recognition that the concrete mixes then in use, in addition to providing more than adequate strength, also provided resistance to atmospheric and other attack.

The many actions which concrete, and particularly reinforced and prestressed concrete, have to resist require that the full range of available material properties, and the reaction and interaction of these properties in use, should be understood and considered by the competent designer. In addition, properly planned maintenance is an essential part of a good design. No designer, or owner, expects that the lifts or the fire-protection system in a building will serve indefinitely without inspection and maintenance but there is still an unspoken assumption that concrete structures will last for ever without any attention. The maintenance plan is made harder by the very nature of reinforced concrete construction in which inspection is often very difficult. Such techniques as potential surveys are beginning to overcome this problem, but it is one that the designer should address right from the beginning. Such details as precast panel connections are especially at fault. The appropriate strategy for maintenance is discussed in Chapter 6. Many problems if found early enough may be contained before they reach epidemic proportions in a structure. For example, the early attention to delamination in concrete facades may stop a problem that could otherwise lead to the complete destruction of the face of a building. That this attention is not often provided reflects not only the lack of communication between designers and owners but also the current scenario concerning liability and insurance. These questions are addressed later in this chapter. There is increasing attention in the building and construction industries to quality assurance and quality control. The extent to which procedures can be adapted to achieving more durable structures is discussed in Chapter 2.

It has often been claimed that the cost of providing long life in buildings is likely to be prohibitive and to require a greater investment of national resources than is often needed. The analogy is drawn between the cost of a Rolls Royce and a Ford automobile. In many instances the analogy and the conclusion are incorrect. Good design, which takes account of environmental hazards, is no more expensive than design which ignores these conditions. Construction by competent operators, properly supervised, has no inherent additional costs and may, as a result of smoother running operations, even be cheaper than haphazard construction. The cost of ingredients for good concrete is no more than the cost of the same ingredients made into poor concrete. It is only in repairs and maintenance that large additional costs can arise, and then when the building is initially of poor quality.

In the remainder of this chapter, we will outline the actions and conditions, other than normal static loading, which have to be resisted by various forms of concrete structure, and examine briefly the current legal responsibilities that the designer has to carry. The ways in which the resistance of the concrete structure can be provided will form the subjects of Chapters 3 to 5. Minor repairs as part of maintenance will require reference to Chapter 7. If failure has occurred then reference will be necessary to Chapters 7 and 8 where repairs are fully discussed. In the examples of repairs in Chapter 8, we note particularly the number of occasions on which previous repairs have been unsuccessful and have required further action.

Reinforced Concrete Buildings

In this section we will consider buildings intended for commercial or residential use, whether low- or high-rise. The special needs of industrial buildings are the subject of the following section.

The foundations of any structure are vitally important yet are the least amenable to inspection after the building is completed or indeed while it is under construction. For this reason the possible attack by groundwater on both concrete and steel must be examined in the design stage. The deleterious conditions most likely to be encountered are sulphate-bearing groundwater, acid groundwater and aggressive industrial wastes. It must be borne in mind that changes of natural growth, such as trees, changes of groundwater level from pumping or drainage changes and future discharge or diversion of industrial wastes can all introduce changed hazards to existing concrete foundations. A general assessment by the designer of such likely changes is an essential part of the design of foundations, whether piled, raft or pier and beam. The impossibility of examining reinforcement for possible corrosion means that special attention is necessary during construction to ensure that steel is properly placed and protected. The possibility of a giant corrosion cell forming in a reinforced concrete building should not be overlooked. Although the more usual situation in such a case is that the steel in the upper parts of the building is anodic while the foundations remain cathodic, the reverse is not impossible where the foundation steel, which is not accessible to inspection, becomes anodic to other metal in the building.

What characteristics are to be expected in the floors of a building? It is common in commercial and residential buildings to provide floor covering which is generally expected to be fixed to the floor. The presence of some cracks in the top surface is therefore often acceptable, provided that the covering does not mirror any cracks below it. The failure of the covering to remain stuck on is however a serious defect. Some curing compounds are therefore to be avoided and in the case of slabs on the ground the penetration of moisture must be prevented. The long-term deflection of slabs, although frequently attributed to creep under load, has been shown to be more often the consequence of differential shrinkage and is therefore a condition to be considered here. The consequences of such deflection are cracking of finishes and partitions and the failure of compactus filing systems.

The external envelope of the building, including the roof and the facade, is under continuous attack from the atmosphere. It has been claimed that the climate of the British Isles is exceptionally detrimental to buildings.[1.7] The climate promotes chemical and physical decay, large dimensional changes, physical wind damage and wind-assisted rain penetration. In other parts of the world conditions even more damaging to concrete and other building materials exist. In the Middle East, daily cycles of extreme dryness and overnight humidity occur. In Australia much construction is on the warm sea-coast. In the south-east of the continent changes of humidity by as much as 50 percentage points accompanied by temperature drops of 20 °C can occur in a few hours.

On the west coast the average humidity remains for long periods at about 50 per cent, the condition which is most encouraging for carbonation to take place. Northern areas are subject to tropical conditions of high temperature and high humidity for long periods, with the added risk of tropical cyclonic winds. In the USA, in addition to the conditions already mentioned, severe and prolonged periods of freezing can occur as can frequent cycles of freezing and thawing. In all countries the incidence of industrial pollution, such as acid rain, is an increasing hazard.

No one building has to stand up to all these forms of attack, but their very variety leads to the conclusion that it will not be adequate to provide exactly the same building envelope for all situations. Both general and local climate have a profound influence on the requirements and performance of external concrete. The common practice of describing the conditions of exposure of the whole external surface of the concrete in a building in a single classification such as 'Mild' or 'Very Severe' ignores the very wide range of micro-climates which can exist around a large building. It has been shown, for example, that the extent of damage from wetting and drying is markedly dependent on how high up the building the damage occurs. The common forms of deficiency seen in building facades are weathering and adhering dirt, corroding reinforcement, bowing resulting from thermal gradients, inadequately protected fixings for precast panels and delamination of veneers. The deficiencies arise from a lack of understanding of the interaction of materials and climate, the failure to ensure that the designer's intentions are achieved in the construction and the lack of adequate, or any, maintenance.

Exposed concrete inside a building must behave so that it does not deflect unacceptably and must be appropriately free of cracks. The extent to which cracking is acceptable depends on the use to which the building is put, the distance from which viewing can occur and the nature of the surface finish. The requirements for the walls on which pictures are displayed in a public gallery are quite different from what is acceptable visually in a plant room. Finally the whole structure of any building must be compatible in terms of movement with finishes and fittings. Brick growth and concrete shrinkage are not compatible without special design provision. Modern buildings are particularly susceptible to dimensional change. In the case of buildings of great height, creep and shrinkage of columns and cores result in large movements and, just as importantly, large differential movements. The solar loads that impinge on buildings and the temperature and humidity gradients that are imposed by air-conditioning in both hot and cold climates can induce differential movements between parts of the structure and between the structure and the finishes. Chemical compatibility is also necessary. For example, bricks which exude sulphates can not be located against concrete or mortar which is susceptible to sulphate attack.

Industrial Structures

In addition to most of the conditions that have to be resisted by reinforced concrete buildings, structures for particular industrial use have further special requirements.

Floors in industrial structures are not commonly covered except possibly by an integral topping. Dusting, crazing and wear of surfaces are therefore important considerations. Industrial floors are likely to be subjected to loading by foot and wheeled traffic which is continually braking and accelerating in short distances. Floors may therefore be classified on the basis of intended use, as for instance proposed by ACI Committee 302.[1.8] Wheeled traffic is especially damaging at joints or cracks in a floor and the control of cracking from all causes including loading and settlement becomes important. Industrial processes such as food processing, the handling of animals and animal products and chemical plants require floors and other parts of the structure to be resistant to acids and organics. Conventional concrete is frequently unsatisfactory in this respect and special processes such as polymer impregnation may have to be considered. Flatness and levelness of floors has increased in importance with the increase in narrow-aisle warehousing, where slight divergences of floor profile may make high stacking plant inoperative. The achievement of satisfactory floor profiles requires a sound understanding of the interaction of construction methods, concrete materials and mix composition. In addition, present specifications, even if met in construction, do not necessarily relate to the real needs.[1.9] As in floors for other uses, cracking may become a major operational hazard.

Many industrial processes subject concrete structures to high temperatures. Examples are aluminium pot-lines, metallurgical furnaces and boilers. Chimneys are almost always expected to operate at above ambient temperature and foundations and supports associated with chimneys may be subject to radiant heat from hot gases. The response of concrete to such elevated temperatures is therefore of fundamental importance in their successful design. The now celebrated English law case (Pirelli General vs. Oscar Faber and Partners) is concerned with a chimney lining which cracked when the furnace was first fired. It was the acceptance of the time at which this cracking occurred by both parties to the dispute that enabled the House of Lords to decide that the matter was out-of-time under the Statute of Limitations. Not only thermal effects but also the chemical reactions possible between lining materials and hot gases may be decisive design factors. When linings are provided to protect a concrete chimney, there is always the possibility that leakage of corrosive gases, followed by their condensation, may occur.

The massive nature of much of the concrete for industrial installations means that the capacity of the concrete to generate heat during the hydration of the cementing materials must be understood and taken into account. Much information in the past has been obtained for the lean concrete mixes typical of dam construction. It is only with the advent of such massive prestressed concrete units as pressure vessels for nuclear power plants that information on the behaviour of higher strength concrete has become readily available. This information, including the use of special cements and admixtures, should now be considered in the design of other massive industrial concrete and even in the design of thick foundation rafts for buildings.

Concrete vessels are often used to provide containment for solids, liquids and gases. The loading on a bin containing solids, such as wheat or coal, depends on the frictional forces between the concrete wall and the granular material and a knowledge of the surface condition is therefore essential. The special needs for vessels containing liquids and gases are freedom from leakage and freedom from chemical attack. Cryogenic vessels used for the storage of liquefied gases require that the designer shall be familiar with the behaviour of concrete at very low temperatures (approaching absolute zero) and with the response of concrete to cycles of low temperature. Blast and impact resistance are necessary for protective structures.

Pavements

Pavements, whether for roads or airfields or other trafficked areas, must be able to support the traffic loading with acceptable riding qualities and without unacceptable deterioration over their design lives.[1.10] A prime function is therefore for the pavement to distribute the wheel loads so that they do not induce conditions in the subgrade which cannot be sustained. A concrete pavement has to endure fatigue loading from traffic and frequent changes of internal stress arising from temperature and moisture changes. During the day solar heating of the top surface tends to warp the slab so that the top is in tension. At night time the reverse occurs. Moisture gradients may reinforce or cancel these effects. Traffic superimposed on these cycles, which may occur many times a day, can lead to cracking which affects both riding quality and durability.

The serviceability of a pavement is determined by the extent to which it fulfils its functions throughout its design life. Riding quality is the requirement most noticed by the public and tends to be a matter of subjective judgement. Measurements of roughness are undertaken but the relation between these measurements and the perceived riding quality of a pavement is not always clear. Skid resistance is an essential requirement for safety of a pavement. The provision of a suitable surface texture that resists wear is the prime requirement for skid resistance. In addition, for high speed surfaces such as freeways and airport runways, the problems of aquaplaning must be addressed. The need is to provide for the rapid discharge of surface water while at the same time making sure that the film of water is broken or interrupted so that contact between the tyre and the pavement is maintained. If an appropriate surface is produced to prevent aquaplaning, it must continue to operate without being affected by subsequent wear.

Roads, airport pavements, and the associated structures represent a large investment of public funds and the public is entitled to expect that they will last and that traffic will not be interrupted by frequent maintenance. Studies in the USA have shown that the average service life of a concrete pavement, described as the average time before the first significant rehabilitation for structural reasons, is 27 years.[1.11] Many pavements may perform for a much longer time than this. For example concrete pavements constructed in Australia over 50 years ago are still performing adequately. Many must therefore have lives less than the average and designers and builders should be paying attention to the cost involved. Joints are often the weakest part of a pavement and the materials used, the geometric design, the construction methods and the maintenance procedures all need careful attention. Pavements may often be restored by the use of overlays. The methods of attachment and the possible use of special concretes, such as fibre reinforced concrete, and of adhesives demand specialist knowledge by repairers.

Bridges

Concrete is especially suitable as a bridge material as evidenced by its ever increasing use in all parts of the world. The reasons for this use are largely functional and economic, but the aesthetic advantages are considerable. Although concrete may suffer various forms of deterioration, the record for well-designed and constructed concrete bridges is good and could, with greater care to detail, be even better.[1.12] In the Federal Republic of Germany it was reported in 1982 that although about 30 000 prestressed concrete bridges, built during the previous 35 years, were giving good service, damage causing public concern had been observed in about 500 or 600 bridges. In 40 the damage was serious. High cost and annoying traffic hold-ups resulted. Although the proportion of badly damaged bridges is small (about 0.14 per cent) any increase would lead to a very heavy financial burden for the next generations.

There is no doubt that engineers must make every effort to prevent damage in existing and future bridges.[1.13]

Bridges combine the hazards of pavements and buildings and in addition are, in cold climates, subject to severe chloride attack from de-icing salt. There has been a dramatic increase in this hazard in the past 30 years with a ten-fold increase in the use of de-icing salt being reported from the USA and an even greater increase in the UK. Immediate action to combat the chloride problem has been proposed by the Cement & Concrete Association (now British Cement Association) as a combination of good design, detailing, material choice and site practice.[1.14] The exposure of the various parts of a bridge to corrosive actions depends very much on the geometry of the whole structure and its relation to salt spray, prevailing winds and tides, and the effectiveness of drainage. The resultant behaviour requires an understanding of the action and reaction of materials and the local and general environment. Even such matters as the prior storage of reinforcing bars within reach of salt spray must not be overlooked. The response of bridge structures to temperature gradients has become in recent years a concern of designers.[1.15] An understanding of the temperature dependent properties of concretes is therefore necessary for satisfactory design in this field.

Hydraulic and Marine Structures

The action of water on concrete takes three distinct forms. First, direct chemical attack may occur. The water itself may be corrosive, as is very soft or 'lime-hungry' water. It may contain salts which react with components of the hydrated cement or aggregate — sulphates are the most common example. The presence of chlorides may lead to corrosion of reinforcement. Secondly, the water may affect the surface of concrete by flow and by carrying suspended and abrasive matter as it flows. High velocity flow can lead to cavitation which has been found to cause rapid and extensive erosion of even the best quality concrete and even to penetrate steel plates. Vapour bubbles will form in running water whenever the pressure in the liquid is reduced to its vapour pressure. These vapour bubbles flow downstream and on entering a zone of higher pressure collapse suddenly and with great impact. The pitting so produced by cavitation is readily recognized from the holes or pits formed, which are distinguished from the smoother worn surface produced by abrasion from solids in the flowing water. If flow is unsteady, fluctuating pressures may act in cracks and openings to dislodge particles of concrete.

The third form of attack, and the one most prevalent in marine structures, involves the action of splash and spray by which salts are built up during drying cycles and subsequently disrupt the concrete and corrode steel. The splash zones are also especially prone to freezing action and consequent disruption. Reinforcement which is subject to fatigue loading in the presence of seawater is more liable to failure than under other fatigue conditions. Special cements may provide additional resistance to chemical attack by seawater and aggressive water but their use does not necessarily overcome the various forms of physical attack or ensure that reinforcement is made free from corrosion.

Alkali—silica reaction requires the presence of moisture and therefore is more likely to be present in hydraulic and marine structures. The reaction is especially encouraged by cycles of wetting and drying. The phenomenon has been known for a long time. Deterioration in the Buck hydroelectric plant in Virginia, USA, was first observed in 1922, 10 years after construction. As early as 1935 it was concluded from petrographic studies that the expansion and cracks were caused by chemical reaction between the cement and the phyllite used as coarse aggregate. In 1940 Stanton published a description of the phenomenon from his experience in California and after 10 years of intensive investigation alkali—aggregate deterioration of concrete had been identified in at least 14 states thoughout the USA.[1.16] The recent upsurge of interest in the phenomenon has probably resulted from changes in the minor constituents of cement brought about by changing patterns of fuel use in cement production. Once again we have a clear example of the need for designers and constructors to understand more than the mechanics of the structure.

Design for Durability

The designer of a concrete structure expects that a structure which is safe, durable and economical will result from his design. If the structure fails in one, or more, of these features the designer will have been less than successful and his self-esteem and professional standing will suffer. This has always been the situation but in recent years complications have arisen. First, it is not always clear who is the designer. In the case of precast concrete units, the engineer may only show the outline of a member and the load to be carried. The details are then provided by the contractor, or more often a specialist sub-contractor, on shop drawings. Who then is the designer? More commonly, in Australia, the engineer provides structural details including concrete strengths, cover and reinforcement. The precaster provides the handling and fixing details. Who is responsible for the long-term durability of such a unit? There are plenty of examples where cover to the main steel is inadequate and plenty more examples where deficiencies of fixings have led to corrosion and deterioration.

Secondly, in recent years, designers have found that they have new and increasing duties and liabilities resulting from changes in the law, especially in Common Law countries in which the legal system is based fundamentally on the common law of England. These countries are the UK, the USA, Canada, Australia and New Zealand. Traditionally the concrete construction industry has been organized on the basis of contracts and sub-contracts. Under the common law, the doctrine of privity of contract was established in the middle of the nineteenth century. Briefly stated, this doctrine is: 'Only parties to a contract may sue for breaches of that contract, notwithstanding that some third party may be damnified by the breach and intentionally so damnified'. The implication of the doctrine in relation to a construction job is that the owner can only recover for unsatisfactory construction from the head contractor and for unsatisfactory design from the architect or other head consultant who has a design contract with the owner. Any claim under a contract must be initiated within a specified time, usually six years, from the date on which the cause of action accrued if the claim is not to be time-barred under the various Limitation Acts. The statutory period begins to run the moment the cause of action accrues, that is to say when the breach of the contract is committed. The fact that the actual damage is not suffered until some date later than the breach does not extend the time available for a plaintiff to start proceedings. Only if a defect is prevented from being found by fraud on the part of the person responsible may the owner be able to extend the time. This was the position until the early 1970s. Since that time both legislatures and courts have been busy extending protection to individuals whom they see as victims of rapacious and negligent operators. The effect has been a wide extension of the liability of all parties involved in the construction and sale of buildings, even when the apparent victims are corporations at least as able to carry the cost of any possible defects as are the builders and designers.

The process began in the English courts in 1972 when the case of Dutton vs Bognor Regis Urban District Council broke new ground. Mrs Dutton had purchased a house from Mr Clark who had himself bought from the builder. The builder had obtained authority from the Council to construct a house with normal foundations on what turned out to be an old rubbish tip. The builder, on realizing that the site was in fact a tip, made no changes to the foundations and the Council's inspector passed the work. A crack had appeared at the time that Mr Clark sold to Mrs Dutton. Further serious defects developed, such as jamming of doors and windows — a situation which was aggravated by Mrs Dutton's inability to find money to prevent deterioration. Judgement was given against the Council for negligence in passing the foundations.

The major change brought about by this decision was that it was held that the council owed a duty of care to subsequent purchasers of the house, a situation which would have been regarded in the past as being too remote. Lord Denning, in the Court of Appeal, put the current point of view which can be seen to be heavily weighted in favour of the consumer. He said:

> Mrs Dutton had suffered a grievous loss through no fault of hers which she was not in a position to bear. In justice it ought to be borne by the builder, owing to his carelessness, the inspector, for failing to do his job, and the council who had failed to protect purchasers and occupiers with the public funds provided for the purpose.

In the UK in 1970, against a background of concern at the lack of protection of home-owners, proposals were published for legislative reform to strengthen the available remedies. When Parliament created in 1972 a restricted remedy in the Defective Premises Act, it was immediately eclipsed by the courts increasing the ambit of the entire law of negligence to encompass almost every aspect of defective building. In retrospect the statutory interventions seem little more than tinkering with an old framework in comparison with the radical changes effected by the judges.[1.17] Cases based on the tort of negligence have been decided in favour of the owner or occupier in successive decisions — Sparham-Souter vs Town and Country Development (Essex) Ltd [1976], Anns and Ors vs Merton London Borough Council [1978], Junior Books Ltd vs Veitchi and Co. Ltd [1982]. In all these cases there has been an extension of the type of damage which may be recovered. In early negligence cases some form of injury to the person was involved, while damage to property was also recognized. Purely economic loss, such as loss of profits, was excluded. But all this has changed. Wallace[1.18] has put this clearly:

> The final and most disturbing development of the law in the Anns field springs from the House of Lords decision in Junior Books. In that case a nominated sub-contractor was held liable in tort to a building owner for defective workmanship, as a result of which a factory floor was of inferior quality, giving rise to a dust problem and requiring more frequent maintenance than normal in order to deal with it. The House expressly discounted any finding that apprehended danger or physical damage was involved — and expressly recognized that the duty must be one to compensate the user for economic loss due to inferior quality ... A powerful and persuasive dissenting opinion from Lord Brandon cannot, unfortunately, prevent the case from representing a damaging four-to-one decision which it may be predicted will haunt the courts of the Commonwealth for many years to come. Its implications, not merely for professionals, but throughout commerce, are potentially devastating.

The essence of the decision is that an economic loss claim in tort now becomes possible, in both manufacturing and construction, not merely in those cases where physical damage is the result (Donoghue and Stevenson) or where apprehended physical damage justifying repairs exists (Anns) but also where no physical damage, actual or apprehended, exists but the end result is unsuitable or of inferior quality (Junior Books).

The most important implication of these changes, brought about by judges rather than legislatures, is that the chain of liability, once quite clear in terms of the various contracts entered into, is now impossibly confused. Any of the parties involved in the construction job can sue any other and, further, later owners and lessees can join in the fray. These new types of claim are all claims in tort and consequently there are two further benefits for the hopeful plaintiff. The first is that the existence of a contract no longer prevents a parallel claim for negligence on the same facts. Secondly, the time-bar for a tort claim, although governed by the same Limitation Acts as apply to contracts, allows a much longer time in which to launch claims. The cause of action in negligence cases accrues on the date when damage is caused. This will often be the date of the defendant's breach of duty. Sometimes, however, the defendant's negligence (or alleged negligence), or its effects, may lie hidden for years. This is particularly so in cases where the defect in question is a lack of durability, such as deterioration of concrete or corrosion of embedded steel. The problem in such cases is to determine when the plaintiff suffers damage so as to start the limitation period running against him. In a building case, for instance, is it the date on which damage to the structure first occurs, or the date on which the plaintiff acquires a negligently designed building, or the date on which he discovers, or could reasonably have discovered, that the building was unsound? After the decisions in Sparham-Souter and Anns it was widely believed that the courts would favour the date of discovery or discoverability; but the point was never entirely free from doubt. However any further developments in the direction of the date of discoverability were halted by the decision of the House of Lords in the Pirelli case, where it was decided that time runs from the date of the 'first physical damage', which in the Pirelli chimneys was agreed as the date when the furnace was first lit up. In most other cases involving deteriorating concrete, the date of first physical damage is almost impossible to determine. Therefore, although the English law is now clearer, the position of people in the concrete industry, and especially designers, is no clearer at all.

The law as it stands has two major defects. First it seems to place an unduly heavy burden on professionals who in theory might be open to claims for negligence for an indefinite period of time. Secondly the law is uncertain as to whether a claim can exist or not, in spite of the Pirelli decision. As the UK Law Reform Committee has pointed out, 'the law has been clarified; but the difficulties of establishing the date of occurrence of latent damage — which are the difficulties arising out of the facts in each and every case — remain as severe as ever. Where there is disagreement neither a plaintiff nor a defendant can be certain as to this date until a court has ruled on it'.[1.19] The situation is little different in the USA, though the time limits set may vary from state to state, ranging from three years or less to 15 years from the date of completion of construction.[1.20] The difficulty about the starting date also exists, with some states holding that the period begins to run only when the defect is first discovered, rather than when construction is completed.

This is not the place to discuss at length the liabilities and responsibilities of designers and reference should be made to specialist texts such as *Design Liability in the Construction Industry*.[1.21] What needs to be said is that, in every project, a good job is more likely to result if everyone involved knows what is his (or her) responsibility. Whoever has the responsibility must have the authority and whoever has the authority must also have the responsibility. Most projects do not have the various positions clearly set out in contracts between owner and designer, owner and head contractor and contractor and sub-contractor. Even the extent to which a sub-contractor for concrete work warrants the quality and fitness of his product is open to question as these matters are often left for the implication of terms into the contract, rather than being spelled out in express terms. The presence of a project manager and fast-track construction may leave responsibilities even more obscure.

Designers, and all other parties to construction, need to establish what are their responsibilities at the time of construction and into the future. Time spent at the beginning of a project on ensuring that everyone is aware of his responsibility and authority can do a lot to avoid problems of both short-term and long-term serviceability and durability and, equally importantly, avoid argument and litigation. If in spite of the best efforts litigation does arise, information and records need to be available, to enable a decision to be made as to whether a defence is possible and then to develop that defence. Most contractors in the concrete industry are concerned with getting things done rather than with keeping meticulous records and when a job is finished and out of the maintenance period, any records there might have been tend to be thrown out. Their retention involves trouble and expense with no apparent return. Our advice, in the face of the law as it now appears to stand, is that records should be better and more complete than those traditional in the industry, and should be kept for 10 to 15 years after the job is finished. The cost involved has to be included in the cost structure of the industry.[1.22]

Nevertheless, the quality of concrete construction will not be improved in any way if designers have to spend their time looking over their shoulders to avoid present or future liability. As an experienced owner and developer has written:

> we gather together responsible design and construction professionals, give them responsible owner support, and the direct result is a project that is completed on time with a minimum of problems upon completion and with the comfort of knowing that future problems should be minimized. An owner expects that these professionals are qualified and responsible, and are therefore prepared to be responsible for their work. Living by this philosophy they are more concerned about doing the job they are educated, trained, and experienced to do than they are about liability.[1.23]

References

1.1 Kong F K, Evans R H 1987 *Reinforced and Prestressed Concrete* 3rd edn Van Nostrand Reinhold 508 pp

1.2 Ferguson P M, Breen J E, Jirsa J O 1988 *Reinforced Concrete Fundamentals* 5th edn John Wiley & Sons, New York 746 pp

1.3 Warner R F, Rangan B V, Hall A S 1989 *Reinforced Concrete* 3rd edn Longman Cheshire 553 pp

1.4 Neville A M, Dilger W H, Brooks J J 1983 *Creep of Plain and Structural Concrete* Construction Press, London 361 pp

1.5 Rusch H, Jungwirth D, Hilsdorf H K 1983 *Creep and Shrinkage — their Effect on the Behaviour of Concrete Structures* Springer-Verlag, New York 284 pp

1.6 CEB-FIP 1970 International recommendations for the design and construction of concrete structures. *Principles and Recommendations* FIP Sixth Congress, Prague June 1970 80 pp

1.7 Allen W 1985 Root causes of degradation. In *Design Life of Buildings* Thomas Telford, London pp 75−85

1.8 ACI Committee 302 Guide for Concrete Floor and Slab Construction *ACI 302.1R−80* American Concrete Institute 1980 46 pp

1.9 Garber G 1988 A new look at floor surface profiles. *Concrete* (London) **22**(1): pp 12−15

1.10 Hodgkinson J R 1982 *Introduction to Concrete Road Pavements* Cement & Concrete Association of Australia, Technical Note, June 8 pp

1.11 Corvi E I, Houghton J U 1971 *Service Lives of Highway Pavements — a Reappraisal* US Office of Highway Planning, Public Roads, Aug

1.12 Liebenberg A G 1983 Bridges. In Kong *et al* (eds) *Handbook of Structural Concrete* Pitman, London Chapter 36 pp 36−1 to 168

1.13 Leonhardt F 1982 Prevention of damage in bridges. *Proceedings of the 9th Congress of Federation Internationale de la Precontrainte Vol. 1: 58−65*

1.14 Somerville G 1985 The interdependence of research, durability and structural design — concrete. In *Design Life of Buildings* Thomas Telford, London pp 233—50

1.15   Priestley M J N 1978 Design of concrete bridges for temperature gradients. *ACI Journal* Proceedings **75**(5): 209—17

1.16   Blanks R F, Kennedy H L 1955 *The Technology of Cement and Concrete Vol 1 Materials* Wiley, New York 422 pp

1.17   Speaight A, Stone G 1982 *The Law of Defective Premises* Pitman, London 205 pp

1.18   Wallace I N Duncan *Conference on Building and Civil Engineering Law* Master Builders' Federation of Australia, Sydney 20/21 Sept 1984

1.19   Law Reform Committee UK 24th report *Latent Damage* HMSO 1984

1.20   Kagan H A, Van der Water J 1986 Design in jeopardy: the expanding legal responsibilities of engineers. *Journal of Professional Issues in Engineering* ASCE **112**(1): 58—67

1.21   Cornes D L 1985 *Design Liability in the Construction Industry* 2nd edn Collins, London 232 pp

1.22   Campbell-Allen D Who pays for repairs and who should do so? *Engineered Repair of Concrete Structures* University of Sydney and Concrete Institute of Australia, 20 Aug 1985 pp 7—1 to 17

1.23   Little W R 1987 Responsibility in concrete construction — the owner's viewpoint. *Concrete International Design & Construction* **9**(8): 37—40

Further suggested reading, p. 352.

2 Quality Assurance For Concrete Construction

Introduction

A quality assurance scheme is a management system which increases confidence that a material, product or service will conform to specified requirements. It outlines the commitments, policies, designated responsibilities and requirements of the owner. These are then implemented through quality assurance programmes to provide a means of controlling, to predetermined requirements, those activities which influence quality. In the manufacture of virtually every complex product a quality assurance scheme of one type or another is used. Depending on the value of the product and methods used in its manufacture, such schemes may themselves become extremely complex and involve individuals who have little empathy for a particular material or process, whilst being very competent in their understanding of others. Under such circumstances some individuals are charged with overall responsibility for acceptance or rejection of the product.

Unfortunately, no matter how complex such a scheme may be, it cannot in all cases guarantee that the functional requirements of the product will be met, particularly when unforeseen circumstances arise. Two recent tragic examples may be quoted which illustrate this point, *viz* the Chernobyl Reactor failure, and the Space Craft Challenger explosion. These examples, quite removed from the problems usually associated with concrete deterioration, have specifically been chosen as they provide lessons to those active in applying quality assurance procedures within the construction industry.

In the case of the Chernobyl Reactor explosion, Franklin[2.1] points out that strict procedural guidelines for use were not followed during a test on the generating plant. He states that the operators not only disregarded standing instructions in relation to operating the reactor, but also disconnected the protection system. In commenting on the suggestion that some blame attaches to the designers for not anticipating the combination of circumstances brought about by the operators in their tests, he notes that it is difficult to legislate for all combinations of folly. It is of importance to recognize that, throughout the active life of any type of structure, guidelines as to permissible loadings, periodic inspections and maintenance cannot be ignored without some possible unanticipated durability, serviceability or functional efficiency problems arising. For this reason such guidelines need to be outlined in some detail at the time of final inspection, and adhered to or upgraded during occupation or use of the structure. Without this post-construction attention, the value of quality assurance during construction is significantly reduced.

In the second example, that of the Space Craft Challenger explosion, material failure under adverse thermal conditions was held to be responsible, despite warnings by some engineers that there was considerable risk in launching the craft at the prevailing temperature. Bell and Esch[2.2] describe a telephone conversation which took place on the night prior to the launch, in which a group of engineers warned their own senior managers, as well as engineers and managers from the National Aeronautics and Space Administration, of the dangers posed to the O-rings by the freezing weather forecast for Florida the next morning. The lesson here is that if a rigorous program of design checking, inspection and testing is instituted, it must be backed by a capable engineer reporting independently to management with authority to reject design details, materials or sub-standard constructed work, which in due course will lead to unacceptable performance. Any attempt to over-rule the opinion of such a capable inspector in his endeavours to establish substantial compliance with specifications, must be resisted unless it is based on strong engineering evidence, and never on the basis of expediency or economic considerations.

The Need for Quality Assurance

All involved with the construction and use of a concrete structure are concerned that the quality necessary to give good performance and appearance throughout its intended life is attained. The client requires it in promoting his next engineering scheme. The designer depends on it for his reputation and professional satisfaction. The materials producer is influenced by the quality of work in his future sales. The building contractor also relies on it to promote his organization in procuring future contracts, but his task is often considerably complicated by the problems of time scheduling and costs. The owner is particularly influenced in trouble-free use and low maintenance costs of such a structure. Finally the user is rewarded by a functionally efficient structure of good appearance. It would seem to follow therefore that since all responsible parties gain by quality it should be automatically achieved. Yet this is not so, and a considerable positive effort must be employed to achieve it. This effort can best be expended by instituting a quality assurance scheme which involves each of the above parties.

In order to ensure that the frequency of faults in new constructions is minimized, any potential faults and their underlying causes must first be recognized. In the UK work by the Building Research Establishment has suggested that most faults in structures are attributable to design errors and poor workmanship on site, with only 10 per cent being due to inadequate materials. A similiar finding was made by Campbell-Allen[2.3] after analysis of data in Australia and by Paterson,[2.4] working on a survey done in France, where less than 5 per cent of defects could be considered as principally caused by materials problems.

The causes of design faults may include:

(*i*)   misinterpretation of the client's needs;
(*ii*)   lack of good communication between members of the design team;
(*iii*) production of, and reference to inadequate and imprecise specifications;
(*iv*)   misinterpretation of design standards or codes of practice;
(*v*)   use of incorrect or out-of-date data;
(*vi*)   failure to appreciate the influence of construction procedures and tolerances on the product at various stages of construction.

The causes of faults in construction may include:

(*i*)   misinterpretation of design drawings or specifications, or deliberate departure therefrom;
(*ii*)   lack of effective communication with suppliers and sub-contractors;
(*iii*) inefficient co-ordination of sub-contracted work;
(*iv*)   inadequate on-site supervision;
(*v*)   poor workmanship due to inadequate skills and experience of the labour force or lack of satisfactory instructions;
(*vi*)   failure to understand the design principle or how this is involved in the construction procedure or sequence.

A report entitled *Structural Failures In Public Facilities* by the House Committee on Science and Technology of the US Government[2.5] came to conclusions which may be summarized as follows: six major factors (and several factors of lesser importance) were identified, which, in the opinion of the sub-committee, contributed most significantly to the occurrence of structural failure. These are:

(*i*)   communication and organization in construction industry;
(*ii*)   inspection of construction by the structural engineer;

(*iii*)   general quality of design;
(*iv*)   structural connection design details and shop drawings;
(*v*)   the selection of architects and engineers;
(*vi*)   timely dissemination of technical data.

Each of the above six factors are to some degree addressed in a well developed quality assurance programme.

It is of importance to note the rather different approach to the classifications of faults arising from premature deterioration of concrete presented by concrete technologists. The ACI Committee 201,[2.6] classifies five groups of mechanisms that can, under unfavourable conditions, produce premature deterioration of concrete. These are freezing and thawing; aggressive chemical exposure; abrasion; corrosion of steel and other embedded materials; and chemical reactions of aggregates. Newman[2.7] notes four problem areas which have done much to tarnish the image of concrete. These include corrosion of reinforcing steel; poor, hurried construction techniques; internal disruption of concrete due to alkali—silica expansions; and conversion of high alumina cement concrete and non-structural cracking during setting and hardening. In a subsequent paper[2.8] the same author discusses these in somewhat greater detail, and highlights certain of the lessons to be learned from these problems. It should be noted that in delineating the causes of concrete deterioration in the manner adopted by concrete technologists, an impression may be established that premature deterioration relates principally to inherent properties of the concrete and steel, and the exposure conditions in service, rather than to unwise choices made by the designer, non-adherence to Codes of Practice or sloppy workmanship. This impression needs to be dispelled, and this can be achieved only by further education at different levels of all those employed in concrete construction, and by rigorous application of presently available technology.

It follows therefore that despite all the valuable work done to date on the materials *per se* both in the laboratory and on site, any approach to construction product quality control,which is based solely on the concrete technologist's concepts of serviceability and durability, is bound to fail, although these concepts must be taken into consideration as part of the whole scheme. The cement manufacturer, the aggregate supplier and the concrete supplier generally consider that their obligations are at an end once their product fulfils the requirements of performance type specifications. Each tends to work within a confined area, unless integrated into the construction team. For this reason therefore any quality assessment scheme must identify all the necessary quality objectives, implement them using the skills of each of the parties, and clearly define individual roles. The importance of human factors in achieving the desired outcome has been stressed by the Comité Euro-International du Beton.[2.9]

A successful quality assurance scheme should give the client greater confidence that his project will be completed to his requirements, on time and within his budget. Because projects vary in size, complexity and the number of inter-related organizations involved, no single quality assurance scheme can be expected to cover all the necessary elements. Several reports and standards offering guidelines for the development of such schemes are available, including one from ACI Committee 121,[2.10] an International Standard, ISO 9000, which was based on the British Standard BS 5750, a series of European standards — from Germany DIN 55355, Switzerland SN 029 100, 1982, Norway NS-ISO 9001 — and an Australian Standard AS3900. The approaches have much in common and define roles for the client/owner/promoter, the designer, the supplier/producer of materials, the construction contractors and the material testing laboratory. The basic mechanisms available for both the development and operation of a quality management system are as follows:

(*i*)   organization, which requires clear definition of responsibilities and relationships for the total construction project;

(*ii*)   auditing, which requires the ability to demonstrate that the tasks defined under responsiblities are continually being executed according to stated methods;

(*iii*)   reviewing, which requires continuous checks on process methods and action procedures adopted if stated requirements are not being met;

(*iv*)   feedback, which requires elucidation in measurable terms, of causes of errors that generate defects, in order that processes can be changed so as to reduce nonconformance and allow the benefit of such change to be demonstrated.

The difference between management quality and product quality certification must be emphasized. For example BS 5750 is used in the UK by assessment agencies to evaluate the quality management systems operated by individual firms. Such firms can demonstrate compliance with that standard by obtaining a certificate from the British Standards Institute or other third party certification body, to show that the firm is capable of producing consistently to a given specification, but this says nothing of the specification *per se*. The possession of such a certification provides the means by which a firm may demonstrate their quality assurance capability to their customers, in much the same way as a laboratory may demonstrate by certification that it can perform a specific Standard Test Procedure after apparatus and personnel have been assessed. It should be noted that such certification is in addition to any certification scheme which may be applied to the individual engineer in order for him to practise in his chosen area of expertise.

In the case of nuclear safety regulated projects, it is an almost world-wide practice that all consultants are required to operate quality management systems appropriate to this type of work, for example to BS 5882 in the United Kingdom, ANSI No 4.5.2 in the USA, and KTA 1401 in Germany. However, as yet, no general move has been made to exclude consultants from work on non-nuclear projects simply on the basis of lack of quality management certification. There exists some considerable fear that if this were to become a mandatory requirement, many smaller consultant groups would be in jeopardy, and competition within the industry would be lost.

Roles of Parties Forming the Construction Process Team

In quality management systems the owner is defined as the organization financially responsible for the project and it bears the ultimate responsibility for the public health, welfare and safety. The owner may, under certain contractual arrangements, also include those organizations or individuals designated by the owner as his agents to perform certain activities (engineering, quality consulting or construction management). The owner must be actively and systematically involved in the project. It follows that he will require contracted parties to be systematic in turn. The owner, in initiating a quality assurance system, precisely defines his needs and documents these together with commitments, policies and designated responsibilities in a quality assurance plan. This quality assurance plan may be developed in conjunction with a quality consultant or engineer. The scope of work, organizational relationships and the quality objectives for the project are fully defined within this document. Stemming from this, a number of quality assurance programme elements are prepared in document form, and these indicate to each participating organization what its individual responsibilities and authorities are in achieving the quality objectives outlined in the plan. These programme elements are best developed in consultation with the concerned organizations. Care must be taken that these are compatible with any standard conditions of contract which may be agreed to between parties.

It is obvious that the principal implications are that the client invests more of his time in the project and is generally forced to bear an additional first cost. This extra first cost must be weighed against the value of reduced risks that the project will be delayed, be completed not to his satisfaction or be found to exceed budget estimates.

The role of the designer is to follow established procedures and document his design in such detail as to permit its correct development, and to permit a qualified engineer to understand and verify the final design. Design bases such as codes, standards, assumptions and other requirements should be identified, and calculations documented, checked and approved. Plans and drawings should be reviewed and approved for correct incorporation of design calculations, materials, processes and in particular constructibility. Responsibility for the overall co-ordination, verification and approvals should be explicitly provided for in the quality assurance programme. Design changes, including site alterations, should be subjected to the same control measures as applied in the original design. It should be the responsibility of the design organization to establish communication links with the other project organizations. A full set of records related to the design, contractor performance and verification documents together with as-built drawings should be provided. The designer's duties should include recommendations with respect to periodical inspections to ascertain the need for maintenance and repair, and to check that no changes in the use of the structure have been made.

In commenting on the role of the designer Newman,[2.7] in a 1986 paper, noted that, although quality control procedures are applied rigorously to materials and construction, they were rarely applied to design. He suggests that the objectives of design quality control systems should be to check whether the design structure has the safety, serviceability and durability required for the use for which it is intended, and to confirm that the technical requirements of the relevant codes are achieved. Furthermore they should ensure that the required structure is clearly defined in the drawings and specifications. With much greater accent now being placed on certification of design groups, his criticism is being addressed.

Structural engineers engaged in the role of design do nevertheless have some reservations with respect to quality assurance, and these are probably best expressed by Campbell and Dougill at a CIRIA Forum[2.11] as follows:

(i)   Recognition that a quality management system cannot compensate for conceptual error or inadequate specifications. The system merely aims for consistent application of procedures to meet the specification. A poor input could produce a consistently unsatisfactory but 'quality assured' result.

(ii) Concern at the cost of introducing and maintaining a management system without reassurance of consequential benefits.

(iii) Doubts on the effectiveness or relevance of a quality management system to design. In particular, doubts that quality assurance procedures for manufacturing processes may not be appropriate for design services. Creativity may be inhibited by forcing designers into a standard production routine conditioned by the management system.

(iv)   Questions on the type of staff needed to operate the system and their influence in a firm operating a quality management system.

(v)   Concern over accreditation procedures and particularly forms of third party certification which put the emphasis on the system and the paperwork without regard to the quality of the engineering.

Despite these concerns Campbell and Dougill note the benefits of such a management tool when applied to the design process as:

(i)   organizing and allocating responsibilities stage by stage within the design process;

(ii)   recording the completion and checking of different stages of design;

(iii)   tracing and correcting errors when they arise;

(iv)   recording and checking the transmission of instructions and information within the design team and to the constructor.

There is an educational need to be met in each step of the development and implementation of quality assurance.

The role of the producer or supplier of materials is probably the most well-defined one in any quality assurance programme. Assessment and testing techniques are well developed, particularly performance as assessed by standard tests. The designer is of course more concerned with performance in use, and may be on occasion misled with respect to the qualities of materials and products if the series of standard tests applied by the producer does not detect problems arising under service conditions.

It is in defining the role of the contractor and having his complete cooperation that some difficulties may arise in a quality assurance scheme. The construction process consists of several different operations occurring in sequence or sometimes overlapping one another. Such operations are often undertaken by persons skilled in different trades and interaction and cooperation is of the essence. Where such operations are repetitive such as in precast works, the application of a quality assurance scheme is relatively straightforward, and follows the same lines as those adopted for most manufactured articles. As a rule, however, very repetitive job details are the exception on most major construction sites, and control becomes much more burdensome.

The role of the contractor in a quality assurance programme is to comply with contract and specification documents, to perform in accordance with an approved internal programme at a level suitable for the project, to require sub-contractors' and suppliers' adherence to the contract documents and to verify that inspections and tests are performed. A guide for concrete inspection is given by ACI Committee 311.[2.12] The contractor is chiefly involved in quality control inspection which helps to assure him that the finished construction will meet all requirements of the project plans and specifications, and thus will be accepted by the owners' representatives. Such in-progress inspection is performed by contractor personnel or others specially hired by the contractor. The contractor may be required by contractual arrangements to provide a specified amount of inspection and testing. In this case inspection performed by the contractor will be much more detailed than is the usual practice for acceptance inspection.

Finally, returning to the owner or user, he has a further responsibilty in the proper use and maintenance of the structure once it has been completed.

The diverse opinions of representatives of the various parties associated with quality assurance in construction are well presented in the Proceedings of a National Quality Assurance Forum for Construction,[2.11] which was organized by the Construction Industry Research and Information Association in the United Kingdom in 1988, and those of a symposium held by the International Association of Bridge and Structural Engineers,[2.13] in Tokyo two years earlier.

Components of Quality Assurance

Available to the parties as a means of ensuring the quality performance desired are the five components of a quality assurance system *viz* standards; production control; compliance control; tasks and responsibilities; and guarantees for users. Each of the parties will use each of these to some degree in formulating the total quality assurance scheme.

Standards or specifications are used to define the important criteria, methods of assessment or testing and levels of acceptance to satisfy the stated requirements. They should, if possible, be expressed in performance terms according to Newman.[2.7] Tuthill,[2.14] however, warns that, 'proposals for "performance type" specifications for concrete work of significance are totally unrealistic and merely reflect the unawareness that they can produce inferior results despite apparently acceptable performance or appearance, unless each step, which will be covered by the next step, is inspected as the work proceeds'. Mather,[2.15] in commenting on specifications, notes that: 'the only good specification is that which requires only those things that need to be done to make concrete suitable for its purpose. A good specification contains no requirements that can be ignored or slighted, and is one that omits no requirements that must be met,' and continues, 'it is not possible to write such a specification; it is only possible to try to do so'.

Production or internal control procedures are required to be done by each of the parties to confirm that its own personnel and operations are conforming to its own quality control standards. Internal control is generally undertaken on a regular basis by the person responsible for the particular operation. For the most part such internal checking has become quite a regular procedure in the materials handling and production side of the concrete industry. Its extension to design and inspection is of importance in a quality assurance scheme.

Compliance or acceptance control procedures are required to be applied to the materials and to the structural and non-structural members at the end of each construction operation. It is often the duty of the person who is to continue work on the resultant product to check such compliance. It may be done at critical stages by an independent authorized body, during regulatory inspections, but such inspections are generally too rare to be fully effective in rectifying errors unless they are either major ones or are repetitive. Compliance or acceptance control may also be undertaken by the design engineers and, in this case, a problem of costs for more regular inspection of work than is usual may arise. Despite the relatively high costs of vigilant inspection procedures applied to all stages of concreting, it is probably here that major scope exists for improvement in the area of quality assurance. Unfortunately no universal scheme for inspection of all such stages of concreting can hope to cover the many variations observed on site. Probably the most convincing evidence of well-conducted inspections lies in the documentation which forms the inspection records. The inspection programme should include written checklists for items of construction and quantitative or qualitative criteria for acceptance in accordance with the project documents. The inspection records should document the date of inspection, area or system inspected, item inspected, inspection results, acceptance criteria, statement of compliance or non-compliance, remarks, inspector's signature and company affiliation. Mere accumulation of such data is, however, useless unless they are constantly and usefully monitored.

ACI Committee 311[2.12] considers that inspection and testing should be undertaken at one of three levels depending on the complexity of the project. The highest level of inspection (Level A) is recommended for such structures as high-rise construction, dam construction, major bridge construction etc. Inspection and testing at Level B for moderate projects relates to structures such as industrial and commercial buildings, low-rise buildings and small bridge construction. Finally inspection activities recommended for minor projects (Level C) require less intense effort and structures of this category include residential construction, catch basins and small drainage structures, kerbs and gutters. A check list is presented by ACI Committee 311 for use as a control for inspection and testing. The tests generally relate back to ASTM Standards: however, these can be substituted by chosen standards of local testing authorities if required.

Any quality assurance scheme must specify actions to be taken if materials or processes are found not to meet project requirements. Non-complying conditions are those of a recurring nature or conditions that could adversely affect the satisfactory performance or appearance of the structure should they remain uncorrected. If the causes of such conditions are determined, appropriate actions can be recommended to preclude them from future work. Evaluation of the consequences of their presence and implementation of corrective actions requires significant integration of purpose between the engineer and other project organizations. Four possible conditions may be considered *viz* acceptance as is: repair; rework; or rejection. Documentation should include the opinions of concerned individuals and the specific instructions to accomplish acceptability. Any repaired or reworked items must be reinspected and criteria for acceptability established, if these are not the same as those outlined in the original requirements.

Definitions of tasks, functions and responsibilities of each party and for each activity need to be established. Tasks and functions should include the total scope as well as any limitations on both technical and organizational roles. It is necessary to recognize in defining individual responsibilities that they may already be covered by legal or regulatory provisions. Provided that under such circumstances no conflicts exist, these arrangements contribute greatly to the success of quality assurance.

Guarantees for the user, including liabilities for faults, should be fully covered by the contract, and in certain cases, by the building control system, or, rather less desirably, by the law of tort. Some aspects of the legal problems facing concrete engineers and building owners have been dealt with in Chapter 1.

Even with these components of quality assurance available to construction engineers, Cornick[2.16] notes that factors inhibit the mechanisms of organization, auditing and feedback being applied to a building project because of traditional methods. In particular he notes the following:

(*i*)   Each project participant 'organizes' their own particular processes according to specialization without necessarily referring to the 'organization' of the other project participants, and consequently no overall 'organization' for the total project ever emerges.

(*ii*)   The documentation used seldom provides evidence of decision-making procedures so tracking is impossible and therefore 'auditing' cannot occur.

(*iii*) 'Reviews' — which are critical during the conceptual phases of a building project — are normally informal, sporadic and carried out against a limited subjective knowledge-base rather than a comprehensive and objective one.

(*iv*)   The identification of errors and their cause is discouraged because of contractual and professional liability and consequently a positive approach to 'feedback' cannot exist.

Cornick further notes that unless these mechanisms are applied throughout a quality management system, it cannot exist, and quality must suffer. He states that a major problem in demonstrating real benefits in applying quality management to building projects is that all methods of procurement still tend to obscure the real cost of not doing things right first time through every phase, i.e. the cost of not getting things completely right in any one phase is usually allowed for in the cost of the subsequent phase or phases.

Examples of Application of Quality Assurance Schemes

Kirkness and Guarre[2.17] described as an example of a comprehensive transportation construction project wherein quality assurance played a major role the Vancouver, British Columbia, Advanced Light Rapid Transit System (ALRT). They stressed three primary objectives of the programme to be fitness for purpose, value for money and confirmatory documentation. As part of the design and specification segment of the project a 'pre-build' 1.1 kilometre segment of the proposed structural system for the elevated guide-way provided data which aided design and cost reviews as well as quality audits. This preliminary phase of the project provided a sound base of quality assurance information. Factors relating to eventual major construction contracts included:

(*a*)   measurement and statistical evaluation of achievable tolerances in:

    (*i*)   column construction and bearing positioning

    (*ii*)   precast concrete beam placement

    (*iii*) survey and setting out

    (*iv*)   running rail position and level

    (*v*)   linear induction motor rail, power rail and other subsystems;

(*b*)   review and revision of beam bearing construction and access;

(c) review and revision of structure durability criteria such as crosshead allowable tensile stresses;

(d) factual data on construction costs;

(e) the preparation of contract specifications, quality control manuals and verification techniques.

During the construction phase the quality assurance programme concept was applied to the manufacturing contract governing precast concrete guideway beams. Some 1040 precast beams were manufactured using specialized formwork which allowed casting of beams that follow precisely the curves and grade changes of the design alignment. The form setting processes were well-suited to quality assurance procedures. Production control and quality control again fitted well into the overall plan. The prime contractor provided a monitoring quality assurance team that developed an inspection checking acceptance procedure, an outline of which is presented in Fig. 2.1.

Inspection check lists and QA/QC sign off sheets were developed for routine production checks. Documentation was aided by use of a microcomputer. The authors, while agreeing that non-conformance is an unwelcome part of a comprehensive quality assurance programme, outline the preplanned and disciplined procedure used to resolve problems. A flowchart indicating the steps followed in the case of non-conformance is presented in Fig. 2.2.

An application of quality management techniques to the construction of both precast and *in-situ* concrete sections of a highly complex structure is decribed by Iguro, Suzuki and Niwa.[2.18] They describe the quality management and organization applied to the construction of the world's first Arctic mobile drilling unit named Super CIDS. The structure was composed of three modular units — the top deck storage barges; the middle concrete basic block of 13.4 m height, constructed primarily of high-strength concrete; and the bottom steel mud base which rests directly on the sea-bed foundation. The basic block consists of a bottom slab, external, shear and internal walls, hollow silos and a top slab. The interior is characterized by a honeycomb structure. Although precast segmental construction was employed to form the silos the remainder of the structure was constructed with cast-in-place concrete to which prestressing was applied. Normal-weight high-strength concrete was used for the internal and shear walls, while the rest was constructed using four different mixes of high-strength light-weight concrete. The overall dimensions of the block are 71.3 m by 71.3 m by 13.4 m. The concrete section of the structure is positioned in the splash zone and designed to resist the severe loads and environmental conditions exerted by this Arctic off-shore locality. High-strength characteristics as well as adequate durability and

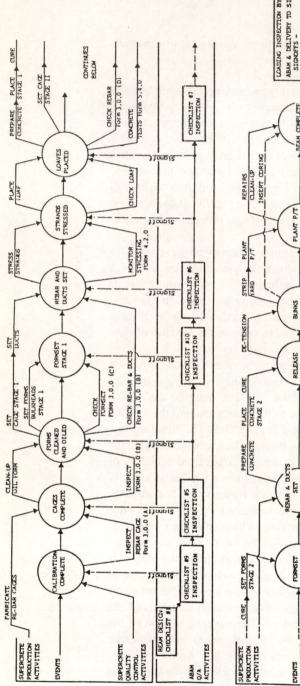

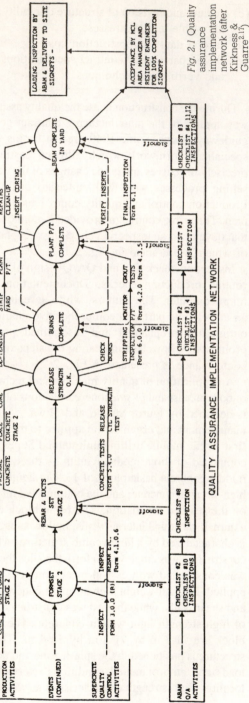

QUALITY ASSURANCE IMPLEMENTATION NETWORK

Fig. 2.1 Quality assurance implementation network (after Kirkness & Guarre[2.1])

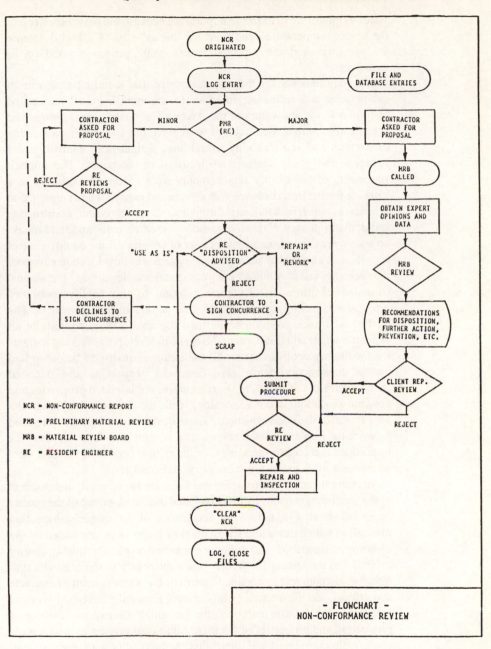

NCR = NON-CONFORMANCE REPORT
PMR = PRELIMINARY MATERIAL REVIEW
MRB = MATERIAL REVIEW BOARD
RE = RESIDENT ENGINEER

- FLOWCHART -
NON-CONFORMANCE REVIEW

Fig. 2.2 Flowchart: non-conformance review (after Kirkness & Guarre[2.17])

water tightness are of extreme importance because the basic block resists the severe temperature conditions (as low as $-50\ °C$), highly intense ice pressures and the repeated freeze—thaw actions exerted on its members.

The requirements for the structure were that it must be capable of submerging and refloating in order to accomplish an all-year-round operation in the icy waters of the Arctic. The quality management and organization which was required to achieve the specifications relating to strength and serviceability, durability, weight/draft control, water tightness and grade controls is discussed in the paper. The implicit statements of the quality requirements were redefined explicitly in a working format in accordance with a proposed materials and construction procedure, which was followed throughout the procurement, construction and delivery phases. Extensive in-house research work and field mock-up tests were carried out to assess a set of criteria for the quality control activities. A site quality management system was developed to carry out the necessary tasks. A quality control manual was developed. The manual consisted of three distinct parts relating to the quality assurance programme, quality control charts and manufacturer's standards. The quality assurance programme defined the quality requirements in an explicit manner and this formed the basis of developing working formats for the quality control charts. All the items of quality to be controlled during construction were classified with respect to the intended performance of the structure. Furthermore, the intended properties and possible deviations were cited along with the factors which rank the importance of their consequences. These quality control charts were used as working formats for the site quality control activities. Criteria for inspections and counter-measures for the critical construction items were established after extensive laboratory and field tests.

An internal quality control team was organized to be totally independent of the construction team. Its tasks included the development of the quality control manual, evaluation and acceptance of the owner's inspection manual, in which items and frequencies of inspections conducted by the owner were described, and in-house inspection work. The quality control extended to the testing for freeze—thaw durability of the high-strength light-weight concrete materials. Control of the water content of concrete was critical, and the internal quality control team enforced quality control measures over the raw materials, the amount of admixtures, the rate of production and properties of the mix. Additional properties which were extensively considered and controlled by the quality management team were thermal cracking, and weight limitations, both properties being of great importance in the functioning of the structure.

Conceptual Bases for Quality Assurance Schemes

In most discussions of quality assurance schemes for concrete construction an impression may be presented that there exists amongst engineers and project owners a general consensus on definitions of concepts such as service life, acceptable appearance and interactions between environment, loadings and time on constructed members. In practice these concepts are not simple and there is still considerable debate as to how they should be handled. This is the reason why building codes tend to avoid any direct reference to expected life of a building or structure. As an example consider an exposed concrete building facade. When designed, no consideration may ever have been given to the fact that concrete repair may be required at some time, or that a surface coating would satisfy aesthetic demand as the structure ages. Yet these two procedures may be responsible for prolonging the working life of the facade. Can or should the application of such procedures be taken into consideration at the design stage? The answer to this type of question is open to debate. One reason for this is the uncertainty associated with future attitudes and economic regimes influencing the structure's history.

Structural concrete in general should be expected to remain serviceable for at least a period of 30 to 50 years. The working life of a concrete structure may be limited either by visible signs of deterioration of the surface or by loss of serviceability, functional efficiency or structural integrity. When individual structures are considered, intended service lives vary considerably. Concrete for military use may have an intended life of days or weeks and a similar period may be sufficient for concrete used in emergency flood-mitigation work. Temporary structures used to house exhibits at a World's Fair or cooking facilities at a Scout Jamboree may need to remain serviceable for only as long as the event is in progress. A dam or sea-wall may be expected to endure almost indefinitely. City buildings of a residential or office type, although expected to have over a 50 year existence, may quickly be demolished because of changes in fashion or rapid increases in the economic valuation of the site. Public buildings, especially of a monumental type such as libraries, churches and law courts, are expected by the public to have the characteristic of longevity and degrees of durability and stability greater than usually exhibited by other city buildings. These expectations result from a psychological reaction which directly relates the function of the building to its physical structure.

Durability or serviceability changes, which may lead to impairment of the ability of one type of structure to perform its intended role, may in no way influence the satisfactory function of another. The classic example of surface finish may be quoted here. Architectural surfaces of concrete, except if off-form finishes are in vogue, require smooth mouldings, unless tooling or other surface treatments are used. Surface defects would never be considered as detrimental in the case of a mine shaft lining. However, although perhaps not aesthetically disturbing, surface defects on a dam spill-way may not be acceptable if they adversely influence its function by increasing its vulnerability to erosion, abrasion and cavitation.

Because of the many variations so far outlined the initiation of a quality assurance programme requires planning which is specifically related to the structure being considered. The quality assurance consultant must depend on experience gained by the study of comparable structures, subject to similiar exposure conditions to those which will affect his planned structure. He must at this stage consider how he can best incorporate his background knowledge into the scheme, and it is here that he encounters some considerable diversity in opinions expressed by experts in the field. The divergence in opinion may be best illustrated by considering the work of a series of authors. In order to minimize distortion of their ideas, direct quotes will be given where considered necessary.

Mather[2.15] argues that differing quality levels for allowable degrees of change over differing periods of intended service, free of excessive change, will be applicable to concretes in differing structures. He bases this on, amongst other arguments, an economic one *viz* 'that what is wanted in any construction is work of the quality that will give the desired performance at the lowest cost that must be paid to get such work.' He states that 'the best concrete for any given purpose is the one that does the job satisfactorily at lowest cost having due regard for both first cost and maintenance and repair'. He suggests therefore that 'the requirements for concrete — concrete materials, concrete mix proportions and concrete construction practices should vary widely.' He suggests that by using computer methods, 'the infinite number of alternatives concerning combinations of materials and methods can be compared with the equally truly infinite number of alternatives regarding the service requirements for the concrete and the environment in which those requirements must be met so that the one proper solution that is most economical can be selected, indicated, and specified.'

Tuthill[2.19] has severe misgivings about this approach and states that

> my concern is that someone may get the idea that it is feasible to put a concrete together that would have a level of resistance to deterioration just sufficient to offset a believed known degree of deteriorating conditions, exposure, and circumstances. Presumably this would be expected to provide lifetime trouble-free service, and in some cases

only for a predetermined number of years. There was also the hint that there may be worthwhile savings achieved by tailoring the concrete to have this just-the-right resistance to deterioration in each exposure condition, and no more, and for just the right length of time, and for no more. But concrete is not like the fabled one-horse shay. Concrete can be very long lasting. It does not come to the end of its life and sericeability like turning off a light. It must be performing its intended service right up to the end without threatening, creeping deterioration.

He urges 'that we not attempt to cut the cloth too close to the seam when there is any question of durability. The greater freedom from maintenance can soon off-set any minor extra costs for putting concrete together the best we know how, rather than something we think will be good enough but may not be in the long run'.

In an important contribution to the concept of design life of concrete structures, Somerville[2.20] noted that their life in service depends not only on the production and placing of durable concrete but also on proper design, detailing and construction methods and on appropriate levels of maintenance. He argues that there is no single condition called 'durability', and that aggressive actions must be identified and quantified prior to dealing with them in a manner similiar to that for the provision of strength, stiffness, stability, and serviceability. He considers that enough information exists to permit the derivation of such procedures to begin, but that a prerequisite to this is the definition of design lives for structures, and uses the analogy of the approach to fire design to illustrate this.

Somerville argues strongly against the opinion that the design life concept is unworkable and impractical simply because a structure does not come to the end of its life suddenly, and that engineers should simply continue to 'do better' in an open-ended way. He presents the following four arguments against adoption of such a procedure:

(*i*)   It is equally bad to produce something that lasts too long as it is to fall short of generally accepted standards. He notes that this is an economic argument, which embraces both first cost and the cost of eventual demolition.

(*ii*) In meeting a particular design-life criterion, the designer will have at his disposal a whole range of options. Under such circumstances he is unable to make the best economic and technical choice, unless he has a predetermined target for which to aim.

(*iii*) From a psychological viewpoint, merely attempting to 'do better' is not helpful. He suggests that what is really required is for the engineer to define what is needed, and more importantly, why it is needed. It follows that once these have been correctly outlined, the chances of a successful outcome are significantly increased.

(*iv*) Attempting to 'do better' at minimum first cost may not lead to maximizing the investment during the complete life-cycle — and may indeed restrict the options for doing so. In presenting this argument Somerville stresses that 'client need' is essential input in setting proper values for 'nominal design lives'.

Newman[2.8] chooses the middle ground in the debate. He states that in economic terms the quality : cost ratio of a building should be expressed in terms of initial cost plus maintenance costs. He then discusses division of the total life-cycle costs into initial costs and costs concerned with maintenance and repair. He quotes results of cost—effect comparisons made on the basis of discounted values over various periods of time which show that for a discount of 6 per cent, maintenance costs after 50 years can be practically disregarded. He notes, however, that the problem of such analyses is that not only do we lack the information to predict durability, but also that to define positively the practical life of a building.

He then defines the functional requirements to obtain the required performance as follows:

(*a*) Adequate strength of concrete in the finished structure, both on completion of the building and throughout its life.
(*b*) Adequate durability of the concrete in the finished structure in terms of:

    (*i*) ability to stand up to the external environment and still maintain a good appearance;
    (*ii*) internal structure of the concrete for chemical stability and paste-pore structure;
    (*iii*) ability to resist corrosion of reinforcing steel.

He then discusses what he considers are the most convenient and appropriate tests for measuring the actual properties and potential performance of concretes in the structure. These include proposals of measurements of actual concrete strength using cores and pulse velocity tests. Maturation of *in-situ* concrete may be measured by estimating the amount of unhydrated cement remaining in the concrete throughout its life. He proposes the use of a model which relates the onset of deterioration due to carbonation and chloride diffusion to concrete properties and cover.

Both Somerville[2.20] and Newman[2.8] propose the use of models, which relate deterioration rates due to different causes to service lives, as aids in design for durability. It is concluded that provided such models, as proposed for use by Newman and Somerville in their individual papers, do accurately portray the phenomena and rates of occurrence, their aid in design of quality assurance systems is of inestimable value. Problems will still arise where, for example, ingress of depassivating ions is not through solid concrete, for which diffusion rates are known, but *via* cracks, or when such ingress is not solely by diffusion, but is hastened by suction. Hence one concludes that the whole system depends on the accuracy of available information, rather than any other factor. Furthermore the experience of the quality assurance consultant in sufficiently narrowing the scope of tests and inspections without excluding any of importance, cannot be under-estimated. He must be capable of choosing his applied models to predict accurately future effects if maximum value of such a scheme is to be achieved.

The task of the design engineer may become more simplified in future with the development of expert systems, which can be applied to these durability problems. An expert system is a computer program that contains and uses the knowledge of experts along with inference procedures to allow a user to solve problems of this sort. The system should accurately represent the knowledge of the experts in a specific field. The knowledge base consists of facts, which are readily available items of information that are widely held to be true, and heuristics, which are rules of thumb, or statements of good judgement, that an expert would use to solve a problem when all the needed facts or theories are not available.

Clifton and Kaetzel[2.21] discuss an expert system entitled Durcon, which relates to four major deterioration processes of concrete, *viz* corrosion of reinforcement; freezing and thawing; sulphate attack; and reactions between cement and aggregates. These four processes were selected because they comprise the most common causes of the deterioration of concrete. The major factors controlling the response of concrete to these deterioration processes and recommendations of measures to minimize their deleterious effects are given in a publication entitled '*Guide to Durable Concrete*' prepared by ACI Committee 201,[2.6] this information was chosen to be the factual component of the knowledge-base. A series of questions were formulated and, depending on the answer to each of these by the system user, a list of recommendations results.

In the above example, only durability as influenced by a few processes has been covered, but it is suggested that through the application of expert systems in concrete construction, a whole series of applications including topics in design, monitoring, planning, control, prediction of remaining service life and repair materials and procedures, will eventually be better comprehended and used to advantage by the quality assurance team. Again the value forthcoming from the use of such expert systems will depend on the knowledge of the expert, and its applicability to the particular structure and available materials and resources.

Conclusion

In response to a recognized need to prevent failures and produce quality in constructed projects, the ASCE in 1988 published a *Manual of Professional Practice*, being a guideline for owners, designers and constructors with regard to quality in the constructed project.[2.22] This document was issued as a trial-use edition, and despite the comprehensive cover of the topic, comment from users was sought until late 1989. Any organization wishing to institute a quality assurance program, or improve an operative one should benefit greatly from referring to this publication.

A well planned and run quality assurance scheme is of great value to all concerned in concrete construction. Its effective functioning depends on the skill of the engineers managing it, both with respect to their background knowledge of tasks and tests, and their capability of influencing the employed workforce. Any quality assurance scheme should lead to greater vigilance on the part of all parties, and hence to an improvement in the finished structure.

References

2.1   Franklin N 1986 The accident at Chernobyl. *The Chemical Engineer* **430**: 17–22

2.2   Bell E, Esch K 1987 The fatal flaw in Flight 51-L. *IEEE Spectrum* **24**(2): 36–51

2.3   Campbell-Allen D 1979 *The Reduction of Cracking in Concrete* University of Sydney/Cement and Concrete Association of Australia 165 pp

2.4   Paterson A C 1984 The structural engineer in context. *The Structural Engineer* **62A**(11): 335–42

2.5   House Committee on Science and Technology 1984 *Structural Failures in Public Facilities* US Government Printing Office, Washington DC

2.6   ACI Committee 201 Guide to durable concrete *ACI 201.2R–77* American Concrete Institute 1977 36 pp

2.7   Newman K 1986 Labcrete, realcrete, and hypocrete — where we can expect the next durability problems *ACI SP 100–64* pp 1259–83

2.8   Newman K 1986 Common quality in concrete construction. *Concrete International Design & Construction* **8**(3): 37–49

2.9   Comité Euro-International du Beton 1983 Quality control and quality assurance for concrete structures. *Bulletin d'Information* No 157 98 pp

2.10 ACI Committee 121 Quality assurance systems for concrete construction *ACI 121R–85* American Concrete Institute 1985 7 pp

2.11 CIRIA 1988 National quality assurance forum for construction. *Special Publication 61* 85 pp

2.12 ACI Committee 311 Guide for concrete inspection. *ACI 311.4R–88* American Concrete Institute 1988 11 pp

2.13 IABSE 1986 *Symposium on Safety and Quality Assurance of Civil Engineering Structures* IABSE Reports 50, 51 and 52: 143, 407, 191 pp

2.14   Tuthill L H 1986 Obtaining quality in concrete construction. *Concrete International Design & Construction* **8**(3): 24−9

2.15   Mather B 1986 Selecting relevant levels of quality. *Concrete International Design & Construction* **8**(3): 30−6

2.16   Cornick T C 1988 Quality management model for building projects. *International Journal of Project Management* **6**(4): 211−16

2.17   Kirkness A J, Guarre J S 1986 Quality assurance for a major transportation construction project *ACI SP 93−23* pp 491−506

2.18   Iguro M, Suzuki T, Niwa M 1986 Quality management for Arctic offshore concrete structures. *IABSE, Safety & Quality Assurance Symposium* pp 309−16

2.19   Tuthill L H 1980 Discussion of 'Concrete Need Not Deteriorate' by Bryant Mather. *Concrete International Design & Construction* **2**(7): 88

2.20   Somerville G 1986 The design life of concrete structures. *The Structural Engineer* **64A**(2): 60−71

2.21   Clifton J R, Kaetzel L J 1988 Expert systems for concrete construction. *Concrete International Design & Construction* **10**(11): 19−24

2.22   *Manual of Professional Practice: Quality in the Constructed Project: a Guideline for Owners, Designers and Contractors — Preliminary Edition for Trial Use and Comment* ACSE, New York 1988 192 pp

Further suggested reading, p. 352.

3 As-built Concrete Properties

Strength

Since the beginning of the use of concrete, strength has been regarded as one of its most significant and important properties. Early systems of reinforced concrete construction advocated limits on both tensile and compressive stress in concrete, with the result that in almost all structures the tensile stress criterion became the one that governed. This criterion resulted in some very durable, and consequently economical, structures as for example the Lamington Bridge at Maryborough, Queensland, Australia, one of the world's first major concrete bridges. Reinforced concrete was selected by the designer, Brady, on the grounds that 'it ensures a structure of very great strength, almost everlasting in character, and the annual expenditure on maintenance is consequently reduced to a minimum'.[3.1] As O'Connor reports,[3.2] 'he has been shown to be correct in these expectations for the structure is still in excellent condition (in 1984)'.

Even in 1890 it was recognized that high strength was allied to many other desirable properties such as durability.

Since the 1960s it has been usual to specify the required quality of concrete on the basis of strength. The strength usually specified is the 'characteristic strength' which is the strength determined by testing at a fixed age samples of the concrete, moulded and treated in a standard way. The characteristic strength is the strength below which only a small proportion — generally 5 per cent — of the population of possible samples will test.

Owing to their simplicity the standard cube and cylinder control tests have attained a unique position as measures of a wide variety of properties of concrete. They have been used for control of uniformity in manufacturing, as the basis for working stresses in concrete design and as a measure of many other remotely related properties such as elasticity, creep, ultimate strain and durability. The results obtained from these tests are not measures of basic properties of concrete. The tests are not even closely reproducible as the results are influenced by many different factors, including the characteristics of the testing machine.

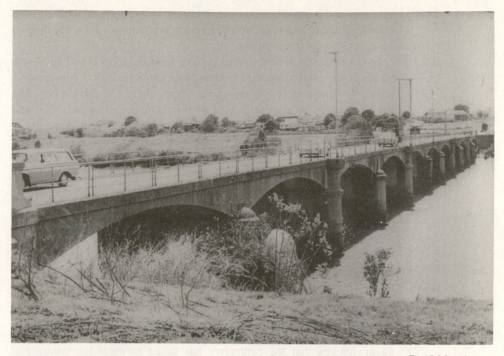

Fig. 3.1 Lamington Bridge, Maryborough, Queensland, Australia

If structural strength is largely dependent on compressive strength, as for instance in axially loaded columns failing in compression, then the results of cube or cylinder tests may be regarded as giving a direct guide towards the member strength. In all other cases, where failure is in tension or under triaxial conditions (as for example in prestressing anchorages) the control can only give indirect guidance towards strength.

In any strength test under a particular stress regime, there is tested the effect of gel strength, gel concentration, degree of saturation, paste— aggregate bond and aggregate strength. All these factors have different inter-relationships as the shape and size of the specimen is changed even when the nominal stress is the same. As moisture conditions change, the inter-relationships become even more complex. Nevertheless, the control test for compressive strength is still the most important piece of information that is available about concrete that is placed in a structure. The capacity of the structure to carry its design loads is related in the mind of the designer to the specified characteristic strength being achieved and being maintained throughout the life of the structure. There is sometimes a hidden hope that the strength in the structure will increase with age beyond that measured. Strength may deteriorate with time as a result of external attack or as a result of internal chemical disruption (see pp. 53—7). Much of this deterioration is associated with internal cracking and is readily observed. Only the conversion of high alumina cement can lead to a reduction of strength without any apparent distress being visible in the concrete.[3.3]

In-situ and non-destructive tests have been developed and in almost all cases are used to provide indirectly an estimate of the compressive strength of the concrete at the age at which the test is carried out. The use of strength as a specification requirement is almost universal and there is no provision in standard specifications in North America or Europe for the acceptance of concrete on the basis of *in-situ* or non-destructive tests. As Malhotra has pointed out, the development of well-documented detailed field test data by testing organizations is a prerequisite before such a procedure could be allowed.[3.4] Non-destructive tests are used to determine early strengths so that prestressing or formwork stripping can be safely carried out.[3.5]

It is now recognized that the most useful purpose of specifying compressive strength is to ensure that the water:cementitious ratio, on which many other properties are primarily dependent, is at an appropriate level. There are, however, situations where somebody wishes to know what is the actual strength of concrete in the structure as built. The exact connection between this strength and that measured by control specimens is still the subject of much debate, especially when the owner, or his agent, alleges that the contractor has failed to meet his contractual obligations.

The immediate question then is: 'How can the strength in the structure be determined?' The most common way and one that is widely respected is by drilling and testing cores. The measured strength obtained from testing a core is affected, however, by many factors. Some of these factors are inherent in the concrete, as for example air-voids, but many more relate to the test method. The major factors that affect the strength of a core as tested are:

(*i*)   the size of the core and its relation to the size of the aggregate;
(*ii*)   the ratio of length to diameter (slenderness ratio) of the core;
(*iii*) the direction of drilling as related to the direction of casting;
(*iv*)   the presence of reinforcement;
(*v*)   the moisture conditioning of the core prior to the test.

Size of core   It has been shown in a number of investigations that the dispersion in individual core strengths increases as the dimensions of the test specimens become smaller. Henzel and Freitag[3.6] showed that the standard deviation for 50 mm diameter cores was about twice that for 150 mm cores. The dispersion also increases markedly if the core diameter is less than three times the maximum aggregate size.[3.7] The practical consequence is that if smaller cores are used, more will have to be drilled to achieve the same reliability in average measured strength. There has been much conflicting evidence on the effect of the core size on the level of strength measured. Some suggest that an increase in strength occurs with smaller size,[3.8] and others that a reduction takes place.[3.9] The size effect in cores is different from that in cast cylinders, where an increase in strength is generally noted for smaller cylinders, owing to the presence of a mortar layer at the curved surface. The most probable situation is that found by Meininger[3.10] and supported by further work by Munday and Dhir,[3.11] namely that 'provided the core diameter is confined to a certain sensible range, the effect of this factor on the measured strength could be ignored for practical purposes. In practice, the maximum aggregate size is generally between 10 and 20 mm and for this range core diameter between 50 and 150 mm could be satisfactorily used'.

Slenderness of core   Evidence on the effects of length:diameter ratio is not wholly consistent[3.12] but if the measured strength of a core of slenderness ratio λ is f_λ, the strength of a core of slenderness ratio 2 can be estimated from

$$f_2 = \frac{2\lambda}{1 + 1.5\lambda} \, f_\lambda \qquad\qquad [3.1]$$

Direction of drilling   A relation of the direction of loading to the direction of casting might be expected to have some effect on strength, particularly if bleeding lenses are present. Petersons[3.13] showed that horizontal cores were 10 per cent lower in strength than vertical cores, but the investigation was limited and he did not repeat these results in his later work on factory-made beams.[3.14] BS1881 Part 120 indicates that on average vertical cores are 8 per cent stronger than those drilled horizontally. However, Keiller[3.15] found that on average horizontal cores were about 2 per cent stronger than vertical cores though he noted a tendency for the vertical cores to be slightly stronger at low strengths and horizontal cores to be stronger at high strengths. These differences were not statistically significant. Once again, the best conclusion seems to be that the measured strength will not be greatly influenced by the direction of casting.

Presence of reinforcement   A matter that sometimes causes arguments is the presence of reinforcement at right angles to the direction of loading. The effect has been shown by Lewandowski[3.16] to be a slight reduction in strength (generally less than 4 per cent) and an increase in dispersion.

Moisture conditioning   The appropriate moisture condition of cores at the time of test is always the subject of arguments. AS1012.14 is typical of standard specifications in that it allows either wet conditioning (three days submerged in water before test) or dry conditioning (seven days in air at 50 ± 10 per cent relative humidity). AS3600 requires that cores shall be tested dry unless the concrete concerned will be more than superficially wet in service. Cores tested dry will usually give a higher strength than those tested after wet conditioning but the relationship between the wet and dry strengths depends on the curing which the concrete has previously received and on the nature of the cementitious material. More reproducible results are obtained by testing wet and this procedure is recommended.

When all these test effects have been brought to account as well as they can be, there is still a question as to the relationship between the strength of control specimens (cubes or cylinders) and the strength shown in the structure by the cores. An extensive investigation is reported by Bloem[3.17] in which pairs of slabs, 150 mm thick, from three different concretes were subjected to good and poor curing. Cores and push-out cylinders were extracted at six different ages, ranging from one day to one year, and tested for strength. Standard-cured and field-cured cylinders were also included. Altogether 216 cores, 216 push-out cylinders and 270 molded cylinders were tested. The results obtained for the concrete using ASTM Type I cement are shown in Fig. 3.2 and those for two mixes using ASTM Type III (high-early-strength) cement are shown in Figs 3.3 and 3.4. It has been suggested that the low strength found in some cores is brought about by the disturbance caused by drilling. In this series the push-out cylinders gave results very close to those for the cores.

Bloem concludes that the strength of cores will be less than that of moist-cured cylinders tested at the same age and will probably never reach the standard 28-day strength even at greater ages. The amount of the deficiency will depend on the field-curing and the type of cementitious material. Field-cured cylinders do not provide a good guide to the core strength. Bloem's investigation was carried out in a climate in which the ambient relative humidity (smoothed over 5 week periods) ranged between 35 and 70 per cent.

In contrast to these results, Keiller[3.15] reported that the strength of concrete *in-situ*, in the UK, is likely to be about 80 to 85 per cent of the standard cube strength at 28 days, if the concrete is fully compacted and 'properly cured'. Both cores and cubes showed a significant increase in strength between 28 days and one year such that the *in-situ* strength at one year is likely to equal or slightly exceed the standard cube strength at 28 days. These results are in line with those found by Murray and

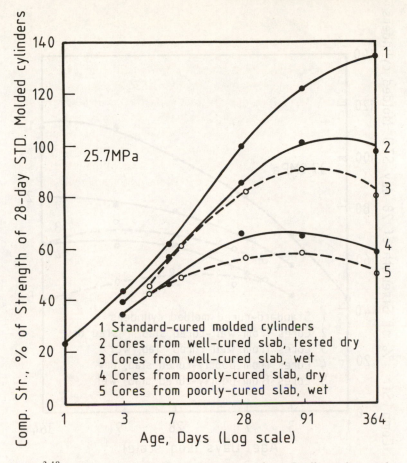

Fig. 3.2 Strength
measures for
Grade 25 concrete
with Type I
cement (after
Bloem[3.17])

1 Standard-cured molded cylinders
2 Cores from well-cured slab, tested dry
3 Cores from well-cured slab, wet
4 Cores from poorly-cured slab, dry
5 Cores from poorly-cured slab, wet

Long[3.18] in Northern Ireland, another damp place. In discussion on the Murray and Long paper, Tomsett[3.19] quoted several papers[3.16, 3.20, 3.21] which all showed a lower strength in site concrete than in standard specimens but most of the papers related to concrete in climates different from that in the UK. Tomsett added that 'this state of affairs may, or may not, continue throughout the life of the structure, depending on service conditions'. He reported that he had observed equivalent cube strengths of over 90 MPa in concrete which had been submerged for its life of 20 years and which probably had a specified cube strength of 30 to 40 MPa when cast. Some writers have even suggested that core tests under-estimate the strength of concrete in real structures. There is no support for this view provided care is taken in assessing core tests so that the results are reduced on a standard basis to an 'equivalent *in-situ* cube strength' or 'equivalent *in-situ* cylinder strength'. The further step of using core tests to deduce an equivalent standard 28-day control strength cannot be justified.

Fig. 3.3 Strength
measures for
Grade 40 concrete
with Type III
cement (after
Bloem[3.17])

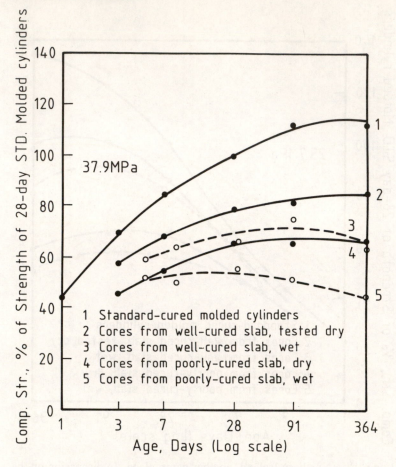

Permeability

Concrete is a permeable and a porous material. The rates at which liquids
and gases can move in the concrete are determined by its permeability.
The porosity determines the quantity of liquid or gas that can be contained
in the concrete. Both properties, and particularly the permeability, affect
the way in which concrete resists external attack and the extent to which
a concrete structure can be free of leaks. Figure 3.5 illustrates the
differences and the inter-relationship between permeability and porosity.
The permeability is much affected by the nature of the pores, both their
size and the extent to which they are inter-connected. There can therefore
be no one measure of porosity which fully describes the way in which
the properties of concrete, or of hardened cement paste, are affected.
As Powers and Brownyard[3.22] pointed out:

To speak of the porosity of hardened paste is likely to be misleading

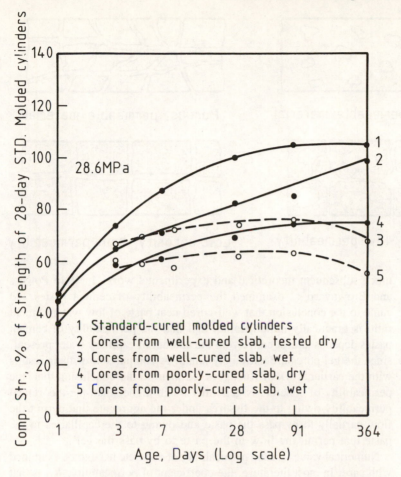

Fig. 3.4 Strength measures for Grade 30 concrete with Type III cement (after Bloem[3.17])

28.6MPa

1 Standard-cured molded cylinders
2 Cores from well-cured slab, tested dry
3 Cores from well-cured slab, wet
4 Cores from poorly-cured slab, dry
5 Cores from poorly-cured slab, wet

Age, Days (Log scale)

unless the term is properly qualified. The word 'porosity' can be interpreted differently according to past experience or to the chosen criterion as to what constitutes porosity. Certainly, the term is misleading here if it calls to mind such materials as felt or sponge.

Whether or not a material is considered to be porous depends in part on the means employed for detecting its porosity. If a material were judged by its perviousness alone, the decision would rest primarily on the choice of medium used for testing its perviousness. For example, vulcanized rubber would be found impervious, and hence non-porous, if tested with mercury, but if tested with hydrogen it would be found to be highly porous. The perviousness of hardened cement paste to water and other fluids is direct evidence of the porosity of the paste.

Early work on the permeability of concrete was generally related to its use in dam construction and the work of Ruettgers *et al.*[3.23] in 1935 on concrete for the Boulder (now Hoover) Dam provided the basis of

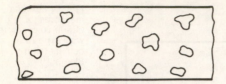

Porous, impermeable material

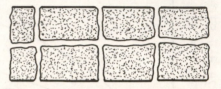

Porous, permeable material

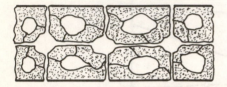

High porosity, low permeability

Low porosity, high permeability

Fig. 3.5
Permeability and
porosity (after
Concrete
Society[3.26])

much subsequent theoretical and experimental work. In 1946 Powers
and Brownyard[3.24] examined the permeability of cement pastes and
came to the conclusion that well-cured neat paste of low water:cement
ratio is practically impermeable, and that the permeability of cement
pastes depends almost entirely on the amount of capillary water present,
since the gel pores are extremely small. A comparison of their results
with the earlier work of Ruettgers led them to the conclusion that 'the
permeability of concrete is generally much higher than the theoretical
permeability owing to the fissures under the aggregate that permit the
flow partially to by-pass the paste and owing to the capillaries in the
paste that permit the flow in the paste to by-pass the gel'.

Numerical values for the permeability of concrete need to be examined
with care. In most literature, the coefficient of permeability, K_1, is that
obtained from applying Darcy's law for low velocity flow:

$$\frac{dq}{dt} \cdot \frac{1}{A} = K_1 \cdot \frac{\Delta h}{L} \qquad\qquad [3.2]$$

where: dq/dt = the rate of volume flow ($m^3\ s^{-1}$)
$\quad\quad\ A$ = area of the porous medium normal to the direction of
flow (m^2)
$\quad\ \Delta h$ = drop in hydraulic head across the thickness of the
medium (m)
$\quad\quad\ L$ = thickness of the medium (m)
$\quad\quad K_1$ = coefficient of permeability depending on the properties
of the medium and of the fluid ($m\ s^{-1}$)

For any set of tests, the value of K_1 depends on both the medium and
the fluid and therefore represents the permeability of the medium to a

specified fluid at a specified temperature. In the case of concrete, the fluid is usually water. An alternative expression in which the effect of the fluid properties is eliminated, by the introduction of the viscosity of the fluid, appears attractive at first sight. Unfortunately, the viscosity of water in concrete is not constant, even at a fixed temperature, as the viscosity is affected by the size of the pore into which the water moves and by the presence of solutes. In addition, further hydration of the cement can occur, especially when the water is under pressure. Values of permeability discussed here will therefore be values of K_1 in eqn [3.2] for water in concrete.

A very wide spread of water-permeability results has been noted by Lawrence.[3.25] As pointed out by the Concrete Society Working Party[3.26]

> a number of factors can account for the wide spread of permeability results for a specific water:cement ratio concrete, due primarily to aspects of the test method, for example:
> (a) varying and continuing hydration of the specimen;
> (b) incomplete and variable initial saturation;
> (c) lack of absolute water cleanliness;
> (d) chemical reaction of specimen with the test fluid;
> (e) effect of dissolved gases where high pressure air is used to pressurise the water;
> (f) silting, due to movement of fines;
> (g) microstructural collapse and macroscopic instability when very high flow pressures are used;
> (h) lack of attainment of steady state condition.

The composition of the water and the presence of dissolved materials can also have a substantial effect.

The permeability of mature hardened paste was found by Powers et al.[3.27] to range from 0.001×10^{-12} to 1.20×10^{-12} m s^{-1} for water:cement ratios ranging from 0.3 to 0.7. These figures related to pastes which had never been allowed to dry. Drying was found to increase the permeability and, for the particular specimens examined, drying at 79 per cent relative humidity increased the permeability about 70-fold. As cement hydration proceeded the permeability was reduced and in 24 days was found to be only one-millionth of its initial value. Table 3.1, based on Powers et al., shows this effect. The cement used was of comparable fineness to modern cements (342 m^2 kg^{-1} as measured by the permeability method) but the compound composition was very different from present day normal Portland cements. C_3S content is quoted as 45 per cent and C_2S as 27.7 per cent with C_3A as 13.4 per cent. Parrott[3.28] has now suggested that with modern cements, containing higher proportions of C_3S, the times to produce complete closure of the capillary pores can be substantially less than those proposed by Powers following his work in the 1940s and 1950s.[3.29]

Table 3.1 Reduction of permeability
by cement hydration (W:C 0.7)
(after Powers *et al.*[3.27])

Age (days)	Permeability coefficient $(\times\ 10^{-12}\ \text{m s}^{-1})$
Fresh	2 000 000
5	400
6	100
8	40
13	5
24	1

The flow tests are appropriate for testing material which has a high permeability, but for concrete of low permeability a method in which the depth of penetration is measured is usually a more practical proposition. Water is applied under pressure to one face for a period of time and the specimen is then split open to determine the depth to which the moisture has penetrated. This method is included in the German Standard DIN1048. Typical results and specification values, using this test under which the pressure is applied for 96 h, are:

dense (50 MPa) concrete with slag and superplasticizer — 5 mm;
specification for good quality concrete to resist aggressive environment — <30 mm;
specification for concrete in water-retaining structures — <50 mm.

Although the results of these penetration tests are expressed as depths, it is possible in some circumstances to derive a value for permeability by using the equation derived by Valenta for uniaxial flow:[3.30]

$$d = (2K_1 Th/v)^{1/2} \tag{3.3}$$

where: T is the time to reach a depth of penetration d under a head h in concrete whose porosity is v.

Some test results for which coefficients of permeability have been calculated from eqn [3.3] were obtained by Browne and Domone.[3.31] For a concrete with a cement content of 405 kg m^{-3} and water:cement ratio of 0.42, a test under a head of 211 m and extending from age 13 days to 28 days gave penetrations between 110 and 125 mm. The calculated permeabilities were 24 and 29 $\times$ 10^{-12} m s^{-1} respectively. The same concrete tested under a head of 100 m from age 60 days to 228 days gave penetrations ranging from 125 mm to 150 mm, with corresponding calculated permeabilities of 5 to 8 $\times$ 10^{-12} m s^{-1}. The permeability of concrete *in-situ* which was suspected of being porous was calculated from tests lasting only four to five hours as 2500 to 3900 $\times$ 10^{-12} m s^{-1}.

For the reasons already discussed, the water-tightness of a concrete structure is not determined by the permeability of the hardened cement paste, or even by the measured permeability of laboratory specimens of the concrete. Additional flow occurs through cracks which can form in the concrete and this flow is generally much larger than that which takes place through the body of the concrete.

However, the permeability of concrete, both to moisture and to gas, is important in relation to the protection afforded to embedded steel. The most practical methods of test for these properties of *in-situ* concrete are suggested as being the Initial Surface Absorption Test (ISAT) and the Figg Test. The ISATest, which is described in BS 1881: Part 5: 1970, measures the rate at which water is absorbed into the surface of the concrete for a brief period under a head of 200 mm. Typical results of ISATests are given in Table 3.2. The Figg Test and subsequent modifications of it measure the permeability of the concrete at the bottom of a fine hole drilled to some depth below the concrete surface. The Figg apparatus has been modified for convenient site use and is available in a portable form. Tentative classifications of concrete are given in Table 3.3.[3.32]

The depth to which water which is absorbed into concrete, under little or no head, has been shown to be initially a linear function of the square-root of time. The slope of this function is called the sorptivity. (In some reports the square of this value is called the sorptivity and the extent of the absorption is measured as the volume of water instead of the depth of wetting. Care must therefore be taken in interpreting any reported results.) The sorptivity has been proposed by Ho and his co-workers[3.33] as a measure for assessing the protection that will be afforded to embedded reinforcing steel, particularly after it has become activated. A possible criterion for durability would be to ensure that the water front from a 24-hour rain period should not reach reinforcement which had

Table 3.2 Typical results of initial surface absorption tests (ISAT)[3.26]

Comment on concrete absorption	ISAT results ml m^{-2} s^{-1}			
	Time after starting test			
	10 min	30 min	1 h	2 h
High	>0.50	>0.35	>0.20	>0.15
Average	0.25–0.50	0.17–0.35	0.10–0.20	0.07–0.15
Low	<0.25	<0.17	<0.10	<0.07

Table 3.3 Tentative classification of protective quality of materials for Figg method. (Method modified by Ove Arup Partnership using air, specimens to constant weight at 50 °C)[3.32]

Quality category	Time (s)	Interpretation	Type of material
0	<30	Poor	Porous mortar
1	30–100	Moderate	20 N mm^{-2} concrete
2	100–300	Fair	30–50 N mm^{-2} concrete densified, well-cured concrete
3	300–1000	Good	
4	>1000	Excellent	Polymer-modified concrete

a cover of 30 mm. This would require that the sorptivity should not exceed 6 mm h$^{-1/2}$. In subsequent work, Ho *et al.* suggested that a prudent value of sorptivity for ensuring durable reinforced concrete was 3.0 mm h$^{-1/2}$.[3.34] In the series reported the only concrete that was found to have a sorptivity that complied with this criterion was a Portland cement concrete with a 28-day strength of 38 MPa. Some unreported

Table 3.4 Typical values of concrete permeability and related properties (after Concrete Society[3.26])

Test method	Units	Concrete permeability/absorption/diffusion		
		Low	Average	High
Intrinsic permeability k	m^2	<10^{-19}	10^{-19}–10^{-17}	>10^{-17}
Coefficient of permeability to water	m s^{-1}	<10^{-12}	10^{-12}–10^{-10}	>10^{-10}
Coefficient of permeability to gas	m s^{-1}	<5 × 10^{-14}	5 × 10^{-14}–5 × 10^{-12}	>5 × 10^{-12}
ISAT 10 min	ml m^{-2} s^{-1}	<0.25	0.25–0.50	>0.50
30 min		<0.17	0.17–0.35	>0.35
1 hour		<0.10	0.10–0.20	>0.20
2 hour		<0.07	0.07–0.15	>0.15
Figg water absorption 50 mm (dry concrete)	s	>200	100–200	<100
Modified Figg air permeability (Ove Arup) −55 to −50 kPA	s	>300	100–300	<100
Water absorption 30 min	%	<3	3–5	>5
DIN 1048 depth of penetration (4 days)	mm	<30	30–60	>60
Oxygen diffusion coefficient 28 day	m^2 s^{-1}	<5 × 10^{-8}	5 × 10^{-8}–5 × 10^{-7}	>5 × 10^{-7}
Apparent chloride diffusion coefficient	m^2 s^{-1}	<1 × 10^{-12}	1–5 × 10^{-12}	>5 × 10^{-12}

tests on a 10 year old concrete from a building facade, made with a water:cement ratio of 0.65 and therefore a much lower strength than 38 MPa, suggested that the sorptivity of surface concrete which had been exposed to the weather was substantially lower than that found by Ho for concrete of comparable strength. The criterion proposed by Ho has a sound practical basis for providing protection to already activated steel, but the low figure of sorptivity may not necessarily be required at an early age, provided it can be achieved as a result of exposure to the elements by the time the steel does become activated. At early ages the resistance of the concrete to carbonation and to the ingress of chlorides is a more important property. Ionic diffusion tests and gas diffusion tests have been used to assess these properties on laboratory concrete and on laboratory specimens cut from actual structures and typical values are given in Table 3.4. But, as the Concrete Society Working Party points out, the figures 'are simply intended to give typical test values. In view of the significant differences which can be obtained by various methods of conditioning specimens and by differences in test technique it is important to note that the values given should not be used for specification purposes or for any contractual arguments on the quality of specific concrete'.[3.26]

Internal Chemical Disruption

Soundness

Soundness of cement is defined as the ablility of the material, after gauging with water and setting *in situ*, not to undergo any appreciable change in volume. The movements which are experienced by a sound cement when either continuously immersed in water, or allowed to dry after setting, are usually small, and can be accommodated without disruption or loss of strength. An unsound cement expands after setting, sometimes after a period of months or years has elapsed. The expansion, during which cracking develops, occurs in mortars and concretes as well as in the neat cement, although the effects in the latter are always more severe. The expansions of unsound cements are due to the slow hydration of certain of its constituents.

The testing of the soundness of cement, so as to ensure such serious expansion does not occur, has been so perfected that cases of problems arising in practice from this cause are almost absent from recent publications. It is nevertheless worthwhile to mention the three errors in cement composition which are known to give rise to unsoundness. These are:

(*i*)   an excess of lime above that which, under the conditions of manufacture, can become combined with the acidic oxides of the cement mix;

(*ii*) an excessive proportion of magnesia;

(*iii*) an excessive proportion of sulphates.

In the first case crystalline calcium oxide may hydrate at a very slow rate and although the volume of the reactants is greater than the reaction product, the particles being formed grow preferentially in one direction producing internal stress and resulting in cracking. The hydration of periclase, MgO, to brucite, $Mg(OH)_2$, is a slow reaction, well known as the cause of unsoundness of dolomitic limes. Due to solid solution and grain size effects, limits on MgO in the meal cannot be used as an estimating parameter of expansion. Autoclave expansion tests have therefore to be used. The expansion associated with excess sulphate in a cement is attributable to the formation of sulphoaluminate hydrates. This process is harmless during the early setting phase, but if excessive amounts of sulphate are present the continuation of this process leads to unsoundness. For this reason the maximum permissible SO_3 content of a cement is governed by the ability of the cement clinker minerals to react with it in the early hardening period. The chosen gypsum content of a cement governs all of its engineering properties (such as strength, elasticity, shrinkage and creep) throughout its life, and hence the necessity of ensuring that this content is close to an optimum value which is governed by the C_3A content, fineness, available alkalis and the temperature at which casting takes place.

Although what has so far been said relates to the effects of composition of the cement alone, if any of the constituents are introduced to the concrete by way of aggregate materials similiar problems arise. Calcium oxide is sometimes present in slags and these have been reported to have led to problems in practice. Periclase is found in metamorphosed dolomites and again in some slag materials. Aggregates containing excessive amounts of gypsum may provide a source of sulphate ions which leads to unsoundness of the concrete manufactured with them. This problem is related to sulphate attack discussed in greater detail elsewhere (see Chapter 4, pp. 128−30), but the source of sulphate ions in that case is, in general, external to the concrete itself. For more details on the chemistry of unsoundness of cements reference can be made to Lea.[3.35]

Phase Changes and Conversion

Phase changes which involve volume changes can cause disruption of concrete, and in particular the phase change of quartz on heating has been studied because of its importance in concrete subjected to extreme temperature changes. The conversion of one hydrate phase to another in concretes under more usual service conditions is less well understood and documented, but such processes are of importance in influencing the durability of concretes. The importance of complete understanding of such conversion processes by civil engineers has been revealed in the severe and costly problems which arose from the use of high-alumina cement in the UK in the 1970s.

High-alumina cement became available in the UK in the 1920s, and although more expensive to produce than Portland cements, it was recommended for use under conditions of high temperature and where sulphate attack was to be combated. The rapid strength gain associated with this cement allowed rapid turn-round times in precasting factories, and use, on this advantage, was made of high-alumina cement concretes from the 1950s until a roof collapse at a school in Stepney in 1974.[3.36] Following on from this event, extensive investigations of structures were carried out to assess the *in-situ* condition of the members manufactured with this cement.

It was known for some time prior to the use of this type of cement in precast structural sections that this cement could produce concretes which lost strength over a period of time due to a process of conversion.[3.37, 3.38] This conversion phenomenon follows the rapid hydration of monocalcium aluminate, the main constituent of high alumina cement, to form the monocalcium deca-hydrate (CAH_{10}). This hydrate slowly decomposes to form tri-calcium aluminate hexa-hydrate (C_3AH_6), aluminium hydroxide (AH_3) and water. All three components are normally present in a partially converted high-alumina cement concrete.

An extensive programme of research on high-alumina cement concretes was commenced in 1964 at the Building Research Establishment in England, and the final results of tests after 20 years of curing have now been published.[3.39] It has been concluded that high-alumina cement concretes with free water:cement ratios in the range 0.25 to 0.67 and cured at 20 °C all continued to lose strength up to the age of 20 years, when testing was discontinued. The strength approached a minimum obtained after three months curing in water at 38 °C, the conversion being accelerated by high temperatures and moist conditions. For structural appraisal purposes a fully converted strength should be taken to be 30 MPa below the one-day compressive strength. Under wet conditions at 20 °C, the degree of conversion may reach 100 per cent before the minimum strength is reached. For determination of strengths of such concretes, accelerated curing methods are recommended. Dry-stored cubes seem to have behaved similarly to the majority of high-alumina cement concretes found in floor joists and roof beams in that minimum strength appears to have occurred before full conversion. Wet concretes of this type have strengths 10 to 15 MPa lower than concretes of identical mixes and ages that have been kept in dry conditions.

Alkali Reactivity

According to research in the USA, alkali−aggregate reaction, which results in excessive expansion of concrete sections and leads to severe cracking thereof, can occur between the hydroxyl ions, present in the pores of hardened pastes of cements containing more than 0.6 per cent soda equivalent, and any reactive aggregate. The soda equivalent is calculated as the sum of the actual Na_2O content plus 0.625 times the K_2O content of the dry cement. The definition of a reactive aggregate is much less precise. Reactive silica occurs as opal, chalcedony and tridymite in certain cherts, siliceous limestones, andesites and rhyolites, but expansive reactions have been observed in a much greater range of rock types. Because of the complex nature of the problem not all workers in the field agree on detailed mechanisms of expansion; however, at this time the three mechanisms causing expansion are considered to be:

(*i*)   alkali−silica reaction — expansion being caused by depolymerization of the silica and uptake of water by alkali−silica products;
(*ii*)   alkali−aggregate reaction — expansion being caused by alkali exfoliation of phyllosilicates;
(*iii*)   alkali−carbonate reaction — expansion being caused by dedolomitization of meta-stable, calcium rich dolomite crystals in close proximity to clay particles.

In the case of the alkali−silica reaction which is most widely spread and best understood, alkali hydroxides in the hardened cement paste liquor attack the silica to form an unlimited swelling gel, which draws in any free water by osmosis and so expands, disrupting the concrete matrix. Expanding gel products exert internal stress within the concrete causing characteristic map cracking of unrestrained surfaces, but cracks may be directionally oriented under conditions of restraint imposed by reinforcement, prestressing or loading.

Cracking resulting from alkali−silica reactions was long held not to pose any structural threat to affected members. However, in the late 1970s examples of structural problems associated with the phenomenon were reported. In Australia at least one major prestressed concrete wharf is suffering from the effects of this problem to the extent that its structural behaviour requires constant monitoring. In Japan columns and crossheads of the pier to the Hashin Expressway in Honshu are causing concern, and in the UK certain structures as different as the A38 March Mills Viaduct and the Val de la Mare Dam in Jersey, have been subject to remedial work.[3.40] In the United States the progression of alkali−silica reaction in the Copper Basin, Gene Wash and Stewart Mountain concrete arch dams has been monitored over a period of 40 years.[3.41]

Structurally, consideration must always be given to the effects of deeply penetrating cracks on the durability of reinforcement and to the self-stress induced by the expansive reactions, which may be of advantage in confined sections of normally reinforced concrete members, but could prove catastrophic in the case of prestressed structures. A systematic procedure for structural assessment has been produced by a working party of the Institution of Structural Engineers (UK).[3.42] The British Cement Association has reviewed the subject of the diagnosis of alkali–silica reaction and has produced a detailed report.[3.43] Approaches outlining procedures may nevertheless need to be modified for certain classes of structure, as indicated by Hammersley for dams and other major water-retaining structures.[3.44]

At the present time avoidance of the problem prior to construction is of the greatest importance. This can be achieved by avoiding the use of reactive aggregates, by the use of low alkali Portland cement (provided that alkalis are not introduced during service), of slag cement or of pozzolanic admixtures. Once the problem does present itself, attempts at control are costly and have generally not been reported as being very successful. Indeed, for the most part, monitoring to ensure that failure is not imminent, rather than restoration, has been the objective of engineers confronted with the problem. As the swelling phenomenon is dependent on water-imbibition, most control measures are geared to decreasing water ingress to the concrete by the use of surface coatings or impregnation materials.

Volume Change

Sources of Volume Change

All structural materials are liable to change in volume as internal and external conditions change but of the commonly used materials only concrete and timber display such large and complex changes. This situation arises because these materials respond to both temperature and humidity effects and are, as a consequence, almost always in a state of dynamic dis-equilibrium with their environments. The effects which may be important in concrete are:

(i)   thermal movement;
(ii)   early shrinkage in plastic concrete;
(iii)   drying shrinkage and cyclic swelling and shrinkage;
(iv)   creep.

Thermal effects will be considered on pp. 72–82.

Early Shrinkage in Plastic Concrete

Early shrinkage is liable to lead to plastic and settlement cracking, but its effects can often be overcome by corrections to mix design and by the adoption of suitable construction practices. Plastic cracking arises when the removal of water from the top surface by evaporation exceeds the rate at which bleed water is coming to the surface. It is most prevalent in slabs. A mix which provides some bleeding is helpful in reducing plastic cracking. Protection of the surface from drying winds by the use of barriers and the earliest possible application of covering to the surface may also help. Plastic cracking is often confused with settlement cracking. Settlement cracking is much more widespread and can even be seen to occur under a film of bleed water, making it quite clear that it is not plastic cracking. The cracks occur because settlement is restrained by one or more mechanisms, usually:

(*i*)   reinforcing bars (see Fig. 3.6);
(*ii*) aggregate particles;
(*iii*) formwork restraint (see Fig. 3.7).

Settlement cracking is much increased if excessive retardation occurs. In a massive bridge crosshead in which an overdose of retarder had delayed setting for three days, cracks were observed over bars and prestressing ducts 1 m and more from the top surface. The reduction in shear strength in the crosshead was a serious consequence. Variations of drainage conditions, particularly in slabs on grade, have been shown to have a marked influence on the settlement of the concrete and hence on the potential for cracking.[3.45] It is common practice to trowel over

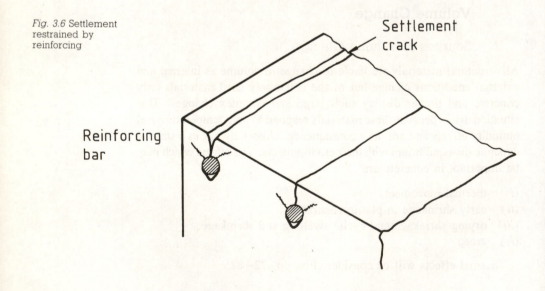

Fig. 3.6 Settlement restrained by reinforcing

Settlement crack

Reinforcing bar

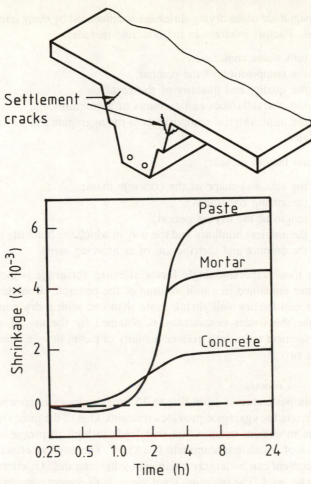

Settlement cracks

Paste

Mortar

Concrete

settlement cracks when they form, but this treatment is not adequate to close the cracks around the top reinforcing bars. These cracks remain as initiators for subsequent drying shrinkage cracks and can only be effectively closed by revibration at the latest possible time. The magnitude of plastic shrinkage may in extreme cases be as large as 10 000 microstrain[3.46] and has been shown by L'Hermite[3.47] as over 6000 for paste and 2000 for concrete (Fig. 3.8) but as the concrete is in a plastic state no great stress is induced and further working of the concrete can generally be applied to eliminate consequential cracks.

Drying Shrinkage

Drying shrinkage of concrete is caused by the contraction of the hydrated hardened calcium silicate gel when its moisture content is decreased.

The magnitude of the drying shrinkage is influenced by many inter-related factors. Factors inherent in the concrete include:

(*i*)   unit water content;
(*ii*)   the composition of the cement;
(*iii*) the quality and quantity of the paste;
(*iv*)   the characteristics and amounts of admixtures;
(*v*)   the mineralogical composition of the aggregate and its maximum size.

External factors include:

(*vi*)   the size and shape of the concrete mass;
(*vii*) the curing conditions;
(*viii*) length of the drying period;
(*ix*)   the ambient humidity and the way in which the humidity changes;
(*x*)   the amount and distribution of reinforcing steel.

The most important single factor affecting shrinkage is the amount of water contained in a unit volume of the concrete. Concrete with a wetter consistency will shrink more than one with a dry consistency because the wetter consistency is obtained by the use of a higher water:cement ratio, by a greater quantity of paste, or by a combination of the two.

Cement

Cement paste shrinks from five to 15 times as much as concrete since in concrete the aggregate provides restraint. One of the most significant factors in cement composition that leads to high shrinkage is a high content of tricalcium aluminate (C_3A).[3.48] The adverse effect of high C_3A content can be largely balanced by keeping the SO_3 content at an optimum level. The optimum level varies with cement composition and fineness and may unfortunately exceed the level permitted by cement specifications. Verbeck[3.49] came to the conclusion that the optimum level of SO_3 is not only dependent on the clinker composition but also on the temperature of early hydration. The higher the early hydration temperature the greater will be the SO_3 demand if minimum shrinkage is to be achieved. Using a particular clinker, Verbeck demonstrated that for an early hydration temperature of 23 °C the optimum SO_3 content for minimum shrinkage was 4 per cent, for 41 °C it was 5 per cent and for 4 °C only 3 per cent was required. The first two cases gave SO_3 contents higher than is permitted by many cement specifications.

It is generally believed among engineers that an increase in fineness of cement leads to an increase in shrinkage. Roper[3.48] and Swamy[3.50] both concluded that fineness does affect the shrinkage of cement paste but there is less evidence of this effect in concrete. No significant relationship between concrete shrinkage and cement fineness was found from tests on 200 commercial cements for different units in France. Nevertheless, in practice finer cements with unchanged composition will lead to more rapid shrinkage than coarser cements simply owing to the more rapid hydration.

Admixtures

As very little concrete is now placed without some admixture, it is important to have some information on the effect of admixtures on shrinkage. Morgan[3.51] found after an extensive investigation that the intrinsic effect of lignosulphonate water-reducing admixtures was to increase the early shrinkage of cement paste and of concrete. The short-term shrinkages (up to three months) were significantly increased by the simple addition of the admixture but the long-term effect was not so noticeable. When accelerators were added, the shrinkage was increased at all ages. These effects are shown in Table 3.5. The results are consistent with those obtained by Collepardi et al.[3.52] for mortars mixed at constant water:cement ratio as shown in Table 3.6. When the concrete mix containing a water-reducing admixture is adjusted to give the same workability and strength as the undosed concrete the apparent increase in shrinkage is much reduced and may be completely offset.

Table 3.5 Effect of admixtures on concrete drying shrinkage (after Morgan[3.51])

Mix	Details	Shrinkage (microstrain) after drying at 50% RH for period:		
		14 days	91 days	203 days
I	Control (cement: 307 kg m^{-3} W:C 0.55)	337	573	630
II	As I, plus lignosulphonate (0.200% × cement wt)	353	633	647
III	As II, plus 0.050% calcium chloride	377	643	653
IV	As II, plus 0.030% triethanolamine	397	673	687

Table 3.6 Drying shrinkage of mortars without and with 0.2% calcium lignosulphonate (CLS). Cured 1 day and dried at 50% relative humidity and 20 °C[3.52]

	Drying shrinkage (millionths)			
	7 days		90 days	
Cement type	Without CLS	With CLS	Without CLS	With CLS
Portland	530	740	910	900
Pozzolanic	630	780	930	940
Slag	500	620	890	890

Superplasticizers when used as water-reducers show similar effects to conventional water-reducers and in general easily meet the requirements of ASTM C494, which permits some increase of shrinkage over the reference concrete of comparable workability and cement content.[3.53] Flowing concrete produced by using superplasticizers also showed little change from conventional concrete.[3.54] The relationship between moisture loss and shrinkage for water-reduced superplasticized concrete is shown in Fig. 3.9, which indicates that for the same moisture loss the superplasticized concrete has larger shrinkage.[3.55] Air-entrained concrete, if made with the same strength and workability as non-air-entrained concrete, will have similar shrinkage.

The influence of cement replacements is complicated. Alexander has pointed out that the effect of fly ash on the drying shrinkage of paste is made up of three different parts.[3.56] Two opposing effects arise from replacing cement by an inert mineral powder of comparable fineness. The drying shrinkage increase associated with increasing the water content is balanced by the stabilizing influence of the mineral powder. If the powder has a shape as favourable as that of fly ash then the water demand will not be increased so much. The third effect from the cementitious or pozzolanic action of the fly ash further reduces the shrinkage, since the effective increase in cementitious material lowers the water:cement ratio. Investigations suggest that shrinkage may be reduced by up to 20 per cent compared with Portland cement concrete of the same strength and workability but the extent to which this reduction occurs (and it may not always occur) is dependent on the morphology, the chemical analysis and particle size of the fly ash and of the Portland cement in use. The shrinkage of concrete containing Portland-blast furnace slag cement has been reported to be somewhat greater than that of comparable concrete made without slag.[3.57] For Australian slag, Hinczak has reported a small increase in shrinkage (up to 11 per cent) in 40 MPa concrete up to 56 days and an even smaller reduction (7 per cent) at 365 days.[3.58] It has been suggested that this variation can be controlled by adjusting the SO_3 content of the blended cement but the influence is not definite. It seems likely that shrinkage of concrete made with Portland-blast furnace slag cement will not be very different from that of concrete of comparable strength and workability made with Portland cement.

Coarse Aggregate

Coarse aggregate has an important influence on the shrinkage of concrete. The aggregate acts as an elastic, and inelastic, restraint to the paste shrinkage and therefore aggregates with a high elastic modulus are more effective. Quartz, granite, feldspar and some basalts are generally regarded as producing low shrinkage while sandstones and shales give

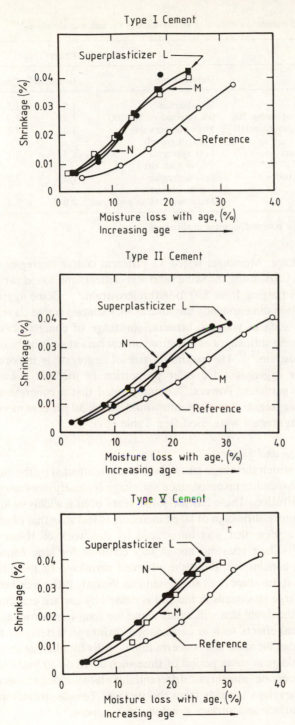

Type I Cement

Type II Cement

Type V Cement

Fig. 3.9 Shrinkage vs moisture loss for superplasticized concretes. L = modified ligno-sulphonate; M = melamine; N = naphthalene type (after Ghosh and Malhotra[3.55])

Table 3.7 Individual and cumulative effects of various factors on concrete shrinkage[3.60]

Factor		Effect*	
Favourable	Unfavourable	Individual	Cumulative
Cement of optimum SO_2	SO_2 deficiency	1.5	1.5
Cement with 15% retained on No 200	0% retained on No 200	1.25	1.9
Less compressible aggregate (quartz)	More compressible (Elgin gravel)	1.25	2.4
More aggregate ($1\frac{1}{2}$ in max size)	Less aggregate ($\frac{1}{4}$ in max size)	1.3	3.1
More aggregate (stiff mixture)	Less aggregate (wet mixture)	1.2	3.7
No clay in aggregate	Much bad clay in aggregate	2.0	7.4

* Multiplication factor for potential increase in shrinkage

high shrinkage. Meininger tested 13 different coarse aggregates from around the USA, using the same sand and cement, and found shrinkage at 180 days ranging from 530 to 990 microstrain.[3.59] Some aggregates are themselves dimensionally unstable as they contain active clays. With aggregates such as volcanic breccia, shrinkage of concrete has been doubled giving shrinkages as measured on test bars at six months of over 1200 microstrain.[3.48] The maximum size of aggregate is important as with larger aggregates a greater proportion of the mix consists of restraining particles. Powers[3.60] demonstrated that a concrete with a maximum aggregate size of 19 mm would shrink 30 per cent more than if 38 mm aggregate were used (see Table 3.7).

Size and Shape

The rate at which shrinkage takes place is much affected by the size and shape of the member involved since shrinkage is closely associated with moisture diffusion. There has for some years been a widely-held view that the ultimate shrinkage of large members is less than that of smaller members, a view that was introduced by the work of Hansen and Mattock.[3.61] Their tests did not in fact continue for long enough to justify any conclusions about the ultimate shrinkage of their largest specimens. It has since been suggested that the only difference between small and large specimens in the end is caused by surface cracking.[3.62] In view of the doubt that still exists about the long-term shrinkage and the additional effects such as carbonation shrinkage, a designer has to decide whether the value that concerns him is really the ultimate shrinkage or the shrinkage at some period of time such as 10 or 20 years hence. In large concrete members, differential volume changes occur as shrinkage develops inwards from the surface. Tensile stresses are set up at the surface and surface cracking may appear.

Curing and Ambient Conditions

Increase in the length of time of moist-curing generally has a rather small effect on the subsequent drying shrinkage of concrete. Prolonged curing may increase shrinkage and the immediate shrinkage that occurs when drying starts is larger after longer curing. Prolonged curing does, however, tend to increase tensile strength and hence reduces the tendency for cracks to be induced by the shrinkage. The rate and magnitude of drying shrinkage is greatly influenced by the local humidity in the surrounding air as illustrated by Helmuth and Turk.[3.63]

Wetting and Drying

Cement paste, and therefore concrete, shows swelling and shrinking when subjected to cycles of wetting and drying. The first drying has an irreversible component and the movements with subsequent wetting and drying are reversible. Measurements on pastes are shown in Figs 3.10 and 3.11.[3.64] These figures show that the irreversible shrinkage and the total first shrinkage are functions of the porosity of the paste, whereas the first swelling is independent of porosity. For concrete similar relations probably hold in that the effects of the first few drying cycles are dependent on the drying conditions but the length changes after that are reversible.

Carbonation Shrinkage

Figure 3.12[3.64] shows that carbonation has a marked effect on the shrinkage of paste and is much affected by the relative humidity of the surroundings. Carbonation proceeds slowly in dense concrete and

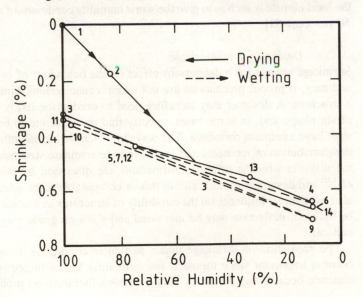

Fig. 3.10 Reversibility of volume changes of pastes with relative humidity after first drying (after Verbeck and Helmuth[3.64])

Fig. 3.11
Dependence of
components of
drying shrinkage
on paste porosity
(after Verbeck and
Helmuth[3.64])

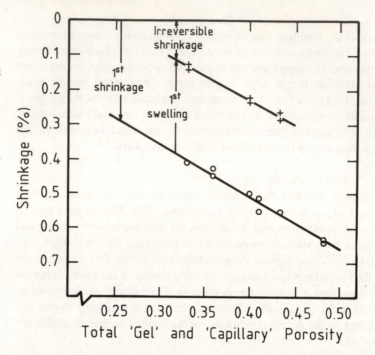

the influence of carbonation on shrinkage would certainly be less than indicated for paste. In porous concrete, where carbonation can proceed rapidly to considerable depths, the additional shrinkage due to carbonation may add to distress already caused by drying shrinkage, particularly if the local climate is such as to give the worst humidity conditions of around 50 per cent RH.

Designing for Shrinkage

Shrinkage always has a deleterious effect on the behaviour of concrete and may, if proper precautions are not taken, cause serious damage to a structure. A designer may therefore need to predict the likely extent of shrinkage and, in some cases, specify that the concrete to be used shall have minimum shrinkage. The ways in which shrinkage influences the distribution of moments and forces in indeterminate structures and the ways in which it affects deformations are discussed by Rusch *et al.*[3.65] and designers should refer to this or comparable texts. Shrinkage also has a direct influence on the durability of structures as cracking may be induced, deflection may be increased and slabs on grade may warp and curl.

The prediction of shrinkage under particular conditions of use has become important since the need for estimating losses in prestressed concrete became significant. A very extensive literature on prediction

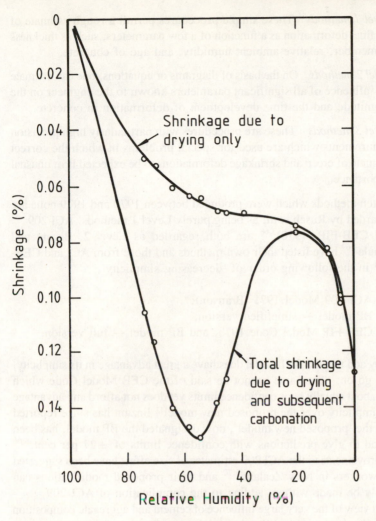

Fig. 3.12 Contributions of drying and carbonation to total mortar shrinkage (after Verbeck and Helmuth[3.64])

methods has developed and many models have been proposed. As pointed out by Bazant et al.,[3.66] 'the model uncertainty, due to the error of the shrinkage formula, is inevitable since without a mathematical model no statistical evaluation is possible. For the formulas used in current codes, this uncertainty is huge. Statistics involving many thousands of data points from the literature have recently been reported, and it has been found that the confidence limits that are exceeded by the errors with a 10 per cent probability are about ±86 per cent of the predicted value for the current ACI 209 model, and ±118 per cent for the current CEB-FIP Model Code.'

Methods of prediction have been conveniently classified by three levels:

Level 1 methods These simple procedures permit a rough estimate of the final deformation as a function of a few parameters, such as thickness of member, relative ambient humidity, and age of concrete.

Level 2 methods On the basis of diagrams or equations, one can estimate the influence of all significant parameters known to the engineer on the magnitude and the time development of deformation in concrete.

Level 3 methods These are procedures with particularly high precision requirements which are used for special problems in which the correct estimate of creep and shrinkage deformations to be expected is of unusual importance.

Of ten methods which were produced between 1970 and 1979 none are regarded by Rusch *et al.* as being purely Level 1 methods. ACI 209[3.67] and CEB-FIP (1978)[3.68] are both regarded as Level 2. Bazant and Panula[3.69] have listed their own methods and those from ACI and CEB-FIP in the following order of 'decreasing simplicity':

1.   ACI 209 Model 1971 (Branson);
2.   BP model — simplified version;
3.   CEB-FIP Model Code 1978, and BP model — full version.

They add that the ACI model does have a great advantage in its simplicity, and go on: 'The same cannot be said of the CEB Model Code which has about equally poor confidence limits yet does not afford any advantage in simplicity over the proposed new model'. Bazant has since reported that the 'proposed new model', now designated the BP model, has been found to give predictions with confidence limits of ± 27 per cent.[3.66] Improvements to the ACI 209 estimates of size effect have been suggested by workers in New Zealand[3.70] and their proposed modifications can easily be made without adding to the complication of ACI 209.

In view of the very large influence of cement and aggregate composition and the lesser, but still substantial, contribution of admixtures and cementitious materials other than Portland cement, none of which are addressed by conventional prediction formulas, there seems little to be gained by using the more complicated expressions. There is always a risk that the designer may be led to believe that because one has used a more elaborate expression, and thereby incurred greater design cost, one finishes with a more precise prediction. There is no real support for this belief. If a designer has no, or only slight, knowledge of the materials that are to be used, the simplest possible prediction is to be recommended and the approach of AS 1481 has much to commend it.[3.71] In that code, the shrinkage is calculated from the expression

$$\epsilon_r = \epsilon_b \cdot k_e \cdot k_h$$

where: k_e is a function of the theoretical thickness and k_h is a function of time and theoretical thickness.

Values to be used for k_e and k_h are shown in Figs 3.13 and 3.14. If the shrinkage is required to be known with a precision of better than ± 40 per cent, tests on the actual materials to be used are prescribed. In other cases, the value of ϵ_b is given for four different environmental conditions as shown in Table 3.8. The values are only appropriate if those aggregates which lead to high shrinkage, such as volcanic breccia, are excluded.

AS 3600, the successor to AS 1480 and AS 1481, has complicated the predictions but it has been shown that they are no more precise than those of its predecessor.[3.72]

If minimum shrinkage is mandatory for any job, the designer can specify shrinkage limits on the concrete when tested in accordance with a standard method. Alternatively one can address those items which could contribute to shrinkage being a minimum, namely:

(*i*)   the use of the lowest cement content, and hence often the lowest compressive strength, compatible with other requirements;

(*ii*) the use of a type of aggregate known to reduce shrinkage;

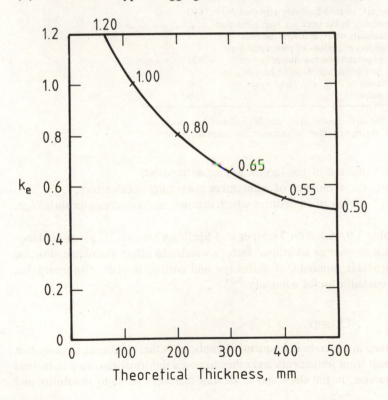

Fig. 3.13 Shrinkage coefficient for theoretical thickness (from AS 1481–1978)

Fig. 3.14 Shrinkage
coefficient related
to time (from
AS 1481–1978)

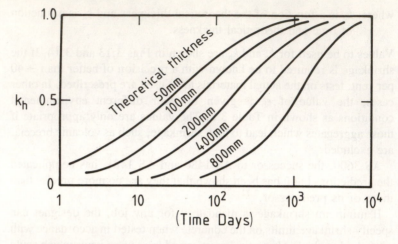

Table 3.8 Shrinkage related to the environment. Values
from AS 1481–1978

Environment*	ϵ_b, Shrinkage ($\times 10^6$)
Dry air, relative humidity less than 50%	600
Generally in the open air with a relative humidity of 50 and 75% and not subject to periods of prolonged high temperature or low humidity	400
In very humid air, relative humidity in excess of 75%, e.g. over water	200
In water	0

* The environment in air conditioned buildings may, in
the absence of other information, be taken as dry air

(*iii*) the use of the largest aggregate possible;
(*iv*) the avoidance of admixtures containing accelerators;
(*v*) the use of admixtures which demonstrate no increase in shrinkage.

Table 3.9, based on Tremper and Spellman's work,[3.73] gives guidance
on a number of additional factors which can affect shrinkage. For the
particular problems of shrinkage and curling in slabs, Ytterberg has
prepared a useful summary.[3.74]

Creep

Creep in concrete is frequently beneficial in that it relieves stresses that
result from settlements and other imposed deformations. In prestressed
concrete, in flat slabs and in slender members liable to instability and

Table 3.9 Cumulative effect of adverse factors on shrinkage[3.73]

Effect of departing from use of best materials and workmanship	Equivalent increase in shrinkage (%)	Cumulative effect
Temperature of concrete at discharge allowed to reach 80 °F (27 °C), whereas with reasonable precautions, temperature of 60 °F (16 °C) could have been maintained	8	$1.00 \times 1.08 = 1.08$
Used 6−7 in (150−180 mm) slump where 3−4 in (75−100 mm) slump could have been used	10	$1.08 \times 1.10 = 1.19$
Excessive haul in transit mixer, too long a waiting period at job site, or too many revolutions at mixing speed	10	$1.19 \times 1.10 = 1.31$
Use of $\frac{3}{4}$ in (19 mm) max size aggregate under conditions where $1\frac{1}{2}$ in (38 mm) could have been used	25	$1.31 \times 1.25 = 1.64$
Use of cement having relatively high shrinkage characteristics	25	$1.64 \times 1.25 = 2.05$
Excessive 'dirt' in aggregate due to insufficient washing or contamination during handling	25	$2.05 \times 1.25 = 2.56$
Use of aggregates of poor inherent quality with respect to shrinkage	50	$2.56 \times 1.50 = 3.84$
Use of admixture that produces high shrinkage	30	$3.84 \times 1.30 = 5.00$
Total increase	Summation = 183%	Cumulative = 400%

buckling, creep may be harmful and the advantages of low creep concrete might be considered by the designer in these circumstances. There is a substantial literature relating to the influences of creep on structural behaviour and on ways of predicting creep and reference should be made to these texts if creep is likely to cause trouble.[3.65, 3.75]

A useful summary of the effects of shrinkage and creep on structural behaviour is given by Rusch et al.[3.65] and the rest of this section is based on that material. The effects of creep can be described in terms of the creep factor, ϕ, which is the ratio of the creep strain under constant load to the elastic strain under the same load. The description of the effects can be separated into those induced by loads (load effects) and those induced by imposed deformations (imposed effects).

For plain concrete, load effects which may be calculated by first-order theory do not change under the effect of creep. Imposed effects are reduced through creep by a factor of about $1/2\phi$ when the effects are generated rapidly; when they develop only gradually — as in shrinkage — this reduction factor is increased by a constant amount of about 0.2.

The deformation caused by a sustained load effect is increased by a factor of $(1+\phi)$ over the course of time, and must be included in any second-order analysis.

In reinforced concrete load-induced effects are the same as for plain concrete. When simplified calculations of uncracked sections are carried out to determine the imposed effects, which may be legitimate if only isolated cracks are expected, one must apply the rules for plain concrete. Imposed effects calculated for cracked sections are considerably reduced by crack formation. Creep causes further reduction. For calculations based on uncracked sections, a factor of about $1/(1+0.3\phi)$ may be used for a rapidly developing imposed action. In the case of an imposed action caused by shrinkage or slow settlement, the factor becomes about $1/(1+0.2\phi)$. For calculations based on cracked sections, the corresponding reduction factors are $1/(1+0.1\phi)$ and $1/(1+0.08\phi)$. Deformations caused by sustained load effects are increased over the course of time by a factor of only $(1+0.3\phi)$. Creep sensitivity is so much lower than in plain concrete that the effect of imposed effects on deformation can be disregarded.

Thermal Properties

Structural Effects of Temperature

In this section attention will be paid to temperature changes and the thermal movements that arise in concrete as a result of temperature change. The response of concrete to high or low external temperatures, which may arise from intentional or accidental usage, is discussed in Chapter 4, pp. 113–22.

A great deal of the literature discussing temperature rises in concrete construction is based on the experience of dam builders.[3.76] Dam construction, however, differs from other forms of massive concrete construction in a number of ways. The strengths, and hence the cement contents, in a dam tend to be low and not to be required until later ages. The blocks of a dam have large dimensions in plan so that the extent of the formed surfaces is comparatively small and the large top surface of each lift is exposed to the ambient conditions. It has now been realized that temperature rises, and particularly those that occur at an early age, may be responsible for a great deal of early cracking in structures which could not be described as 'mass concrete' structures. Nevertheless, the only section of the ACI Committee 224 report of 1980 which considers temperature effects appears in the chapter entitled 'Control of cracking in mass concrete' where the authors state that 'although a large amount of the data for this chapter has been obtained by experience gained from the use of mass concrete in dams, it applies equally well in mass concrete used in other structures such as steam power plants, powerhouses, bridge and building foundations, navigation locks, etc.'[3.77] BS 8007 'The design of concrete structures for containing aqueous liquids' and its predecessor BS 5337 provide for the existence and control of early thermal strain and cracking in immature concrete. The code considers the thermal gradient that occurs in a surface zone which is 150 mm thick and therefore applies to structures which would certainly not normally be regarded as mass concrete.[3.78]

The process of hydration of cementitious materials releases heat which raises the temperature of the concrete. This heat must eventually be lost to the atmosphere and the concrete temperature has to reach equilibrium with the long-term atmospheric conditions. During this period, two structural effects, which are effectively independent, may occur. At early ages, say one to three days, temperature gradients develop in the concrete as the internal temperature is raised above the surface temperature of the concrete member. This surface temperature is dependent on the material in contact, which may be older concrete, soil, formwork, insulation and curing materials or air. The resulting temperature gradient will produce tensions in the surface and may be sufficient to cause cracking. The application of cold curing water and the use of wet hessian, which is cooled by evaporation, may exacerbate the gradient. After some days, the second effect begins to operate as the mean temperature of the member may remain above that of connecting members and the subsequent cooling will induce tensions which are similar to those induced by restrained shrinkage.

Heat Generation

Heat generation from the hydration of cementitious materials has been commonly reported as the total heat in the first three days and the first seven days, as for example by Bogue.[3.79] The maximum temperature in a concrete member such as a slab may occur at 20 hours or even earlier and therefore calculations of thermal response require information on the rate of temperature rise and its variation during the first day after adding water. An immediate difficulty arises in that the rate of reaction is dependent on the temperature at which the reaction is occurring. Mather,[3.80] discussing the observation by Ragan that higher temperatures produced a noticeable slump loss, recorded this conversation:

> Ragan said 'I don't know how to explain this.' I said: 'I do, the cement hydrates faster when the concrete is warmer.' He said: 'That much, that quick?' I said: 'Yes, there's a great spurt of reaction almost instantly upon wetting the cement with water — see Powers, (Fig. 3.15).[3.81] Also all your instantaneous values of slump were at least 8 min after the water hit the cement.'

Fig. 3.15 Within
the first few
minutes after
cement is mixed
with water, the
hydration process
begins (from
Mather[3.80] after
Powers[3.81])

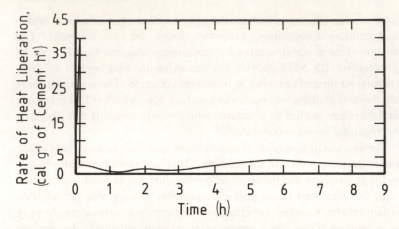

Fig. 3.15 Within the first few minutes after cement is mixed with water, the hydration process begins (from Mather[3.80] after Powers[3.81])

Some curves showing the rate of heat liberation from Portland cements and from Portland-blast furnace slag blends have been obtained by Roy and Idorn[3.82] and are shown in Figs 3.16−19. Samples were prepared by blending Type I/II Atlantic Portland cement and granulated blast furnace slag from Sparrows Point, USA. The figures show the rate of heat evolution of the mixtures hydrated at 15, 27, 38 and 60 °C. The first thermal peak which occurs in the first few minutes is not shown but is included in the value for the total heat evolved. The authors point out that 'it is obvious that the second peak consists of a single peak in the Portland cement, while the slag cements show two peaks, particularly when hydrated at 27 and 38 °C'. They comment that the data seem to present clear evidence of substantial slag hydration already within the first 24 hours. Similar early hydration reactions have been deduced from temperature measurements in test blocks containing Portland cement and fly ash concrete.[3.83] These observations are not consistent with the widely held view, reported by Mehta,[3.84] that mineral admixtures such as natural pozzolans, fly ash and blast furnace slag have the potential of reducing the temperature rise almost in proportion to the amount used as Portland cement replacement, on the grounds that they generally do not react to produce any heat of hydration for several days. This anomaly is explained by Roy and Idorn as arising from the predominant use of low-temperature isothermal conditions in research and testing, which inevitably cause slag cements to be characterized as low-heat, slow strength-developing cements compared with ordinary Portland cement. Under conditions which are conducive to high temperature rises, such as rich mixes, high ambient temperatures, thick sections, or insulated formwork, with many mineral admixtures there may be only minimal reduction in temperature possible.

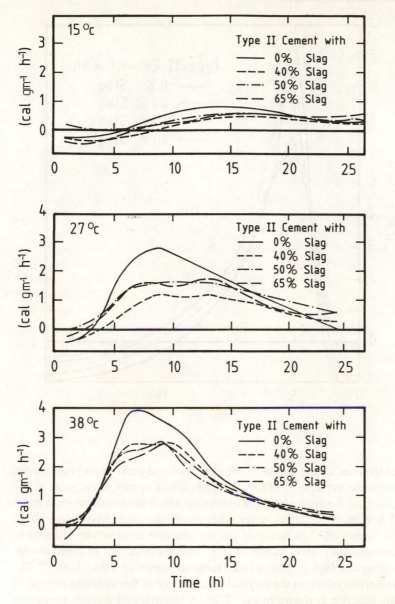

Fig. 3.16 Rate of heat liberation of cements hydrated at 15 °C (after Roy and Idorn[3.82])

Fig. 3.17 Rate of heat liberation of cements hydrated at 27 °C (after Roy and Idorn[3.82])

Fig. 3.18 Rate of heat liberation of cements hydrated at 38 °C (after Roy and Idorn[3.82])

Conductivity and Diffusivity

If calculations of temperature distribution in a concrete mass are to be carried out, it is necessary to obtain values for the thermal conductivity and specific heat (or thermal diffusivity) of the concrete mass. The thermal conductivity of hardened concrete is dependent on the richness

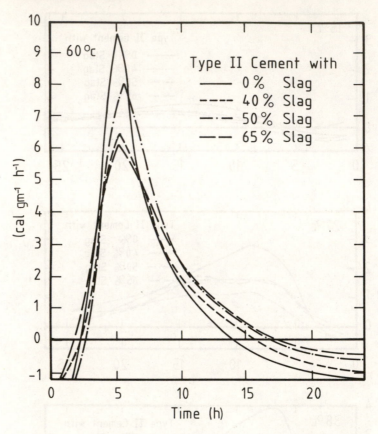

Fig. 3.19 Rate of
heat liberation of
cements hydrated
at 60 °C (after Roy
and Idorn[3.82])

of the mix, on the nature of the aggregate and most importantly on the
moisture content.[3.85] The aggregate effect is not very great as by
changing the aggregate from diorite, of which the conductivity is about
$2.3 \ W \ m^{-1} \ K^{-1}$, to steel shot with a conductivity of about
$52 \ W \ m^{-1} \ K^{-1}$ (a factor of 20) the concrete conductivity is found to
increase only by a factor of 3.7. The conductivity of conventional
aggregates falls in quite a small range as shown in Table 3.10.[3.86] The
important effect on the thermal conductivity of the moisture content of
the concrete is shown in Fig. 3.20. A summary of thermal properties
of concrete was prepared by Zoldners[3.87] but the large number of
different units used in the various references quoted, and in texts in
general, make it difficult for the reader to make use of the data provided.
Table 3.11 has been prepared to help to overcome this difficulty.

The thermal conductivity of any concrete can be calculated from the
conductivity of the coarse aggregate, the mix proportions and the
conductivity of the mortar using the expression:

Table 3.10 Thermal conductivities of
some common materials
(after Birch *et al.*[3.86])

Material	Thermal conductivity (Wm^{-1} K^{-1})
Basalt	1.4
Diorite	2.2
Granite	3.5
Quartzite	3.1–5.4
Shale	0.9
Limestone	0.9–2.6
Mercury	6.2
Cast iron	45
Mild steel	61
Copper	382
Pure water	0.6
Impure water	0.5

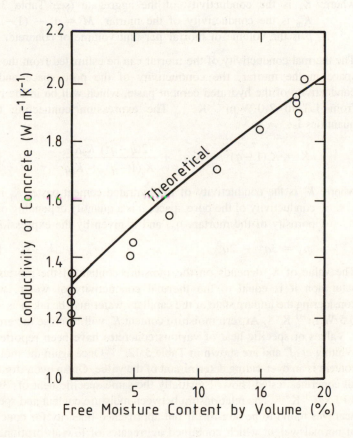

Fig. 3.20
Theoretical and
experimental
variation of
conductivity of
dolerite concrete
with moisture
content (after
Campbell-Allen
and Thorne[3.85])

Table 3.11 Units for thermal properties

Property	SI unit	To convert SI unit to:	Multiply by:
Heat generation	$kJ\ kg^{-1}$	cal/gm	$\times\ 0.239$
Rate of heat generation	$W\ kg^{-1}$	$kJ\ kg^{-1}\ h^{-1}$	$\times\ 3.6$
		cal/gm/hr	$\times\ 0.860$
Thermal conductivity	$W\ m^{-1}\ K^{-1}$	Btu/ft/hr/deg F	$\times\ 0.578$
		kg-cal/m/hr/deg C	$\times\ 0.860$
		cal/cm/sec/deg C	$\times\ 0.239$
Specific heat	$kJ\ kg^{-1}\ K^{-1}$	Btu/lb/deg F	$\times\ 0.239$
Thermal coefficient of expansion	K^{-1}	per deg F	$\times\ 0.556$
Density	$kg\ m^{-3}$	lb/cu ft	$\times\ 0.0634$
Thermal diffusivity	$m^2\ s^{-1}$	sq ft/hr	$\times\ 38.7\ \times\ 10^3$

$$K = K_m(2M - M^2) + \frac{K_m K_a (1-M)^2}{K_a M + K_m(1-M)} \qquad [3.4]$$

where: K_a is the conductivity of the aggregate (see Table 3.10), K_m is the conductivity of the mortar, $M = 1 - (1-p)^{1/3}$, p is the volume of mortar per unit volume of concrete.

The thermal conductivity of the mortar can be estimated from the pore space in the mortar, the conductivity of the pore space and the conductivity of the hydrated cement paste, which will be in the range from 1.3 to 2.0 W m^{-1} K^{-1}. The expression connecting these quantities is:

$$K = K_c(1-q)^2 + K_s q^2 + \frac{2(q-q^2)\ K_c K_s}{K_s(1-q) + K_c q} \qquad [3.5]$$

where: K_c is the conductivity of the hydrated cement paste, K_s is the conductivity of the pore space, q is a quantity depending on the porosity of the mortar, p_0, and is given by the expression:

$$p_0 = 3q^2 - 2q^3 \qquad [3.6]$$

The value of K_s depends on the moisture content of the mortar. At saturation it is equal to the thermal conductivity of water, which considering the impure state of the capillary water may be taken as about 0.5 W m^{-1} K^{-1}. At zero moisture content K_s will be close to zero.[3.85]

Values of specific heat of various concretes have been reported by Whiting *et al.* and are shown in Table 3.12.[3.88] Once again the moisture content is an over-riding determinant of the value. On an oven-dry basis all concretes tested showed virtually the same specific heat of 795 ± 40 J kg^{-1} K^{-1}. The relationship between moisture content and specific heat was found to be non-linear but approached linearity for concretes of normal weight which contained aggregates of low absorption.

Table 3.12 Specific heats of concretes (kJ kg^{-1} K^{-1})[3.88]

Description of mix	SSD	Oven-dry	Normally-dry
Normal weight — siliceous sand/gravel	0.925	0.757	0.791
Normal weight — dolomitic crushed stone	1.075	0.845	0.895
Structural lightweight — expanded shale/sand	1.151	0.833	0.904
Lightweight — expanded shale coarse and fine	1.301	0.849	0.950

The thermal diffusivity is calculated from the thermal conductivity, K, the specific heat, c, and the density, ρ, by the expression:

$$\text{Diffusivity} = K/c\rho \qquad [3.7]$$

For temperature calculations at early ages the concrete will be saturated and the appropriate values of thermal properties should be used.

Thermal Movement

There is a considerable variation in the values of the coefficient of linear thermal expansion of concrete, α, reported by different investigators and consequently standards and codes of practice differ in the values to be used in design. Mitchell[3.89] reports values of α ranging from 6.3 to 11.7 $\times$ 10^{-6} °C^{-1}. Dettling[3.90] reports α as 5.5×10^{-6} °C^{-1} for a water-saturated lean limestone concrete and 14×10^{-6} for an air-dry concrete made with quartzose aggregate and a high cement content. When a precise value is not required, α is often taken as 10×10^{-6} °C^{-1}.

The wide divergence is due firstly to the influence of the coefficient of expansion of the coarse aggregate. The main factor influencing this quantity is the amount of quartz present in the rock. Rocks with a high quartz content, such as quartzite and sandstone, have the highest coefficients averaging about 12×10^{-6}.

Rocks containing little or no quartz, such as limestone, have coefficients of about 5×10^{-6}.[3.87] Furthermore the coefficients for particular rocks need not be isotropic and depend on the crystallographic orientations of the minerals, which themselves may show preferred orientations in the rock. The coefficient of expansion of mortar is dependent on the mineralogical composition of the sand, the ratio of sand to cement, on the type and composition of the cement used, on the water:cement ratio and on the degree of saturation of the paste. Typical influences of the first two factors are shown in Fig. 3.21 due to Harada.[3.91] The coefficients of expansion of hydrated cement pastes vary with the type and composition of cement used and are generally rather higher than

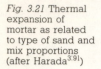

Fig. 3.21 Thermal
expansion of
mortar as related
to type of sand and
mix proportions
(after Harada[3.91])

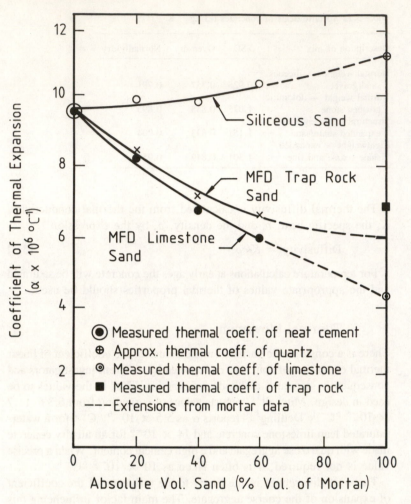

those of common aggregates. Verbeck and Helmuth[3.64] showed that thermal movements in hardened pastes are affected by moisture redistribution, a process that depends very much on the rate at which the temperature change takes place. A paste with a value of α of 13.7 $\times 10^{-6}$ in equilibrium conditions had a contraction coefficient of 29 $\times 10^{-6}$ when rapid cooling was applied. When the cooling ceased, some expansion occurred at the constant temperature reached, as shown in Fig. 3.22. Subsequent heating of the paste reversed the process and faster rates caused greater lags in length change. The value of α for cement paste depends therefore on the rate of heating or cooling and is much lower when the temperature change rates are slower. Hardened cement pastes that have been completely dried show no time-dependent

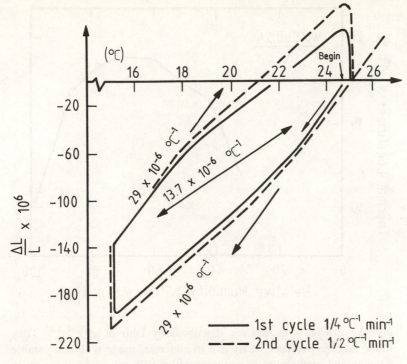

Fig. 3.22 Effect of redistribution of moisture on paste volume after rapid temperature changes (after Verbeck and Helmuth[3.64])

cooling or heating effects. Values of α between 10.5 and 12.4 $\times 10^{-6}$ were found with the higher value corresponding to more completely hydrated cement paste.

Three different coefficients may be considered. The instantaneous coefficient α_i is used to measure a process which is too rapid for any appreciable redistribution of moisture to take place. A coefficient α_w measures a process that is slow enough for continuous redistribution of moisture to occur, but without change in the total moisture content. The value of α_w depends on the ambient humidity and is higher at intermediate relative humidities than at either saturation or extreme dryness, as shown in Fig. 3.23. For a process in which hygral equilibrium is maintained in a paste that is saturated and has access to water a coefficient α_{sat} is proposed. Zoldners[3.87] shows values of the coefficient α_{sat} ranging from 9.0 to 21.6 $\times 10^{-6}$ depending on cement fineness, cement composition, water:cement ratio and moisture content. He also suggests that the value becomes smaller with time as the cement gel ages, a conclusion which is contrary to that of Verbeck and Helmuth. Zoldners' results, drawn from Mitchell[3.92] and Meyers[3.93], are shown in Fig. 3.24.

The influence of the addition of fly ash and of superplasticizers to

Fig. 3.23 The thermal expansion coefficient varies with the relative humidity within the paste (after Verbeck and Helmuth[3.64])

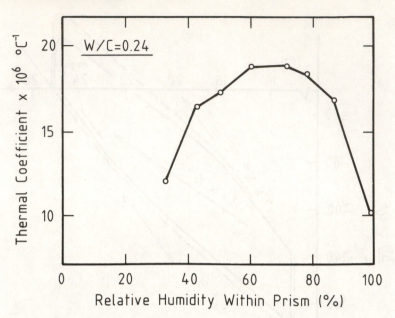

produce flowing concrete are discussed by Dhir *et al.*[3.94, 3.95] They found that the presence of fly ash in concrete, made with comparable strength and workability to concrete without fly ash, did not seriously affect the thermal expansion. For concrete at very low temperatures refer to work by Planas and Corres.[3.96–98] Values of α for light-weight and insulating concretes, cast in place, are given in Table 3.13.[3.99]

Cracking

All reinforced structural concrete contains cracks; indeed the reinforcement cannot work effectively otherwise. This is the natural condition. Concrete is also liable to crack in both the plastic and hardened states owing to the internal stresses that arise from the response of the constituent materials to their environment. The construction industry does not readily accept that intrinsic cracks are equally part of the nature of concrete and consequently concrete containing harmless cracks may be unjustifiably condemned. Hill[3.100] in an entertaining and relevant article points out that 'cracks in concrete are a highly emotive subject'. He quotes a case where 'the client's representative had resorted to saturating the concrete with water and, as it dried out, marking the locations of the 'cracks' before they became invisible to the naked eye. Their positions were then drawn with a firm pencil line on a sketch and presented as necessary remedial work. The pencil line was many times thicker than the actual cracks.'

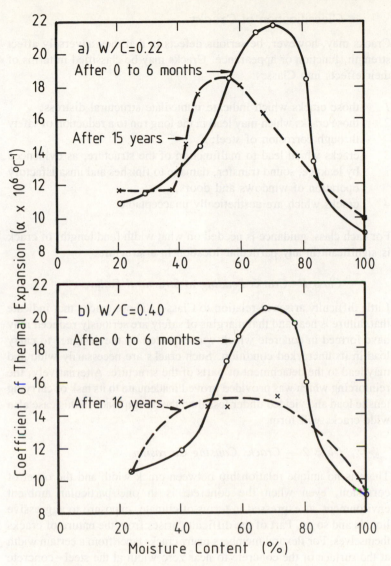

Fig. 3.24 Effect of age and moisture content on thermal expansion coefficient of hardened paste (after Zoldners[3.87])

Table 3.13 Thermal expansion ranges for low density concretes (after *ACI 523.1R*[3.99])

Concrete type	Oven-dry unit weight (kg m^{-3})	Coefficient of thermal expansion ($\times 10^6$ °C^{-1})
Perlite	320–640	7.7–11.0
Vermiculite	240–640	8.3–14.2
Cellular	240–640	9.0–12.6
Polystyrene beads	545–770	9.0–14.2

Classification of Cracks

Cracks may, however, be serious defects in that they adversely affect strength, function or appearance. Cracks may be classified in terms of their effects into Classes:

1   those cracks which indicate immediate structural distress;
2   those cracks which may lead in the long run to a reduction of safety through corrosion of steel;
3   cracks which lead to malfunction of the structure, as evidenced by leakage, sound transfer, damage to finishes and unsatisfactory operation of windows and doors;
4   cracks which are aesthetically unacceptable.

For each class, guidance is needed on what width (and length) of crack is significant in any particular location in a structure.

Class 1 - Cracks Leading to Structural Failure

Little difficulty arises in relation to Class 1. Those cracks that indicate that failure is near and that margins of safety are seriously reduced may have formed in concrete which was expected by the designer to carry load in its uncracked condition. Such cracks are necessarily wide and may lead to the detachment of parts of the structure. Alternatively, the reinforcing which was provided proved inadequate to its task of carrying tensile load and yielded under less than the design load. In this case also wide cracks will form.

Class 2 — Cracks Causing Corrosion

There is no unique relationship between crack width and the onset of corrosion, even when the concrete is in one particular ambient environment, as expressed in terms of climate, exposure to aggressive liquids and so on. Part of the difficulty arises from the nature of cracks themselves. For flexural members many cracks taper from a certain width at the surface of the concrete to near zero width at the steel—concrete interface. This hypothesis is supported by the extensive measurements made by Base and his colleagues on beams and may possibly apply to other flexural members such as slabs.[3.101] However, flexural cracks that are controlled by the overall depth of the beam are not of the tapered shape and it is likely that cracks due to temperature and shrinkage are nearer to being parallel sided.[3.102] It has been assumed for many years that, since wider cracks would give easier access to aggressive substances, corrosion could be controlled by controlling crack widths and that permissible widths should be a function of how aggressive the environment was. Many complicated formulas for the calculation of crack widths in flexural members have been devised with the object of controlling corrosion. But extensive tests on beams in which the cracks are normal to the axis of the bars (that is, the bars run through the cracks) show no evidence of any relationship between corrosion damage and crack width.[3.103, 3.104] As Beeby points out, the nature of the corrosion process and the nature of cracking give strong theoretical reasons for believing that such a relationship would be unlikely.[3.105] Halvorsen has reviewed a large number of codes of practice and suggests that, for many, corrosion control consists of unsubstantiated limitation and blind reliance on flexural crack width calculations. He recommends removing the link between exposure and limiting crack width so that any crack width limit would then only be concerned with appearance.[3.106] When cracks run along a bar, much more of the bar is in an exposed position and it might be expected that there would be a closer relationship between crack width and corrosion in this situation. There is little evidence however that cracks, whether transverse to the bars or running along the bars, pose any great risk of increased corrosion if they are less than 0.3 mm in width (see Chapter 5, pp. 179−86, 198−9).

Some cracks which are parallel to a bar may have been caused by the corrosion of that bar. These cracks will widen as corrosion proceeds and will eventually lead to spalling and exposure of the corroded bar. A crack of any width, which is judged to be brought about by corrosion, is an indication of a deteriorating structure and therefore no minimum width, below which the crack is not significant, can be set. There are of course some structures from which reinforcement can be safely removed at local points without endangering the structural capacity. The general reinforcement in a shell structure is such a case. For the majority of designed reinforced concrete structures, however, bars are placed for a specific purpose and the removal of one bar will reduce the safety of the structure. A crack that indicates that a bar has started to corrode is therefore a portent that the corrosion process will continue unless positive means are used to check it. Merely filling the crack will not achieve this result.

Class 3 — Cracks Affecting Function

The cracks in Class 3 which have the most serious consequences are those that allow liquid-retaining structures to leak or that occur in roofs or other structures intended to be waterproof. BS 8007 prescribes limiting crack widths and details methods of predicting the widths. The maximum design surface crack width for direct tension and flexure or restrained temperature and moisture effects are:

severe or very severe exposure — 0.2 mm
critical aesthetic appearance — 0.1 mm.

There are only limited test data available on what constitutes the limiting crack for preventing leakage. Flow through a parallel-sided smooth crack can be calculated in terms of head, crack width, crack length and fluid viscosity. The difficulty with concrete is that the cracks are not smooth or parallel-sided. Clear has suggested that an effective width can be used in the formulae for flow and that this effective width is about one-half the measured width.[3.107] He also found that if the effective width was less than 0.2 mm the cracks would seal themselves and leakage would stop in a few days. Anchor *et al.*[3.108] point out that the requirements for cracking in BS 8007 represent a very significant relaxation of the requirements contained in the previous code (BS 5337) and assume in effect that cracks which do not exceed 0.2 mm can heal autogenously — given favourable conditions — before the structure is brought into service. For many years, concrete tanks designed by the Engineering and Water Supply Department of South Australia have been designed with comparatively high steel stresses, which allow early leakage. These cracks subsequently heal and although they may be regarded by some as aesthetically objectionable the tanks perform their function of retaining water successfully. The bonding material formed during this 'autogenous healing' was found to be calcium carbonate and calcium hydroxide crystals, with no amorphous hydrated products of cement present. Water is essential for the development of maximum bond and it appears that the concrete should remain continually damp for the strength that is reached to be maintained.[3.109] Healing will not occur if the crack moves during the healing period and it will not occur if the flow of water is sufficiently great to wash out the deposits as they occur and before they can carbonate.[3.110]

Bridge-decks that crack may suffer from poor riding quality or increased vibration. In the case of deck slabs supported on steel girders, cracks may lead to corrosion of the girders as a result of leakage, particularly when accompanied by de-icing salts. There are indications that through-cracks that reach a width of 0.4 mm are likely to leak.[3.111] However, on bridge-decks, cracks once formed tend to grow in length and width under the influence of traffic and penetrating dirt. In those structures in which cracks are eventually going to be deleterious, a quite narrow crack, perhaps only 0.2 mm or 0.1 mm in width, may well be significant.

Class 4 — Cracks Affecting Appearance
For Class 4 cracks, it has been suggested that crack widths up to 0.3 mm in width are acceptable aesthetically but there are no good guidelines.

Various attempts have been made to establish what constitutes an acceptable crack on an aesthetic basis, but in the end there is no rational basis for aesthetic decisions. The aesthetic objection to cracks may be summarized as:

(*i*)   cracks cause alarm about the safety of the structure;

(*ii*)   cracks lower the visual acceptance of the structure (a) by modifying surface textures and damaging the visual effect intended by the designer, and (b) by giving an appearance of cheapness or bad building.

The response of any individual to any crack pattern will depend on the position that such an individual occupies in relation to the structure. The acceptability of a perceived crack to an engineer may well be different from its acceptability to a building owner or to a member of the public renting or using the building. A minimum width related to viewing distance and to the prestige of the structure has been proposed as shown in Fig. 3.25.[3.112]

Causes of Cracking

Cracking may occur in plastic concrete and in hardened concrete. The major sources of plastic cracking are discussed on pp. 58−9. Cracks in hardened concrete may be separated into:

(*i*)   those due to structural response to applied loads and external displacements;

(*ii*)   those due to the intrinsic nature of the concrete and its constituent materials.

Structural cracks are discussed in Chapter 4, pp. 139−55. Intrinsic cracks can be classified simply into three groups, which relate not only to the cause of the crack but to the time scale in which a crack can be expected to appear. These groups are:

(*i*)   plastic cracking, appearing in the first few hours;

(*ii*)   early thermal cracking, appearing at a time between one day and two to three weeks;

(*iii*)   drying shrinkage cracking, appearing after several weeks or even months.

A great deal of the cracking observed in concrete structures is of the intrinsic sort and much of the literature discussing cracks is devoted to this aspect.[3.110, 3.113] Examples of various forms of intrinsic cracking (and a few structural cracks) are shown in Fig. 3.26.

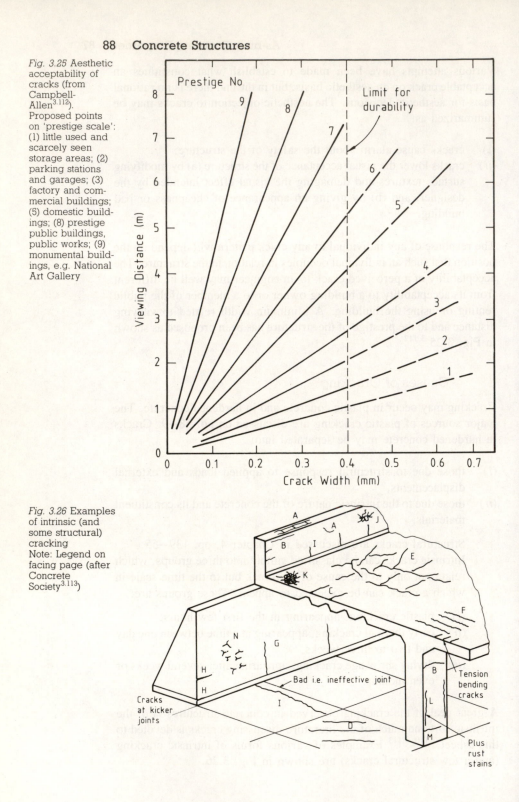

Fig. 3.25 Aesthetic acceptability of cracks (from Campbell-Allen[3.112]). Proposed points on 'prestige scale': (1) little used and scarcely seen storage areas; (2) parking stations and garages; (3) factory and commercial buildings; (5) domestic buildings; (8) prestige public buildings, public works; (9) monumental buildings, e.g. National Art Gallery

Fig. 3.26 Examples of intrinsic (and some structural) cracking Note: Legend on facing page (after Concrete Society[3.113])

Type of cracking	Letter (see Fig. 3.26	Subdivision	Most common location	Primary cause (excluding restraint)	Secondary causes/factors	Time of appearance
Plastic settlement	A	Over reinforcement	Deep sections	Excess bleeding	Rapid early drying conditions	Ten minutes to three hours
	B	Arching	Top of columns			
	C	Change of depth	Trough and waffle slabs			
Plastic shrinkage	D	Diagonal	Roads and slabs	Rapid early drying	Low rate of bleeding	Thirty minutes to six hours
	E	Random	Reinforced concrete slabs			
	F	Over reinforcement	Reinforced concrete slabs	Ditto plus steel near surface		
Early thermal contraction	G	External restraint	Thick walls	Excess heat generation	Rapid cooling	One day to two or three weeks
	H	Internal restraint	Thick slabs	Excess temperature gradients		
Long-term drying shrinkage	I		Thin slabs (and walls)	Inefficient joints	Excess shrinkage Inefficient curing	Several weeks or months
Crazing	J	Against formwork	'Fair faced' concrete	Impermeable formwork	Rich mixes Poor curing	One to seven days, sometimes much later
	K	Floated concrete	Slabs	Over-trowelling		
Corrosion of reinforcement	L	Natural	Columns and beams	Lack of cover	Poor quality concrete	More than two years
	M	Calcium chloride	Precast concrete	Excess calcium chloride		
Alkali-aggregate reaction	N		(Damp locations)	Reactive aggregate plus high-alkali cement		More than five years

The causes of plastic cracking and early thermal cracking are discussed earlier in earlier sections of this chapter (see pp. 57—82).

Temperature changes and drying shrinkage can only cause cracking if there is sufficient restraint so that the tensile stress induced is greater than the strength of the concrete. Restraint may be provided by:

(*i*)   external end restraint;
(*ii*)   external continuous restraint;
(*iii*) internal restraint arising from non-uniform volume changes;
(*iv*)   internal restraint from embedded steel.

Examples of external restraint are shown in Fig. 3.27. The most common internal restraint arises when a shrinking surface layer is restrained by an internal mass of concrete in which drying has not begun or which is at a higher temperature than the surface.

If any of these restraints is sufficient to suppress a potential strain of 100 to 200 microstrain a crack can be expected to form. If the potential shrinkage occurs slowly a greater strain is necessary to cause a crack. The crack once formed will widen as further volume reduction occurs. The only way in which such cracks can be certainly prevented is by multi-axial prestressing.[3.114] For most purposes, however, we can be satisfied if the cracks do not open in an uncontrolled manner but can be contained to an acceptably small width. This implies that sufficient reinforcement must be included to ensure that the force required to widen an existing crack is greater than the force needed to form a new crack.

A particular form of cracking formed by internal restraint of shrinkage

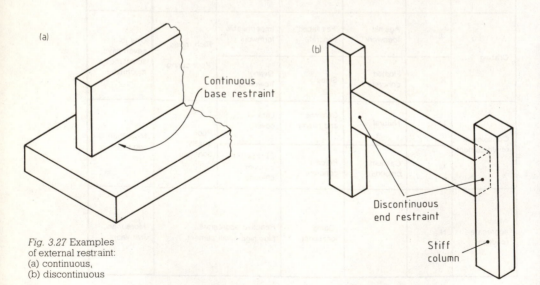

(a)

Continuous base restraint

(b)

Discontinuous end restraint

Stiff column

Fig. 3.27 Examples of external restraint: (a) continuous, (b) discontinuous

Fig. 3.28 Crazing
in a floor slab

is known as crazing or pattern cracking (see Fig. 3.28). Crazing is characterized by irregular, often near hexagonal, areas typically between 5 and 80 mm across. The cracks are usually only a few millimetres deep and in themselves do not affect the structural integrity of the concrete and should not be the cause of further deterioration except in wearing surfaces. The cracks become more noticeable as they fill with dirt and for this reason it is often suggested that concrete containing white cement is more susceptible to crazing.[3.113] Crazing is caused by differential moisture movement due to either or both (*i*) a high moisture gradient or (*ii*) a discontinuity in composition near the exposed surface. It follows from (*i*) that neither a very porous material nor a very impermeable material is likely to craze. From (*ii*) it follows that a concrete on which driers have been used or one which has been over-trowelled to bring laitance to the surface is very likely to craze. Crazing is often apparent within a week of casting, but it may occur much later if sufficiently severe drying conditions occur at a later age.

Measuring and Reporting of Cracks

Cracks should be reported by location, direction, width and depth. For reporting direction, the following adjectives are useful: longitudinal,

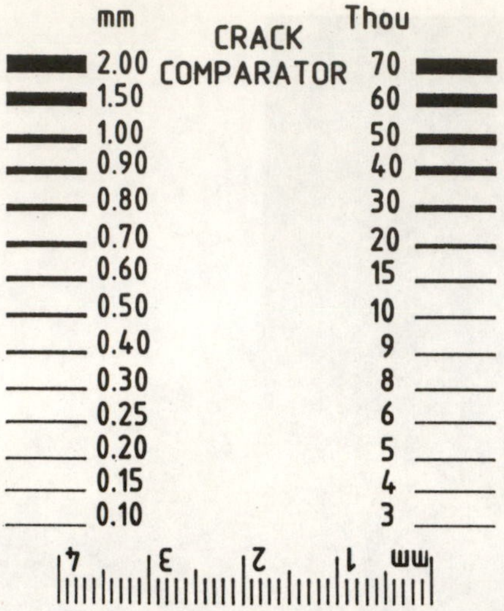

Fig. 3.29 Crack
comparator (full
scale)

CRACK COMPARATOR

mm	Thou
2.00	70
1.50	60
1.00	50
0.90	40
0.80	30
0.70	20
0.60	15
0.50	10
0.40	9
0.30	8
0.25	6
0.20	5
0.15	4
0.10	3

transverse, diagonal (all related to a major structural axis, to be defined
if necessary); vertical, horizontal, north-south, etc. (all related to
topographical axes); random. The widths of cracks should be measured
by some form of comparator, rather than by using estimated judgements.
The simple form, shown in Fig. 3.29 (reproduced full-scale), is valuable
in giving a feel for crack widths to those not very experienced in searching
for cracks. As crack widths vary from place to place and as cracks are
not commonly straight lines, the reporting of widths by the divisions
shown on the comparator is adequate for most purposes. *ACI
201.1R*[3.115] suggests the use of width ranges, described as: fine —
generally less than 1 mm; medium — between 1 and 2 mm; wide —
over 2 mm. In documents which may eventually be used by non-technical
people, and in legal proceedings, these classifications could be confusing
since even cracks in the 'fine' category are far wider than the 0.1 mm
to 0.3 mm cracks considered as the maximum acceptable under design
conditions. If the classifications are used it would be better to use only
the limits of measurement and not the accompanying titles. Depths are
only accurately measurable if a sawn or drilled surface is available.
Sketches and photographs add valuable information for subsequent
analysis. Side-lit photographs are particularly useful for revealing any
offset that may have occurred at a crack.

If the cracking is being assessed as a possible contribution to lack of
durability, rather than on its effect on appearance, the relationship of
the cracks to underlying reinforcement and metal embedments is
important. A method of classifying cracks for this purpose is shown in
Fig. 3.30.[3.116] For each class of crack related to reinforcing bars, the

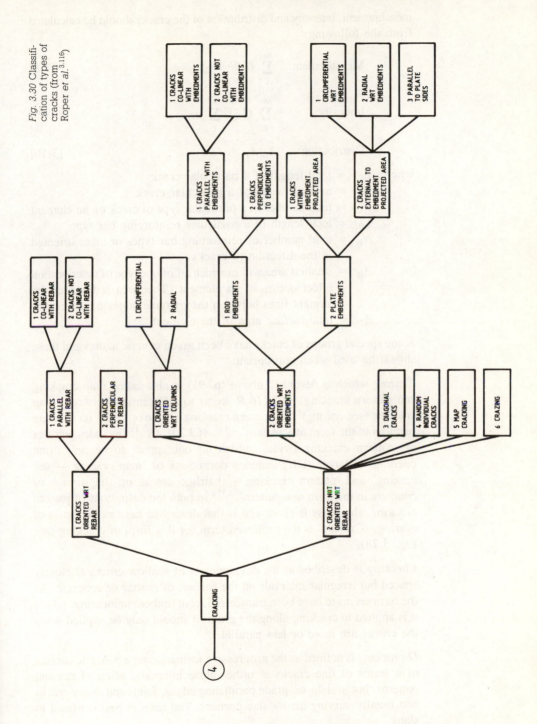

Fig. 3.30 Classification of types of cracks (from Roper et al.[3.116])

measurement, intensity and distribution of the cracks should be calculated from the following:

$$\text{Measurement:} \quad \sum_{1}^{N_C} (L_C \cdot W_C) \quad\quad\quad [3.8]$$

$$\text{Intensity:} \quad \sum_{1}^{N_C} (L_C) \Big/ \sum_{1}^{N_B} (L_B) \quad\quad\quad [3.9]$$

$$\text{Distribution:} \quad A_B / A_T \quad\quad\quad [3.10]$$

where: L_C = total length of a particular crack

$\quad\quad\quad W_C$ = average width of a particular crack

$\quad\quad\quad N_C$ = total number of a particular type of crack on an element

$\quad\quad\quad L_B$ = total length of a particular reinforcing bar type

$\quad\quad\quad N_B$ = total number of reinforcing bar types or sizes oriented in the direction of cracks

$\quad\quad\quad A_B$ = smallest area within which all of one type of deterioration effect occurs in an element (This area is bounded by straight lines between the extreme edges of the effect)

$\quad\quad\quad A_T$ = total surface area of the element

Some special groups of cracks have been given generic names and these should be used when appropriate.

Crazing which is described above (p. 91) is also called map cracking and pattern cracking. *ACI 116.R* seems to differentiate between 'map cracking (see crazing)' and 'pattern cracking — fine openings on concrete surfaces in the form of a pattern'.[3.117] *ACI 201.1R*[3.115] includes pictures of 'pattern cracking (wide)' which do not appear to be only 'fine openings'. Barker (UK) includes definitions of 'map crazing — see crazing' and 'pattern cracking — hairline cracks on the surface of concrete in the form of a pattern'.[3.118] In both the definitions of pattern cracking, the cause is given and is that described here as the cause of crazing. 'Crazing' is the preferred term for this form of cracking (see Fig. 3.28).

Checking is described as the development of shallow cracks at closely spaced but irregular intervals on the surface of mortar or concrete. As the term seems to have been transferred from timber engineering, where it is applied to cracking along the grain, it should only be applied when the cracks are more or less parallel.

D-cracking is defined as the progressive formation on a concrete surface of a series of fine cracks at rather close intervals, often of random patterns, but in slabs-on-grade paralleling edges, joints and major cracks and usually curving across slab corners. The term is best confined to slabs.

When cracks are recorded and reported, it is important in all cases to describe the crack by its nature and appearance rather than by an assumed cause, which may on further investigation turn out not to be the true cause.

References

3.1 Brady A B 1899 Low-level concrete bridge over Mary River, Maryborough, Queensland. *ICE Minutes of Proceedings* **141**: 246—57

3.2 O'Connor C 1985 *Spanning Two Centuries* University of Queensland Press 255 pp

3.3 Bate S C C 1983 Lessons from the past. In Kong *et al* (eds) *Handbook of Structural Concrete* Pitman, London Chapter 2 pp 2—34

3.4 Malhotra V M 1984 In situ/non-destructive testing of concrete — a global review *ACI SP—82* pp 1—16

3.5 Carette G G Malhotra V M 1984 In situ tests: variability and strength prediction of concrete at early ages *ACI SP—82* pp 111—41

3.6 Henzel J, Freitag W 1969 Zur Ermittlung der Betondruckfestigkeit im Bauwerk mit Hilfe von Bohrkernen kleineren Durchmessers. *Beton* No 4

3.7 Rechardt T 1965 *Betonin Lujunden Toteaminen Porausnaytten* The State Institute of Technical Research, Finland

3.8 Bungey J G 1980 Determining concrete strength by using small diameter cores. *Magazine of Concrete Research* **32**(111): 124

3.9 Campbell R M, Tobin R E 1967 Core and cylinder strengths of natural and lightweight concrete. *ACI Journal* Proceedings **64**(4): 190—5

3.10 Meininger R C 1968 Effect of core diameter on measured strength. *Journal of Materials* **3**(2): 320—36

3.11 Munday J G L, Dhir R K 1984 Assessment of in situ concrete quality by core testing *ACI SP—82* pp 393—410

3.12 Murphy W E 1984 The interpretation of tests on the strength of concrete in structures *ACI SP—82* pp 377—92

3.13 Petersons N 1964 Strength of concrete in finished structures. *Transactions No 232* Swedish Cement & Concrete Research Institute, Stockholm

3.14 Petersons N 1970 Betonghallfasthet hos element (Concrete strength in precast elements). *Report No 7014* Inspection Department, Swedish Cement & Concrete Research Institute, Stockholm

3.15 Keiller A P 1984 An investigation of the effects of test procedure and curing history on the measured strength of concrete. *ACI SP—82* pp 441—58

3.16 Lewandowski R 1970 Einfluss von Bewehrungabschnitten auf das Festigkeit-sergebnis von Beton-Bohrkernen. *Betonstein-Zeitung* No 12

3.17 Bloem D L 1968 Concrete strength in structures. *ACI Journal* Proceedings **65**(3): 176—87

3.18 Murray A McC, Long A E 1987 A study of the in situ variability of concrete using the pull-off method. *Proceedings of the Institution of Civil Engineers* Part 2 **83**: 731—45

3.19 Tomsett H 1988 Discussion of Murray & Long. *Proceedings of the Institution of Civil Engineers* Part 2 **85**: 583—6

3.20 Kaplan M F 1958 Compressive strength and ultrasonic pulse velocity relationships for concrete in columns. *ACI Journal* Proceedings **54**(8): 675–88

3.21 Lewandowski R 1976 Comparison strength of concrete in structures and structural members. *Beton Herstellung und Verwendung* **24**(12): 469–75

3.22 Powers T C, Brownyard T L 1946 Studies of the physical properties of hardened Portland cement paste — Part 2: Studies of water fixation. *ACI Journal* Proceedings **43**(3): 249–336

3.23 Ruettgers A, Vidal E N, Wing S P 1935 An investigation of the permeability of mass concrete with particular reference to Boulder Dam. *ACI Journal* Proceedings **31**: 382–416

3.24 Powers T C, Brownyard T L 1947 Studies of the physical properties of hardened Portland cement paste — Parts 6 and 7. *ACI Journal* Proceedings **43**(7): 845–80

3.25 Lawrence C D 1985 Water permeability of concrete. *Concrete Society Materials Research Seminar on the Serviceability of Concrete* Slough July 1985

3.26 Concrete Society Working Party, Permeability testing of site concrete — a review of methods and experience. Final Draft, Nov 1985, Permeability of concrete and its control, Papers, London, 12 December 1985, pp 1–68

3.27 Powers T C, Copeland L E, Hayes J C, Mann H M 1954 Permeability of Portland cement paste. *ACI Journal* Proceedings **51**(3): 285–98

3.28 Parrott L J 1985 Effect of changes in UK cements upon strength and recommended curing times. *Concrete* (London) **19**(9): 22–4

3.29 Powers T C, Copeland L E, Mann H M 1959 Capillary continuity or discontinuity in cement pastes. *Journal of the Portland Cement Association Research and Development Laboratory* **1**(2): 38–48

3.30 Valenta O 1970 The permeability and durability of concrete in aggressive conditions. *Proceedings of the Xth International Congress on Large Dams* Montreal Paper Q3 9R6 pp 103–17

3.31 Browne R D, Domone P L 1975 Permeability and fatigue properties of structural marine concrete at continental shelf depths. *Proceedings of the International Conference on Underwater Construction Technology* University College Cardiff p 39

3.32 Cather R, Figg J W, Marsden A F, O'Brien T P 1984 Improvements to the Figg method of determining the air permeability of concrete. *Magazine of Concrete Research* **36**(129): 241–5

3.33 Ho D W S, Lewis R K 1984 Concrete quality as measured by water sorptivity. *Civil Engineering Transactions* Institution of Engineers Australia **CE26**(4): 306–13

3.34 Ho D W S, Hinczak I, Conroy J J, Lewis R K 1986 Influence of slag cement on the water sorptivity of concrete *ACI SP–91* pp 1463–73

3.35 Lea F M 1970 *The Chemistry of Cement and Concrete* 3rd edn Edward Arnold 366 pp

3.36 Bate S C C 1974 Report on the failure of roof beams at Sir John Cass's Foundation and Red Coat Church of England Secondary School, Stepney. *Building Research Establishment Current Paper CP 58/74* 18 pp

3.37 Bate S C C 1984 High alumina cement concrete in existing building superstructures. *Building Research Establishment Report S040* HMSO, London

3.38 Midgley H G, Midgley A 1975 The conversion of high alumina cement. *Magazine of Concrete Research* **27**(91): 59—77

3.39 Collins R J, Gutt W 1988 Research on long-term properties of high alumina cement concrete. *Magazine of Concrete Research* **40**(145): 195—208

3.40 Cole R G, Horswill P 1988 Alkali—silica reaction: Val de la Mare dam, Jersey, case history. *Proceedings of the Institution of Civil Engineers* Part 1 **84**: 1237—59

3.41 Tuthill L H 1982 Alkali—silica reaction — 40 years later. *Concrete International Design & Construction* **4**(4): 32—6

3.42 Institution of Structural Engineers 1990 *Structural Effects of Alkali—Silica Reaction*

3.43 British Cement Association 1988 *The Diagnosis of Alkali—Silica Reaction* Report of working party, BCA, London

3.44 Hammersley G P 1988 Alkali—silica reaction in dams and other major water retaining structures: diagnosis and assessment. *Proceedings of the Institution of Civil Engineers* Part 1 **84**: 1193—211

3.45 Anderson T A , Roper H 1977 Influence of an impervious membrane beneath concrete slabs on grade. *Symposium, Concrete for Engineering: Engineering for Concrete* Institution of Engineers, Australia, Brisbane

3.46 Troxell G E, Davis H E, Kelly J W 1968 *Composition and Properties of Concrete* 2nd edn McGraw-Hill 292 pp

3.47 L'Hermite 1960 *Proceedings of the IVth International Symposium on the Chemistry of Cement* Washington DC Vol 2 pp 659—94

3.48 Roper H 1974 The influence of cement composition and fineness on concrete shrinkage, tensile creep and cracking tendency. *Proceedings of the 1st Australian Conference on Engineering Materials* The University of New South Wales pp 45—71

3.49 Verbeck G J 1968 *Mechanism of Shrinkage and Shrinkage-Compensating Cements* Portland Cement Association Research and Development Laboratories — Mill session papers — M—144 Dec 1968

3.50 Swamy R N 1974 Shrinkage characteristics of ultra-rapid hardening cement. *Indian Concrete Journal* Apr: 127—31

3.51 Morgan D R 1973 *The Effects of Lignin Based Admixtures on Time Dependent Volume Changes in Concrete* PhD Thesis University of New South Wales 353 pp

3.52 Collepardi M, Marcialis A, Solinas V 1973 The influence of calcium lignosulphonate on the hydration of cements. *Il Cemento* **70**: 3—14

3.53 Lane R O, Best J F 1978 Laboratory studies on the effects of superplasticizers on the engineering properties of plain and fly ash concretes. *Proceedings of the International Symposium on Superplasticized Concrete* CANMET, Ottawa Vol 1 pp 379—402

3.54 Kishitani K, Kasami H, Lizuka M, Ikeda T, Kazama Y, Hattori K 1981 Engineering properties of superplasticized concretes *ACI SP—68* pp 233—52

3.55 Ghosh R S, Malhotra V M 1978 Use of superplasticizers as water reducers. *CANMET Div Report MRP/MRL 78—189(J)* CANMET, Ottawa

3.56 Alexander K M, Wardlaw J, Ivanusec I 1984 Fly ash: the manner of its contribution to strength and the magnitude of its effect on creep and related properties of high-quality concrete. *Civil Engineering Transactions* Institution of Engineers Australia **CE26**(4): 296–305

3.57 Hogan F J, Meusel J W 1981 Evaluation for durability and strength development of a ground granulated blast furnace slag. *Cements, Concrete & Aggregates* **3**(1): 40–52

3.58 Hinczak I 1988 Properties of slag concrete — an Australian experience. In Ryan W G (ed) *Papers, Concrete 88 Workshop* Concrete Institute of Australia July 1988 pp 199–229

3.59 Meininger R C 1966 *Drying Shrinkage of Concrete* Engineering Report No RD3 National Ready Mixed Concrete Association, Silver Spring June 1966 22 pp

3.60 Powers T C 1959 Causes and control of volume change. *Journal of the Portland Cement Association Research and Development Laboratories* **1**(1): 29–39

3.61 Hansen T C, Mattock A H 1966 Influence of size and shape of member on the shrinkage and creep of concrete. *ACI Journal* Proceedings **63**(2): 267–90

3.62 Campbell-Allen D, Rogers D F 1976 Shrinkage of concrete as affected by size. *Materiaux et Constructions* **8**(45): 193–202

3.63 Helmuth R A, Turk D H 1967 The reversible and irreversible drying shrinkage of hardened Portland cement and tricalcium silicate pastes. *Journal of the Portland Cement Association Research and Development Laboratories* **7**(2)

3.64 Verbeck G J, Helmuth R H 1968 Structures and physical properties of cement paste, Principal Paper Session III-1. *Proceedings of the Vth International Symposium on Chemistry of Cements* Tokyo pp 1–44

3.65 Rusch H, Jungwirth D, Hilsdorf H K 1983 *Creep and Shrinkage — Their Effect on the Behaviour of Concrete Structures* Springer-Verlag, New York 284 pp

3.66 Bazant Z P, Wittmann F H, Kim J K, Alou F 1987 Statistical extrapolation of shrinkage data — Part I: Regression. *ACI Materials Journal* **84**(1): 20–34

3.67 ACI Committee 209 Prediction of creep, shrinkage, and temperature effects in concrete structures *ACI 209R–82* 1982 92 pp

3.68 CEB-FIP *Model Code for Concrete Structures* Comité Euro-International du Beton, Paris Apr 1978

3.69 Bazant Z P, Panula L 1982 New model for practical prediction of creep and shrinkage *ACI SP–76* pp 7–23

3.70 Bryant A H, Vadhanavikkit C 1987 Creep, shrinkage-size, and age at loading effects. *ACI Materials Journal* **84**(2): 117–23

3.71 Campbell-Allen D 1973 The prediction of shrinkage for Australian concrete. *Civil Engineering Transactions* Institution of Engineers Australia **CE15**(1 and 2): 53–7 and 62

3.72 McDonald D, Roper H, Samarin A 1988 Prediction accuracy of creep and shrinkage models for Australian concrete. *Proceedings of the 14th ARRB Conference* **14**(7): 66–78

3.73 Tremper B, Spellman D L 1963 Shrinkage of concrete — comparison of laboratory and field performance. *Highway Research Record No 3* Highway Research Board pp 30–61

3.74 Ytterberg R F 1987 Shrinkage and curling of slabs on grade Part I — Drying shrinkage; Part II — Warping and curling; Part III — Additional suggestions. *Concrete International Design & Construction* **9**(4): 22–31; (5): 54–61; (6): 72–81

3.75 Neville A M, Dilger W H, Brooks J J 1983 *Creep of Plain and Structural Concrete* Construction Press, London 361 pp

3.76 ACI Committee 207 Mass concrete for dams and other massive structures *ACI 207.1R–70* American Concrete Institute (reaffirmed 1980) 1980 37 pp

3.77 ACI Committee 224 Control of cracking in concrete structures *ACI 224R–80* American Concrete Institute 1980 42 pp

3.78 Hughes B P, Mahmood A T 1988 An investigation of early thermal cracking in concrete and the recommendations in BS 8007. *The Structural Engineer* **66**(4): 61–9

3.79 Bogue R H 1947 *The Chemistry of Portland Cement* Van Nostrand Reinhold, New York pp 434 *et seq*

3.80 Mather B 1987 The warmer the concrete the faster the cement hydrates. *Concrete International Design & Construction* **9**(8): 29–33

3.81 Powers T C 1968 *The Properties of Fresh Concrete* John Wiley & Sons, New York 440 pp

3.82 Roy D M, Idorn G M 1982 Hydration, structure, and properties of blast furnace slag cements, mortars and concrete. *ACI Journal* Proceedings **79**(6): 444–57

3.83 Symons M G 1989 Unpublished report, South Australian Institute of Technology

3.84 Mehta P K 1984 Mineral admixtures. In Ramachandran V S (ed) *Concrete Admixtures Handbook* Noyes New Jersey Chapter 6 p 331

3.85 Campbell-Allen D, Thorne C P 1963 The thermal conductivity of concrete. *Magazine of Concrete Research* **15**(43): 39–48

3.86 Birch F, Schairer J F, Spicer H C 1942 *Handbook of Physical Constants* Washington, Geological Society of America, Special Paper No 36 Jan 1942 325 pp

3.87 Zoldners N G 1971 Thermal properties of concrete under sustained elevated temperatures *ACI SP–25* pp 1–31

3.88 Whiting D, Litvin A, Goodwin S E 1978 Specific heat of selected concretes. *ACI Journal* Proceedings **75**(7): 299–305

3.89 Mitchell L J 1966 Hardened concrete — thermal properties. *ASTM STP 169–A* p 202

3.90 Dettling H 1964 Die Warmedehnung des Zementsteines, der Gesteine und der Betone. *Deutscher Ausschuss für Stahlbeton*, Heft 164 pp 1–64

3.91 Harada T 1949 Thermal expansion of concrete at high temperatures. *Transactions of the Architectural Institute of Japan* No 39 Nov 1949 No 42 Feb 1951 No 46 Mar 1953

3.92 Mitchell L J 1953 Thermal expansion tests on aggregates, neat cements and concretes. *Proceedings of ASTM* **53**: p 963

3.93 Meyers S L 1950 Thermal expansion characteristics of hardened cement paste and of concrete. *Proceedings of the Highway Research Board* **30**: 193–203

3.94 Dhir R K, Munday J G L, Ong L T 1986 Investigation of the engineering properties of OPC/pulverised-fuel ash concrete: deformation properties. *The Structural Engineer* **64B**(2): 36—42

3.95 Dhir R K, Yap A W F 1984 Superplasticized flowing concrete: strength and deformation properties. *Magazine of Concrete Research* **36**(129): 203—15

3.96 Planas J, Corres H, Elices M, Chueca R 1984 Thermal deformation of loaded concrete during thermal cycles from 20 °C to − 165 °C. *Cement & Concrete Research* **14**(5): 639—44

3.97 Elices M, Planas J, Corres H 1986 Thermal deformation of loaded concrete at low temperatures, Part 2: Transverse deformation. *Cement & Concrete Research* **16**(5): 741—8

3.98 Corres H, Elices M, Planas J 1986 Thermal deformation of loaded concrete at low temperatures, Part 3: Lightweight concrete. *Cement & Concrete Research* **16**(6): 845—52

3.99 ACI Committee 523 Guide for cast-in-place low-density concrete. *ACI 523.1R—86* American Concrete Institute 1986 8 pp

3.100 Hill J 1987 Cracks in structures. *Concrete* (London) **21**(7): 36—8

3.101 Base G D, Read J B, Beeby A W, Taylor H P J 1966 *An Investigation of the Crack Control Characteristics of Various Types of Bar in Reinforced Concrete Beams* Research Report 18, Parts I and II Cement & Concrete Association, London Dec 1966

3.102 Base G D 1971 Causes, control and consequences of cracking in reinforced concrete. *Australian Road Research* **4**(7): 3—13

3.103 Beeby A W 1978 *Cracking and Corrosion, Concrete in the Oceans* Report No 1, CIRIA/UEG

3.104 Schiessl P 1986 Einfluss von Rissen auf die Dauerhaftigkeit von Stahlbeton- und Spannbetonbauteilen, Berlin, Ernst and Sohn. *Schriftenreihe des Deutschen Ausschusses fur Stahlbeton* Nr 370 pp 10—52

3.105 Beeby A W 1985 Cracking of concrete — prediction and control of flexural and direct tension cracking. Concrete Institute of Australia *Cracking of Concrete: Its Importance Prediction and Control* Nov 4 1985

3.106 Halvorsen G T 1987 Code requirements for crack control *ACI SP—104* pp 275—322

3.107 Clear C A 1985 *The Effects of Autogenous Healing Upon the Leakage of Water Through Cracks in Concrete* Cement and Concrete Association Technical Report 42.559 May 1985

3.108 Ancher R D, Allen A H, Hughes B P, Quinian D W, Thorpe E H 1988 BS 8007: the new code. *The Structural Engineer* **66**(3): 40—4

3.109 Lauer K R, Slate F O 1956 Autogenous healing of cement paste. *ACI Journal* Proceedings **27**(10): 1083—98

3.110 ACI Committee 224 Causes, evaluation, and repair of cracks in concrete structures. *ACI 224.1R—84* American Concrete Institute 1984 20 pp

3.111 Csagoly P, Holowka N 1975 *Cracking in Voided Post-tensioned Concrete Bridge Decks* Ministry of Transport and Communications, Ontario June 1975

3.112 Campbell-Allen D 1979 *The Reduction of Cracking in Concrete* University of Sydney/Cement and Concrete Association of Australia 165 pp

3.113 Concrete Society Working Party 1982 *Non-Structural Cracks in Concrete* Concrete Society Technical Report No 22 Dec 1982 39 pp

3.114 Leonhardt F 1976 Rissebeschrankung. *Beton- und Stahlbetonbau* No 1: 14—20

3.115 ACI Committee 201 Guide for making a condition survey of concrete in service. *ACI 201.1R—68* American Concrete Institute (revised 1984) 1984 14 pp

3.116 Roper H, Baweja D, Kirkby G 1984 Towards a quantitative measure of durability of concrete structural members. *ACI SP—82* pp 639—58

3.117 ACI Committee 116 Cement and concrete terminology. *ACI 116R—78* American Concrete Institute (reaffirmed 1982) 1982 50 pp

3.118 Barker J A 1983 *Dictionary of Concrete* Construction Press, London 111 pp

Further suggested reading, p 352

4 Influences on Serviceability and Durability

The Durability Information System

In a paper on durability of a group of building materials, including concrete, Wright and Frohnsdorff[4.1] highlighted the fact that at all stages of the building cycle an input of information on the durability of building materials is called for, and, in turn, several of these stages contribute data to the durability information system. A diagram of this interaction is presented in Fig. 4.1. The data gathering and its useful application is a formidable task, which is probably best done using computer databases developed by individual groups and relating to the specific areas of practice of concern to them. Much premature deterioration of concrete is preventable merely by the application, during the design and construction stages, of existing knowledge of concrete behaviour.

A paper entitled 'Concrete durability — A multibillion-dollar opportunity'[4.2] summarizes the findings of a Report of the National Materials Advisory Board of the National Research Council, USA. This report suggests that there is a lack of proper application of the durability knowledge-base by practitioners, as evidenced by the continuing problem of inadequate durability of concrete in service. It notes that the magnitude of the problem is great enough to merit national action, and seeks reasons for the lack of proper implementation of available data by engineers. It finds that available concrete technology is inadequately used because of the fragmentation of the industry, confused responsibility for the training and skill of the workers, lack of financial or other incentives for the industry, and poor management and dissemination of available technical information. It further finds that two institutional factors have significant influence on the durable lifespans of structures. The first is the system's failure to make durability the responsibility of the organization which can most directly provide it — the contractor. The system in no way rewards the contractor for good durability, which is in any case assumed by the specification writer. The second institutional factor, applicable especially to buildings, is a product of economic management and tax structure. Most developers have no intention of maintaining the ownership of new buildings for the life of those buildings,

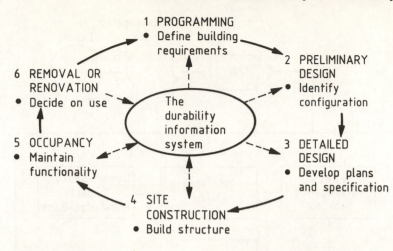

Fig. 4.1 Durability information system (after Wright and Frohnsdorff[4.1])

because it is advantageous to sell them within a few years of construction. The initial owner has therefore little incentive to spend money to ensure continued durability when the benefits will not be realized during the period of his ownership. The report argues that for much construction neither the owner nor the contractor is motivated to ensure long-term durability of the structures, yet there is potentially a severe financial loss to the community.

One of the several recommendations of the committee responsible for the report is that knowledge-based systems should be developed to facilitate, at affordable costs, effective use of available knowledge related to concrete durability by persons not themselves expert in the field of concrete durability. Amongst those non-experts may even be structural designers, whose interests have not led them to intensive study of the chemical and physical changes affecting concrete with time. A diagram of a possible configuration of such a system for use by a non-expert is given in Fig. 4.2.

At the present time few expert systems have been developed for durable concrete. Clifton and Kaetzel,[4.3, 4.4] while working at the National Bureau of Standards in Washington, DC, developed an expert system entitled Durcon [DURable CONcrete]. This contained a knowledge base for freeze−thaw, sulphate, alkali−aggregate and corrosion aspects of concrete durability, formed from the American Concrete Institute's Committee 201, Guide to Concrete Durability.

Although of considerable value, the Durcon programme is not directly related to the format of a design code. For that reason Roper *et al.*[4.5] have developed an expert system entitled Auscon [AUStralian CONcrete], which may be directly interrogated as part of the structural design process. Auscon is divided into three sections: the first relates to the type of

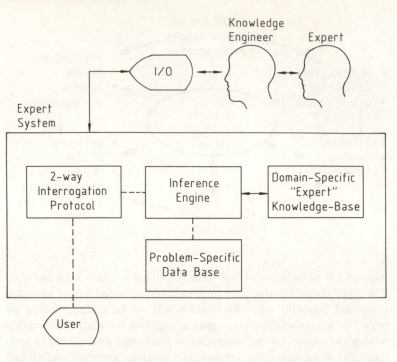

Fig. 4.2 Expert
system architec-
ture. The expert-
head silhouette
symbolizes the
source of
knowledge, which
is converted by
the knowledge
engineer into
production rules
that are contained
in the domain-
specific
knowledge-base.
The information on
a specific problem
provided by the
user is placed in
the problem-
specific data-base.
Based on the
problem-specific
data, the inference
engine selects the
appropriate rules
to be fired (from
Clifton and
Kaetzel[4.4])

member being designed and its exposure conditions; the second is based
on the durability recommendations outlined in the Australian Standard
'Concrete Structures' (AS 3600), but this section can be simply modified
to fit any given code which covers durability requirements; the third
section is concerned with special structures and exposure conditions which
are not covered by the code recommendations, and draws the user's
attention to other aspects of durability that do not get a specific mention
in the code. Apart from the durability aspects covered by Durcon, Auscon
also includes knowledge bases relating to such aspects as fire protection,
high temperature and mass concrete.

It is considered that, in the future, more and more the designer and
maintenance engineer will successfully use such knowledge-based
programmes and data bases to improve durability of concrete, but their
use will not reduce the necessity for engineers in general to be acquainted
with the many individual influences on serviceability and durability of
concrete in service. Indeed even with fairly detailed explanations within
the expert systems, some backgound knowledge of the concrete system
must be taken for granted. Certain of the more important influences on
concrete durability will hence be considered in some detail in this and
the next chapter.

Climatic and Weathering Effects

Freezing and Thawing

The most severe climatic attack on concrete occurs when concrete, containing moisture, is subjected to cycles of freezing and thawing. The capillary pores in the cement paste are of such a size that water in them will freeze when the ambient temperature is below 0 °C. The gel pores are so small that water in them does not freeze at normal winter temperatures. As water when freezing expands by 9 per cent of its volume, excess water in the capillaries has to move. Since the cement paste is relatively impermeable, high pressures are necessary to move the excess water even over quite small distances. For normal strength concrete it has been found that movement of the order of 0.2 mm is sufficient to require pressures which approach the tensile strength of the paste.[4.6] Concrete can be protected from freeze–thaw damage by the entrainment of appropriate quantities of air distributed through the cement paste with a spacing between bubbles of no more than about 0.4 mm. The air bubbles must remain partially empty so that they can accommodate the excess water moved to them. This will generally be the case since the bubbles constitute the coarsest pore system and are therefore the first to lose moisture as the concrete dries. Fully saturated concrete, if permanently submerged, will not need protection against freezing, but concrete which has been saturated and is exposed to freezing, as for example in the tidal range, may not be effectively protected by air-entrainment.

For effective protection an air entraining agent (AEA) must be added to the mix to entrain the appropriate amount of air and to induce a bubble system with an appropriate spacing. When AEA is used, it is only the amount of air entrained which can be measured in the wet concrete. The amount of air required is between 4 and 8 per cent, depending on the maximum size of aggregate, figures which correspond to about 9 or 10 per cent air in the mortar fraction. Air is entrained during the mixing action even when no AEA is added. The effect of AEA is to stabilize the air bubbles in the form desired. More air is entrained with a larger dose of AEA but the effect is not linear and with most agents levels off with larger doses. For mixes with higher slump more air is entrained. It is difficult to entrain air in very stiff mixes. The grading and nature of the particles in the fine aggregate have a very marked effect on the amount of air entrained. Hollon and Prior claimed that sand was the most important single factor in air-entrainment.[4.7] At higher temperatures, less air is entrained and this temperature effect is most noticeable in higher slump mixes. At 180 mm slump, a rise in temperature of 15 °C has been shown to reduce the air content by one percentage point.[4.8] Finely divided materials such as fly ash reduce the amount of entrained air and the dose of AEA for a mix containing blended cement will generally be larger than the dose when the cementitious material is only Portland cement. A useful summary of the action of AEA is given by Dolch.[4.9]

Because of the large number of variables which affect the air content it is necessary to adjust the dose rate at frequent intervals and to measure the air content at least as frequently as measuring slump. The common habit of adding AEA in accordance with the manufacturer's suggestion on the label only introduces a further random variable into the mix control process.

When AEA is used with other admixtures, and particularly with high range water-reducing admixtures (superplasticizers), the air void system is changed. With superplasticizers a larger bubble spacing has generally been found.[4.10, 4.11] However, this larger spacing has not always been found to indicate unsatisfactory freeze—thaw resistance. Nevertheless the results produced by Kobayashi et al.[4.12] show a clear relationship between bubble spacing and durability as can be seen in Fig. 4.3.

It has been suggested that if concrete can be made so dense that there are no inter-connected capillary pores then resistance to freeze—thaw deterioration will exist without the need for air-entrainment. The use of high cement content and low water:cement ratio will lead in this direction as will the introduction of silica fume but there is as yet no firm evidence to show that it would be wise to dispense with air-entrainment if freeze—thaw resistance is wanted.

The particular form of deterioration due to frost action known as D-cracking or durability-cracking is caused by failures in the coarse aggregate[4.13, 4.14] and air-entrainment does not therefore provide protection. The absorption and pore size in the aggregate particles have a marked effect on the behaviour. It is claimed that only sedimentary or poorly metamorphosed sedimentary rocks are prone to D-cracking but no entirely satisfactory methods of predicting performance appear to exist. The only certain method of preventing D-cracking is to use a coarse aggregate with a history of satisfactory service. Available information on the topic has been summarized by Sawan.[4.15]

Carbonation

The reaction between calcium hydroxide in hydrated cement and atmospheric carbon dioxide is not one which is directly damaging to the concrete itself. In fact carbonated concrete is generally found to be less permeable than the same concrete which has not carbonated and is therefore more resistant to the ingress of aggressive fluids. Unfortunately the alkalinity of the concrete is much reduced when carbonation takes place and therefore, if the carbonation front approaches reinforcement

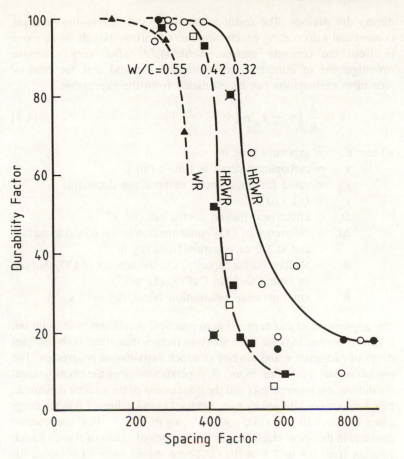

Fig. 4.3 Freeze–
thaw durability
related to bubble
spacing (after
Kobayashi et al.[4.12])

or other embedded metal, protection against corrosion is much reduced
and the steel becomes activated. Corrosion of the steel may subsequently
follow (see Chapter 5). It is therefore important to ensure that the quality
and thickness of the concrete cover is sufficient to prevent the steel ever
becoming activated during the life of the structure.

It is often accepted that carbonation penetration can be described by
a 'square-root of time' law and such an expression has been found to
be reasonably accurate for constant environmental conditions at a low
relative humidity such as in indoor conditions, particularly in air-
conditioned premises. For outdoor exposure the penetration is less than
predicted by a square-root law and it appears that at infinite time the
penetration will still remain finite. The most important contribution to
the departure from the square-root expression is the changing moisture
content of the outer layer of concrete. This concrete takes up water much
more rapidly during wetting periods than it loses water by evaporation

during dry periods. The result is that we get an increasing moisture content and a decreasing penetration rate of carbon dioxide as we move in from the concrete surface. Schiessl,[4.16] after very extensive investigations of concrete carbonation, has found that for outdoor structures carbonation can be estimated from the expression

$$t = \frac{A}{B}\left(y + y_\infty \ln\left(1 - \frac{y}{y_\infty}\right)\right)$$ [4.1]

where: t = exposure time (s)
y = carbonation depth at time t (m)
y_∞ = value for the ultimate carbonation depth (m)
y_∞ = $D_{eff} \cdot \Delta C / B$
D_{eff} = effective diffusion coefficient ($m^2\ s^{-1}$)
ΔC = difference of CO_2 concentration at the concrete surface and at the carbonation front (kg m^{-3})
A = alkaline buffer capacity, i.e. the amount of CO_2 that can be bound to form $CaCO_3$ (kg m^{-3})
B = environmental retardation factor (kg $m^{-2}\ s^{-1}$)

The application of this expression to practical conditions is difficult but it can be used to indicate the important factors that affect both the final depth of carbonation and the rate at which carbonation progresses. The environmental retardation factor, B, depends mainly on the environmental conditions, the concrete mix and the dimensions of the structural element. For concrete not affected by rain, Schiessl found values of B in the range from 0.8×10^{-9} to 1.5×10^{-9} kg $m^{-2}\ s^{-1}$. For unprotected concrete in the open, much greater fluctuations of values of B were found, ranging from 0.8 to 2.8×10^{-9}. These values were all obtained for exposure in central Europe. Subsequent observations have indicated the following influences on the value of B.

Environment   The value of B increases with increasing number and intensity of wetting and drying cycles; with increasing duration of wet periods; increasing relative humidity of the environment.

Concrete   The value of B increases with decreasing water:cement ratio and with increasing duration of moist curing; with increasing cement content.

Structure   The value of B increases with increasing thickness of the structural element.

The important influence of water:cement ratio is shown by the ratios of the depths of carbonation measured after 10 years for three different water:cement ratios with the results averaged over four different Portland cements.

W:C	Relative depth of carbonation
0.45	0.43
0.60	1.00
0.80	2.25

Concrete containing fly ash or blast furnace slag cement has been shown to carbonate to greater depths at an early age when compared with concrete containing only Portland cement. This is particularly the case when only a brief period of moist-curing (less than 7 days) has been provided.[4.17, 4.18] However, long-term measurements of the depth of carbonation do not show conclusively that concretes containing blended cements eventually carbonate to any greater depth than those containing only Portland cement. One of the critical factors determining the depth of carbonation is the ratio of water to Portland cement (not the water:cementitious ratio).[4.19]

Acid Precipitation

The existence of highly acidic rain in the vicinity of industrial plants has been known for many years. Indeed the term 'acid rain' was used by the Scottish chemist, Robert Angus Smith, in 1872 after a long study of the chemistry of rain around Manchester, England.[4.20] Such rain may have a pH of 4 or lower and as such is likely to be highly damaging to alkaline building materials. In recent years the existence of acid precipitation has been noted over large tracts of North America and Europe, even in areas well away from industrial activity. Damage to forests, lakes, agricultural crops and buildings and monuments has been ascribed to this source. All rain is somewhat acid, but the term 'acid precipitation' is now applied when the pH is below 5.6. The major acids present are sulphuric and nitric acid, resulting from those combustion products consisting of the oxides of sulphur and nitrogen. From the mid 1960s the contribution of sulphuric acid has been found to decrease and that of nitric acid has increased as a fraction of the acid in precipitation. The level of acidity has not decreased and in the northeast USA and in eastern Canada average pH levels between 4.0 and 4.5 have been recorded.

Kong and Orbison[4.21] in introducing an accelerated laboratory program found very little information in the literature on the effect of acid rain on Portland cement concrete or mortar but they did note evidence of high levels of atmospheric pollution which greatly accelerated the deterioration of statues and structures made with calcareous stone in and around industrialized regions of Europe.[4.22] They referred also to some qualitative indications of increasing rates of deterioration in reinforced concrete structures caused by acid deposition.[4.23, 4.24] Their laboratory experiments involved immersing concrete cylinders in solutions containing sulphuric and nitric acids with pH levels of 2, 3, 4 and 5. The concretes contained type I Portland cement and a mixture of calcareous and non-calcareous aggregates and air-entrainment. The concretes were made in seven grades ranging from 20 MPa to 62 MPa (3000 to 9000 psi). Loss of surface material increased with increasing acidity. It was particularly interesting to note that the high-strength concretes (with strengths of 48 MPa and above) experienced the greatest deterioration in the form of a relatively uniform erosion of surface material. The lower strength concretes were somewhat less susceptible to surface erosion but suffered a moderate degree of sulphate attack, as shown by expansion, cracking and spalling. All concretes suffered discoloration. Clearly more information on the resistance of Portland cement and blended cement concretes to acid precipitation is still needed as there is as yet no chance of acid precipitation not remaining very general for many years.

Microbiological Attack

Lichens

Concrete which no longer has an alkaline surface can support the growth of lichens and other micro-organisms. Lichen growths can contribute to surface erosion as they excrete acid products and the rootlet growth may penetrate a millimetre or two into the surface and add to mechanical disruption of the surface. Fine surface cracks encourage this growth. As surface is lost, the central part of the lichen colony destabilizes and further surface material is removed. A detailed examination of lichen growth on a reservoir wave wall in Lincolnshire, UK, is reported by Figg et al.[4.25] There was no sign of lichen growth in the first four to six years after casting, but by 10 years substantial colonies of three groups of lichens were in place. Yellow species are common early colonizers on inorganic building materials in coastal regions as they have a greater salt-tolerance. Grey-green species were the most widespread after 10 years, probably because by this time there had been sufficient build-up of bird droppings to provide the nitrogenous nutrition that these species demand. Black species were associated with the most severe surface deterioration.

Microbiological growth is linked with the availability of moisture as well as with loss of alkalinity and the most absorptive surfaces were the first to become colonized by lichens. A reduction of water:cement ratio (by the use of a water-reducing admixture) was found to be the most help in delaying the onset of colonization, an increase in cement content was less effective and air entrainment, without any other modification to the mix proportions, did not assist at all.

Sewage

The corrosion of concrete sewers has been the subject of extensive research.[4.26] Corrosion of concrete sewers typically occurs only above the water line, the severest attack being immediately above the average daily level. Along the sides above this point some corrosion may be evident, increasing in severity until the crown is reached. In severe cases the crown may be perforated. The first evidence is the appearance of white powdery efflorescence. Initial deterioration is slow but as it progresses it becomes more rapid, the initial deposits flaking off the concrete. At its final development the surface concrete is reduced to a soft putty-like material from which the aggregate falls.

Sulphur compounds present in sewage are reduced to suphides by micro-organisms in those parts of the system where the dissolved oxygen in the sewage becomes depleted as a result of biological action. Some of the sulphides thus formed escape to the sewer atmosphere in the form of hydogen sulphide gas. This dissolves in the condensed moisture on the pipe walls above the sewage flow line and on the walls of manholes and sumps, where it is converted by sulphur-oxidizing bacteria in the presence of oxygen, to suphuric acid. The acid causes deterioration of the pipe material, the severity of the corrosion depending on conditions in the sewer and on the degree of resistance of the pipe material to acid attack. This process is depicted in Fig. 4.4. The optimum activity of sulphide-producing organisms and hence of production of sulphides, depends on the quantities of sulphates, organic sulphur and other organic material available, the oxidation-reduction potential, the hydrogen ion concentration, the temperature and habitat conditions.

Possible methods for prevention or control of attack are as follows:

(i)   methods of preventing or minimising the generation of hydrogen sulphide;
(ii)   methods of preventing the release of hydrogen sulphide into the sewer atmosphere;
(iii) methods of minimizing the oxidation of hydrogen sulphide to sulphuric acid;
(iv)   selection of materials for construction of sewers that are resistant to sulphuric acid attack.

Sewer corrosion is likely to be severe in the following conditions:

(i)   A rising main or a sewer flowing under pressure discharges into a gravity sewer.

Fig. 4.4 Processes
of concrete attack
in a sewer

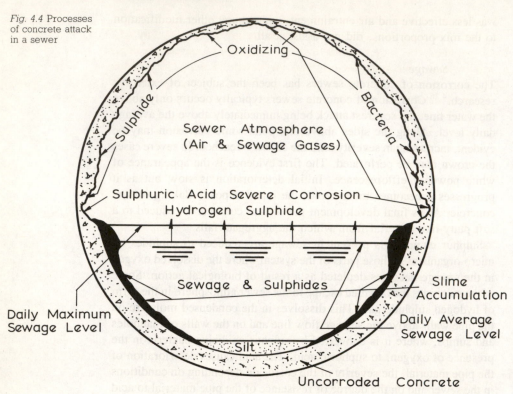

(*ii*)   Night soil or conservancy tank contents or other carbonaceous
wastes of high strengths are discharged into the sewer.

(*iii*) Certain industrial effluents, especially those with low pH, are
discharged into the sewer.

(*iv*)   Main sewers have been laid at such a flat grade that the velocity
of the sewage is less than the minimum required for scouring the
sewer. In such sewers, sewage turns septic and slime accumulates
on the sewer walls under the water level. The slime produces an
excellent breeding ground for sulphur producing organisms.

(*v*)   The retention system time in the system is long and the sewage
becomes septic in the lower reaches.

It has been demonstrated that corrosion could be eliminated or
substantially reduced by good design and good sewer practices. Corrosion
in an existing sewage system can to some extent be controlled but this
is likely to be much more effective when control measures are provided
for in the design of new systems. The system should be designed so that
the likelihood of corrosion occurring is minimized, by providing non-
turbulent flow with sufficient velocity and by providing proper

ventilation. If H_2S production can not be eliminated by these means, chemical or other control methods may be needed, as indicated by ACI Committee 210.[4.27] If chlorine or lead salts are considered, the legal requirements concerning the production of carcinogens and the addition of heavy metals to waste water should be checked. Corrosion resistant materials should be chosen for those parts of the system where corrosive conditions are likely to occur. Also it is essential that the influence of design on the incidence of corrosion should be understood so that corrosive conditions can be predicted under specific circumstances. It is further necessary to have an accurate conception of the economic implication of various design practices and of the properties of the materials used in the construction of those sewers where these practices are to be put into effect.

It has been suggested that siliceous aggregate concrete may be used without reservation in instances where the likelihood of corrosion can fairly confidently be ruled out. Nevertheless, in view of the small difference in price, good calcareous aggregate concrete pipes should be used wherever possible. In this way a substantial safety factor is ensured at little increase in cost. Calcareous aggregates are also recommended for concrete sewers where some form of protection of the concrete is contemplated. The calcareous aggregate concrete will then serve as a second line of defence if protective linings fail or when additional corrosion control measures are inadequate. There is some disagreement with the use of Portland cement concrete for sewers unless linings of plastic material are used. The alternative is to use pipes spun using polymer concrete. Bares[4.28] discusses use of furane resin concrete sewer pipes of large dimension in Czechoslovakia.

Aspects of existing sewer evaluation and rehabilitation methods are given in a publication by the American Society of Civil Engineers.[4.29]

Exposure to Extremes of Temperature

High Temperatures

Concrete at High Temperatures

In some industrial applications, such as aluminium plants and brickworks, the concrete may be occasionally or frequently subject to temperatures well above ambient. These temperatures are likely to be applied cyclically, generally with rather a long period. Jet aircraft and vertical take-off aircraft may subject the pavement to very high temperatures.[4.30] Temperatures of concrete, other than special refractory concrete, have to be kept below 300 °C and for these conditions designs are perfectly feasible provided due account is taken of the changing properties of concrete. Heat may affect concrete as a result of:

(*i*)   the removal of evaporable water;
(*ii*)   the removal of combined water;
(*iii*) alteration of cement paste;
(*iv*)   disruption from disparity of expansion, and resulting thermal stress;
(*v*)   alteration of aggregate;
(*vi*)   change of the bond between aggregate and paste.

The last three of these factors involve the properties of the aggregate used and for this reason it is not possible to reach conclusions without taking account of the petrographic nature of the local aggregates. Abrams[4.31] showed that specimens of concrete made with carbonate aggregate retained more than 75 per cent of their original strengths at temperatures up to 650 °C when heated without load and tested hot. For siliceous aggregates the corresponding temperature was only 425 °C. At 575 °C quartz undergoes a phase transformation, accompanied by a 0.4 per cent increase in volume. This phase change is often associated with spalling of concrete containing siliceous aggregates. Strengths of specimens heated and then tested cold were lower than the companions tested hot. Light-weight aggregates produced by a process involving heating do not generally show such a reduction in strength. The cement paste, and the paste containing cementitious materials other than Portland cement, are also affected by heat. Dry heat applied to mortar removes moisture and may at temperatures up to 100 °C slightly increase strength. At about 400 °C dehydration of $Ca(OH)_2$ occurs and produces a major reduction of mortar strength. There is evidence that some changes in the cement paste occur at quite low temperatures and that these changes are independent of loss of water. X-ray diffraction analysis, reported by Campbell-Allen and Desai, showed changes in cement pastes occurring at 65 °C.[4.32] The effects of moisture content on the strength and other structural properties of concrete have been discussed in detail by Lankard *et al.*[4.33] They conclude that if the moisture can evaporate as heating begins then little or no change will take place in the compressive strength up to 260 °C. If free moisture is retained in the concrete during heating, deterioration in the structural properties can be expected, possibly due to the build up of steam pressure.

In spite of the many different effects which are operating, an approximation to the effect of temperature on compressive strength can be obtained from Fig. 4.5 for Portland cement concretes.[4.34] High-alumina cement concretes lose strength at much lower temperatures and lose a much larger proportion of their strength.

Cycles of temperature can have a progressive effect on the reduction of strength. Campbell-Allen *et al.* reported investigations using dolerite aggregate and Portland cement.[4.35] One cycle to 250 °C showed no loss of compressive strength. One cycle to 300 °C produced 15 per cent

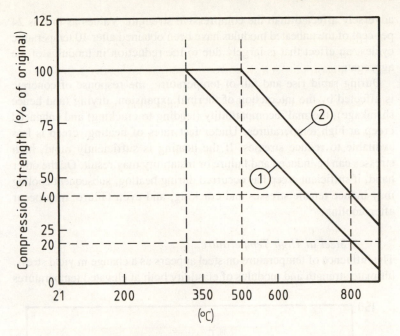

Fig. 4.5 Concrete strength as affected by temperature — idealized curves:[4.34] (1) dense concrete; (2) lightweight concrete

strength loss. Longer curing did not improve this loss. After 5 cycles, 35 per cent loss had occurred and a further 4 per cent loss occurred after 10 cycles. Cycles to 200 °C had a similar effect. The aggregate, a crushed dolerite widely used for concrete construction, consisted of laths of plagioclase felspar and ferromagnesian minerals, olivine and clinopyroxene. Changes brought about by heating could be seen in hand specimens. The rock, which is greenish-grey, assumed a pink hue when heated to 300 °C in air, a result of the oxidation of iron. Twenty-five per cent of the iron present changed to a higher oxidation state in a period of nine hours at elevated temperature. The major effect on the physical properties of the aggregate was to reduce its elastic modulus by about one-third. In concrete the bond between aggregate and paste was noticeably changed.

Tensile strength is more affected by heat than is compressive strength. With dolerite aggregate one cycle to 300 °C produced about 15 per cent loss and 10 cycles over 50 per cent loss of tensile strength. A similar loss of tensile strength was observed with limestone aggregate but with fireclay (and to a lesser extent with expanded shale aggregates) the reduction in tensile strength was nearer to that of mortar. It was apparent that the bond between aggregate and paste was reduced for both the dolerite and the limestone by the oxidation of iron compounds, even though, in the case of the limestone, the iron content was very small. In the range of temperature up to 300 °C, the elastic modulus is more adversely affected than the compressive strength. Values as low as 24 per cent of the unheated modulus have been obtained after 10 temperature cycles, an effect that is largely due to the reduction in modulus of the aggregate.

During rapid rise and fall of temperature, the response of concrete is affected by the interaction of thermal expansion, drying (and hence shrinkage), thermal incompatibility (leading to cracking) and enhanced creep at high temperatures. Under fast rates of heating, creep is less available to reduce stresses. If the heating is sufficiently rapid, high stresses can be induced and failure or instability may result. On the other hand, if sufficient creep has occurred during heating, subsequent cooling may induce tensile stresses and cracking, and even failure, may occur after cooling.[4.36]

Steel at High Temperature

The influence of temperature on steel appears as a change in yield stress, ultimate strength and modulus of elasticity both at elevated temperatures

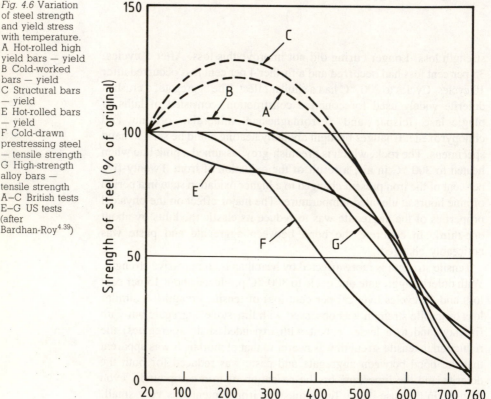

Fig. 4.6 Variation of steel strength and yield stress with temperature.
A Hot-rolled high yield bars — yield
B Cold-worked bars — yield
C Structural bars — yield
E Hot-rolled bars — yield
F Cold-drawn prestressing steel — tensile strength
G High-strength alloy bars — tensile strength
A–C British tests
E–G US tests
(after Bardhan-Roy[4.39])

and after cooling. The changes depend on the type of steel and are greater in cold-worked steels. The strength of hot-rolled reinforcement is not reduced if the temperature does not reach 300 °C, but at temperatures of 500–600 °C, the yield stress is reduced to the order of the working stress and the elastic modulus is reduced by one-third.[4.37, 4.38] Bars heated to these temperatures virtually recover their normal temperature strength when returned to room temperature. Bars heated to 800 °C have a lower residual strength after cooling to room temperature. Prestressing wire and strand starts to lose strength at 150 °C and may have only 50 per cent of its room-temperature strength when heated to about 400 °C. Figure 4.6 gives a summary of test results on different sorts of steel.[4.39] These results can be idealized as in Fig. 4.7.

Behaviour in a Fire

Failure in a fire occurs either through the spread of fire from the compartment or through structural failure of a member or assembly of members. Structural failure of a member most frequently occurs when the temperature of the steel reduces the yield stress to the working stress.

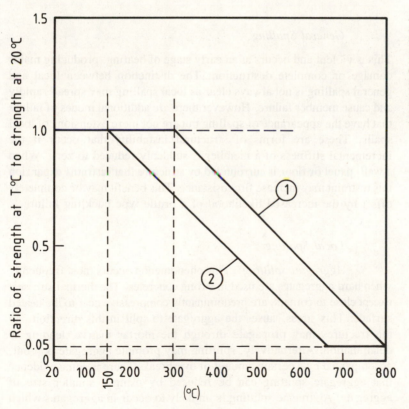

Fig. 4.7 Steel strength and yield stress as affected by temperature — idealized curves:[4.34] (1) high yield and structural bars, high-strength alloy bars; (2) prestressing strands and wires

The length of time for this to occur depends on the severity of the fire, the thermal conductivity of the protecting concrete, and whether spalling of the protection occurs. Provided that the concrete remains in place, the temperature of the steel can be calculated. Lefter[4.40] describes a heat flow computer program which is particularly adapted to short duration high intensity fires as it uses constant values of the thermal properties of the concrete. Salse and Lin refer to finite difference calculations using values of conductivity and specific heat that vary through the course of the fire.[4.41] Spalling of the concrete cover immediately exposes the steel to the fire temperatures. If it occurs violently, it may lead to immediate failure of the entire member. Spalling can be classified[4.42] as:

(a)   general or destructive spalling;
(b)   local spalling, subdivided into

    (i)   aggregate splitting,
    (ii)   corner separation;
    (iii)   surface spalling;
    (iv)   sloughing off.

General Spalling

This is violent and occurs at an early stage of heating, producing major damage or complete destruction. The distinction between local and general spalling is not always clear as local spalling may spread rapidly and cause member failure. However there are additional modes of failure that have the appearance of spalling but are not mere extensions of local spalls. These are forms of structural instability that occur if the incremental stiffness of a member is suddenly reduced to zero. When a wall, panel or floor is surrounded by structure that restrains expansion this restraint may increase fire resistance. This benefit may be completely offset by the increased likelihood of a brittle type buckling failure.

Local Spalling

Aggregate splitting   This phenomenon occurs most frequently when hard aggregates are used in strong concretes. The thermal stresses, except close to corners, are predominantly compressive near to the heated surface. This stress causes the aggregate to split in this direction and the fractures may propagate through the mortar matrix, leading to delamination. Alternatively, splitting may promote aggregate pop-out, accompanied by a certain amount of overbreak. There is some evidence that aggregate splitting can be reduced by using a smaller size of aggregate. Aggregate splitting is unlikely to occur in aggregates which have lower values of Young's modulus, such as light-weight aggregates.

Corner separation This is a very common occurrence and appears to be due to a component of tensile stress causing splitting across a corner. In fire tests, corner separation occurs most often in beams and columns made of quartz aggregates and only infrequently with limestone or light-weight aggregates. If a more rapid rate of heating occurs, it is possible that any type of concrete is at risk.

Surface spalling Spalling not associated with aggregate splitting may be due to thermal stress or to steam pressure in the pores. The influence of moisture is plainly demonstrated in tests by Meyer-Ottens[4.43] in which dramatic spalling occurred with high water contents but none with moisture contents less than 2 per cent. Surface spalling is less likely to occur in poor-quality porous concrete than in a high-quality material.

Sloughing off This is a gradual and progressive form of breakdown, involving partial displacement of surface layers, separated from the main member by long irregular cracks and cavities. It can be expected to occur for most quartz aggregates. The loosely adhering layers do not contribute to the strength or stiffness of the member, but there is an advantage in keeping the damaged material in place, by means of mesh, so that it can continue to act as insulation for the steel.

After the Fire

'The first reaction as you survey the smoke blackened scene of a fire is usually one of despair that anything can be saved. A cool headed investigation will usually show that a lot of the structure can be saved and the damage repaired'.[4.44] Any reinstatement of fire-damaged concrete must be preceded by a careful and painstaking appraisal of the fire characteristics and the effects of fire exposure on the materials and components. A simple guide to the temperature that has been reached by the concrete is given by the colour change that occurs.[4.45] Concrete made with non-igneous aggregate that has been heated above 300 °C, and hence will have lost substantial strength, turns a pinkish tinge. From 300 °C to 600 °C the pink hue deepens to red and at still higher temperatures the colour change goes back through grey to buff.[4.46] The change depends presumably on both cement and aggregate characteristics and is not therefore a certain measure of the temperature reached. Harmathy[4.47] has suggested measurement techniques on samples taken from the structure. These include thermogravimetric and dilatometric tests to indicate the changes that occurred during the fire, particularly in relation to moisture loss. If these tests are to be reliable, the samples must be taken within a day or two of the fire. Further guidance can be obtained from the fusion temperature of ancillary materials such as brass door-knobs, copper water-stops and lead flashings.

The special problem of a fire which affects very immature concrete is discussed by Smith in his description of the fire which destroyed a large part of the formwork and falsework being used for the construction of two additional arches alongside an existing three-span bridge.[4.44]

Cryogenic Temperatures

Reinforced and prestressed concrete have been accepted as useful components in the construction of storage systems for liquefied natural gas (LNG) and may therefore be required to operate from room temperature down to −165 °C. As in so many other situations, the water content of concrete has a marked effect on its behaviour. Rostasy and his co-workers[4.48, 4.49] have shown that water-saturated concrete, when cooled in this temperature range, goes through three distinct phases. From +20 to −15 °C, contraction occurs. Between about −20 and −60 °C, expansion is observed, as a result of water in the small pores changing to ice. This expansion is more pronounced with larger water contents and with higher water:cement ratios. Below −60 °C almost completely reversible contraction takes place (Fig. 4.8). Reheating above −60 °C produces irreversible strains and with each cycle of cooling and reheating the expansion and the irreversible strain both increase. Dried concrete shows almost linear and reversible contraction over the whole temperature range. Prestressing steel behaves similarly.

The interaction of concrete and steel during cooling and heating gives

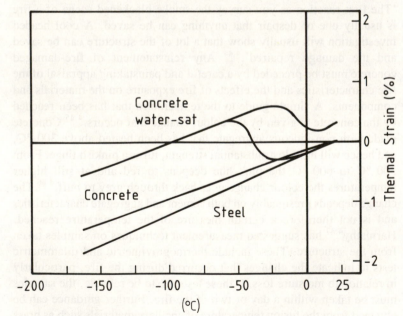

Fig. 4.8 Thermal strains related to temperature for (a) water-saturated concrete; (b) normally stored concrete; (c) prestressing steel (after Planas *et al.*[4.50])

rise to additional internal stresses, with possible overstressing of the steel and additional compression in the concrete. Some of the effect of this interaction has been studied by Planas *et al.*[4.50] They measured the thermal deformation of water-saturated concrete with a normal temperature strength of about 40 MPa under a constant compressive stress of 15 MPa, with the results shown in Fig. 4.9. The behaviour of the loaded specimens differs significantly from that of the unloaded ones. The irreversible expansion exhibited by the unloaded specimens after cycling becomes a contraction for the loaded ones. In the transition range (-20 to $-60°$ C) the loaded specimens become much more like the dry unloaded ones. The departures from linearity are explained by the effects of micro-cracking. In a subsequent paper, Elices *et al.*[4.51] showed that the volumetric strain is independent of the applied load, indicating that a fixed volume increase resulting from freezing in the pores has to be accommodated (see Fig. 4.10).

Design of cryogenic structures must take account of the total interaction of the various components throughout progressive temperature cycles if damage during these cycles is to be avoided. The effects of low temperature on typical properties of concrete and reinforcement are shown in Tables 4.1 and 4.2. The FIP Guide to Good Practice, from which these tables are drawn, summarises useful advice for the designer, pointing out that no direct credit should be taken for any enhancement of properties which may occur at low temperatures.[4.52] The behaviour has been found to be substantially affected by aggregate type. The use

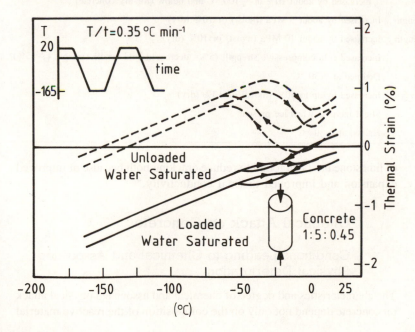

Fig. 4.9 Thermal strains related to temperature for loaded water-saturated concrete (after Planas *et al.*[4.50])

Fig. 4.10 Volumetric
thermal strains
(after Elices
et al.[4.51])

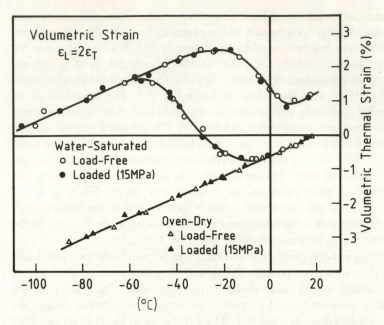

Table 4.1 Typical properties of concrete at low temperatures (compared with properties at +20°C)
(after FIP[4.52])

Compressive strength	Increased by around 70 MPa at −100 °C and below (for moist concrete). Increased by about 10% at −100 °C and below (for dry concrete)
Tensile splitting strength	Increased by about 5 MPa (moist) or 10% (dry)
Tensile bending strength	Increased by about 10 MPa (moist) or 10% (dry)
Thermal cycling	Increased cold compressive strength (5%), decreased cold tensile strength (−15%)
Impact strength	Doubled at −30 °C
Modulus of elasticity	Increased by up to 50% (moist) or 10% (dry)
Thermal conductivity	Slight increase when ice present
Creep	Halved at −30 °C

of limestone has been found to reduce cracking both because of improved
expansion and improved thermal conductivity.[4.53]

Chemical Attack on Concrete

Conditions Leading to Chemical and Associated Physical Deterioration

The characteristics and degree of chemical and associated physical attack
on concrete depend not only on the composition of the reactive material

Table 4.2 Typical properties of prestressing steels and reinforcement at low temperatures (after FIP[4.52])

Property	Prestressing wires	Strands	High alloy prestressing bars	Reinforcing bars
Tensile strength	Slight increase (8–10%)	Slight increase (8–10%)	Slight increase (8–10%)	Slight increase (8–10%)
Yield strength	Slight increase (8–10%)	Slight increase (8–10%)	Slight increase (8–10%)	Slight increase (8–10%)
Modulus of elasticity	6% increase	6–8% increase	Slight increase	5–10% increase
Extension at failure	2–4% (50 mm gauge length)	About 7%	1–6%	
Notch effects	Impact value reduced by 25%	Low impact value. Notches can cause premature failure	Impact value reduced by 80%	Very low impact value
Fatigue	Limit halved approximately	Not known	Not known	Not known
Relaxation	Much reduced	Not known	Not known	—

acting on the concrete, but also on the chemical composition and physical structure of the paste and aggregate. For the most part the paste is the more vulnerable of these constituents, but preferential leaching of carbonate aggregate may sometimes occur. It is generally accepted that the introduced reactants must be present in liquid or vapour form in order for the reactions to proceed; however, stored solid materials in contact with concrete may release reactants to the concrete pore solution under certain conditions. The attacking species must also be above a certain concentration limit to have a significant effect on the concrete, and mass transfer factors are important with respect to both the rate of introduction of the attacking species, and the continued exposure of the vulnerable paste phase by leaching or other phenomena such as cracking and spalling.

Some of the factors which increase concrete vulnerability to external chemical attack are:

(i)   high porosity, and hence high permeability and absorption, resulting from too high a water:cement ratio, unsatisfactory grading of aggregate, inadequate proportioning or poor compaction, or a combination of these;

(ii) improper choice of cement type for the conditions of exposure;

(iii) inadequate curing prior to exposure;

(iv) exposure to alternate cycles of wetting and drying and to a lesser extent heating and cooling, with allowance for the fact that higher temperatures increase reactivity;

(*v*)   increased fluid velocities, which may bring about both replenishment of the aggressive species, and increases in the rate of leaching;

(*vi*)   expansive reactions of any sort which may cause cracking, and any other physical phenomena which lead to greater exposure of reactant surfaces;

(*vii*)   suction forces which may be caused by drying on one or more faces of a section, while at least one other face is exposed to the aggressive liquid or vapour;

(*viii*)   unsatisfactory choice of shape and surface-to-volume ratios of concrete section; in particular massive elements are more capable of withstanding attack than are thin shells, or for that matter small prisms so often used for laboratory testing.

Popovics,[4.54] in a *Corrosion and Chemical Resistant Masonry Materials Handbook*[4.55] edited by Sheppard, has classified various types of deterioration due to fluids into six classes, and his scheme, somewhat modified and extended, will serve as a general basis of further discussion as follows:

Class I   Deterioration is brought about by leaching of concrete constituents by water containing carbonic acid or having a low carbonate hardness. Very pure water with a low dissolved calcium content, that is water of low temporary hardness, attacks hardened concrete. Such soft water may also often contain aggressive carbon dioxide, and in this case solvent action is increased. The action commences by leaching of the calcium hydroxide, and as the local supply of portlandite becomes exhausted, the soluble ions are removed from the silicate hydrates, thus leading to decomposition of practically all the hardened cement.

Class II   Deterioration results from non-acidic reactions between calcium minerals in the hardened cement paste and externally introduced ions. The reaction products are in this case weaker than the original reaction products, are not capable of any cementing action, have a tendency to be leached, or suffer from a combination of these deficiencies. The process weakens the concrete, and heightens its vulnerability to further attack by increasing its permeability. There are various mechanisms active in this class. One mechanism is base exchange as in the case of the substitution of magnesium for calcium in the hydrate mineral lattices. This may be associated with sulphate attack when the aggressive solution contains magnesium sulphate. Another mechanism is the saponification reaction between animal fats or oils and the calcium ions of the paste. Reactions between sugars and calcium to form calcium saccharates have been responsible for degradation of warehouse floors.

Class III Deterioration is a consequence of disruption due to reactions leading to the formation of products which occupy a greater volume than the reactant material. These reactions are either due to the slow hydration of a mineral present in the cement, such as periclase (MgO), or reactions between calcium aluminates and salts, in particular sulphates. In the case of the formation of brucite from periclase, the resulting product is not only disruptive because of associated expansion, but it possesses no cementing properties. Calcium sulpho-aluminate hydrates on the other hand are capable of cementitous action, but continuous growth of these crystals induces excessive internal stresses which lead to cracking and eventual disintegration of the concrete.

Class IV Acid attack, which typically results in decomposition of the calcium hydrate compounds of the cement phase, results in rapid deterioration of concrete. The hardened paste is progressively attacked, and even if the aggregate particles are acid resistant, the paste is unable to support them. Both organic and mineral acids are aggressive to concrete. Furthermore acid-producing substances, which may not initially appear to be aggressive, can lead to problems. Examples of these materials include some industrial wastes, animal wastes, silage, fruit juices, milk products and salts of weak bases.

Class V Disruptive crystallization within the concrete, or more often just beneath the concrete surface, leads to this type of attack. The process is set in train when a salt solution in sufficient supply penetrates the concrete, becomes concentrated at a face due to evaporation, and salt crystallization occurs. The crystals may appear on the surface as efflorescence and/or fill pores in the concrete, developing disruptive pressures as they crystallize. This process is akin to frost damage, where the penetrating water freezes and pressures are generated by the ice.

Class VI Deterioration is due to mechanical abrasion, erosion and cavitation. These factors, which may accompany the chemical corrosion actions so far described, are dealt with on pp. 131—9, and so will not be considered here except to say that they invariably exasperate the chemical processes by removing products of reaction, thus exposing greater surface areas for further reaction and allowing the more rapid ingress of reactants.

Class VII Deterioration is due to the change of properties caused by pore penetration of the chemical, the presence of which reduces the strength of concrete by surface-action. This type of process has been noted when oils enter a concrete, and lead to a reduction in the compressive strength, particularly if the oil is hot, and the concrete is dry. This type of phenomenon needs to be considered in designing storage structures.[4.56]

Much has been published on the corrosion and chemical resistance of concrete, and for details on the action of specific chemical species, reference can be made to textbooks by Sheppard,[4.55] Biczok,[4.57] Kleinlogel,[4.58] Kuhl[4.59] and Lea and Desch.[4.60] In most of these publications tables of reactants are given, and the consequences of their presence in contact with concrete are listed. Rather than adopt this type of approach, a few of the more common active aqueous solutions which influence concrete in practice will be considered in somewhat more detail than was presented in the classification system above.

Attack by Soft Waters

The action is a Class I attack in which very pure water, which may or may not carry aggressive carbon dioxide, attacks hardened concrete by a leaching mechanism. Provided that the concrete is dense, the action is generally superficial and, although the concrete surface is leached and presents a sandy appearance with coarse aggregate becoming more obvious as the process advances, the bulk of the concrete remains sound. If on the other hand, lean, pervious concrete is subjected to this type of attack, its useful life may be considerably reduced. Attack of this type is most serious in the case of thin sections such as concrete conduits, flumes, canal linings and low quality concrete pipes or mortar linings of steel pipes. An example of complete destruction of a series of rain water tanks on a Pacific Island was recently noted by the authors. In this case the pure rain water was able rapidly to leach the cement-poor mortars, because the single sized beach sand, which was used, led to a very permeable wall.

Lea and Desch[4.60] discuss the types of soft water and their potential for solvent action on concrete in detail, but Fulton[4.61] provides two test procedures of different levels of sophistication which will probably provide sufficient information to satisfy the requirements of most engineers.

The first of these tests is done as follows:

(a)   determine either the alkalinity or pH value of a sample of the water;
(b)   to another portion of the same water add an excess of chemically pure washed precipitated calcium carbonate (a teaspoonful to 150 ml). Stir for a few minutes and allow to settle and filter. Determine the alkalinity or pH as in (a).

Interpretation of the results is based on the following three possible outcomes:

(*i*)   if the alkalinity or the pH value in (*b*) is greater than in (*a*), the water is not saturated with calcium carbonate;

(*ii*) if the alkalinity or the pH value in (*a*) and (*b*) are equal, the water is in chemical balance with respect to calcium carbonate;

(*iii*) if the alkalinity or pH value in (*b*) is less than in (*a*), the water is supersaturated with calcium carbonate.

The second test procedure, which is used in water engineering practice to control carbonate deposition in pipelines, is based on an expression derived by Langelier:[4.62]

$$pH_s = (pK_2' - pK_s') + pCa^{++} + pAlk \qquad [4.2]$$

where: pH_s is the pH that the water should have in order to be in equilibrium with $CaCO_3$;

pK_2', pK_s' are the negative logarithm of the second dissociation constant for carbonic acid and the activity product of $CaCO_3$ respectively;

pCa^{++} is the negative logarithm of the molal concentration of calcium;

pAlk is the negative logarithm of the equivalent concentration of titratable base (methyl orange indicator).

The Langelier (saturation) index is obtained by subtracting pH_s from the actual pH of the water. A positive index indicates that calcium carbonate will be deposited, while a negative one indicates that the water is aggressive, and will leach the cement paste. A graphical method of obtaining a solution to the above equation is possible using Fig. 4.11. An example taken directly from Fulton and based on a hypothetical water analysis follows:

Temperature 60 °F	pH 7.4
Total dissolved solids	20 parts per 100 000
Ca (as $CaCO_3$)	9 parts per 100 000
Alkalinity (as $CaCO_3$)	10 parts per 100 000

Then $pK_2' - pK_s'$	=	2.37
pCa^{++}	=	3.04
pAlk	=	2.70
Total = pH_s	=	8.11
pH $- pH_s$	=	-0.71

Consequently the water will pick up $CaCO_3$.

Should the engineer be responsible for sampling and testing water, warnings on contamination and changes due to prolonged storage and even handling and exposure should be heeded.

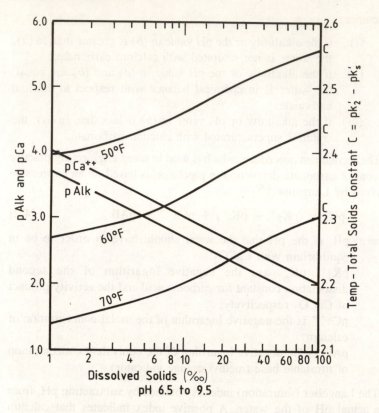

Fig. 4.11 Graphical determination of the Langelier index (after Fulton[4.61])

Probably the most simple yet effective method of reducing the attack of soft water is to bring about exhaustion of the aggression by leading the water over or through beds of crushed limestone before it comes in contact with the concrete.

Sulphate Action

The sulphate ion, present in most Portland and blended cements in the form of gypsum, probably has more effect in modifying the properties of cement paste during all stages of its history than any other anion. Its presence in controlled concentration is required to retard the cement during the plastic phase, and at optimum levels it maximizes strength, and minimzes shrinkage and creep. On the other hand in uncontrolled concentrations, it may cause severe deterioration of the concrete with which it comes into contact.

The reactions between the sulphate ion and hydration products of cements are complex, and are still the subject of detailed studies by cement chemists using more and more elaborate study techniques. Another approach, typified by the work of Mather,[4.63] evaluates the resistance of concretes and concrete making materials to sulphate attack, with the aim of selecting the least susceptible of these for use under adverse conditions.

Broadly considered, the deterioration of Portland cement concretes exposed to sulphates may be ascribed to the following reactions.

(*i*)   The conversion of calcium hydroxide derived from cement hydration reactions to calcium sulphate, and the crystallization of this compound with resulting expansion and disruption.

(*ii*)   The conversion of hydrated calcium aluminates and ferrites to calcium-sulpho-aluminates and sulpho-ferrites or the sulphate enrichment of the latter minerals. The products of these reactions occupy a greater volume than the original hydrates, and their formation tends to result in expansion and disruption.

(*iii*) The decomposition of hydrated calcium silicates.

In the presence of calcium sulphate, only reaction (*ii*), can occur, but with sodium sulphate, both (*i*) and (*ii*) may proceed. With magnesium sulphate all three may occur. For this reason not all sulphates are equally aggressive to hydrated cements, and not only are ion concentrations important, but so are the cation species present. The sulphate salts which can attack concrete are present in soils or crusts on soils in desert areas, and in ground waters. The most important of these salts are gypsum (calcium sulphate), Epsom salt (magnesium sulphate), and Glauber's salt (sodium sulphate). Oxidation of sulphide minerals produces sulphuric acid, which on neutralization produces aggressive sulphate salts. Other sulphate sources may be derived from trade wastes or even waste gases.

Not only do different sulphates attack in different ways, but different cements show very different resistance to such attack. It has long been known that as far as Portland cement performance is concerned, the resistance to attack depends primarily on two factors, *viz* the cement content of the concrete, and the content of tri-calcium aluminate in the cement. For rich mixes, the sulphate attack is slow, and little difference in the resistance of various cements are observed. For lean mixes, the attack is more rapid, and the clear distinction between cements having less than about 7 per cent of tri-calcium aluminate and less resistant cements having higher proportions becomes evident.

Enhanced resistance to sulphate attack is generally displayed by blended cements as discussed by Mehta.[4.64] He suggests that most fly ashes and slags reduce the water demand of the concrete and therefore a less permeable concrete results, even before any chemical effects are considered. Permeability plays a major role in determining the severity of sulphate attack. It should be noted, however, that many fly ashes used in the manufacture of concrete do have a higher water demand than equivalent amounts of Portland cement. As hydration proceeds, a finer pore structure is developed than is usually found in Portland cement pastes. He notes that it is the mineralogy rather than the chemical composition which determines the resistance to attack, and empirical guidelines based on chemical composition of a mineral additive do not prove to be reliable. Nevertheless he notes that as a first approximation, a blended cement will be sulphate resisting when made with a highly siliceous natural pozzolan or a low-alumina fly ash or slag, provided that the proportion of the blending material is such that most of the free calcium hydroxide can be used up during the course of cement hydration.

Under severe conditions, hydraulic cements other than Portland cements should be used, for example supersulphated cement and high-alumina cement. It should be noted here that high-alumina cement should not be used in continuously warm, damp conditions, or in mass construction from which the heat of hydration cannot be easily dissipated. Finally, inert coatings may need to be applied under conditions when even these cements are vulnerable.

Sea Water Exposure

Mehta[4.65, 4.66] in two papers on durability of concrete exposed to sea-water again stresses that of all chemical and physical properties, permeability of concrete is the most important factor influencing performance. He notes that concretes containing even high tri-calcium aluminate cements have excellent service lives if the permeability is sufficiently low. Such concretes are achieved by using mixes having high cement contents and low water:cement ratios, thorough consolidation and control of thermal and shrinkage cracking, and limiting cracks due to mechanical loading. Pozzolanic materials, particularly high proportions of blast furnace slag, improve the impermeability of the concrete by reducing the volume of large pores in the paste fraction of concretes.

Popovics[4.54] notes that sulphate attack from sea water is less than would be expected from its ion concentration. He proposes that chlorides, reacting with the tri-calcium aluminates, inhibit the expansive phenomena associated with sulphate reactions. He also notes the advantage of the formation of tri-calcium chloro-aluminate hydrates which act as absorbers of some of the chlorides which would otherwise be free to attack the reinforcing steel.

Regourd[4.67] summarized her study of the physico-chemical effects of sea water on hydrated cement as follows:

(*i*)   chemical attack by sea-water on cement only occurs in the case of permeable concretes;

(*ii*) C_4AF, in contrast to C_3A, has no deleterious effects;

(*iii*) Portland cements with C_3A contents lower than 10 per cent resist chemical attack in sea-water;

(*iv*) cements containing more than 65 per cent slag are most resistant to sea-water attack;

(*v*) The effects of pozzolan depend on their mineralogical composition and reactivity;

(*vi*) compressive or flexural strengths are not a good basis for assessing durability once reactions commence; a much better basis is the measurement of expansions as they continue.

Processes such as the corrosion of steel, freeze—thaw erosion and cavitation of concrete occur at the same time as concrete attack, and always increase vulnerability of the concrete.

These processes are dealt with in other chapters.

Wear and Erosion

Floors and Slabs

Concrete floors can be produced to have a high degree of resistance to traffic, both wheeled and foot, with foot traffic arising from both people and animals. The resistance of a floor slab is dependent on achieving a hard and durable surface which is plane and free from cracks. The need for avoiding wear has become more acute since the introduction of high-level stacking in warehouses, as slight variations of surface level are magnified at the top of tall stackers to the extent that interference can occur between the stacker and the shelves. The properties of the surface layer are determined largely by the quality of the concreting operations. Furthermore, the timing of these operations, and particularly the timing of finishing and curing, is particularly critical. If errors are made in timing, undesirable conditions occur in the wearing surface which may lead to soft or dusting surfaces, permeable concrete, cracking and generally poor durability.

Without special precautions, the top surface of the concrete can easily be of lower quality than the remainder. The usual practice of specifying a minimum concrete strength and hoping thus to achieve a satisfactory floor can easily be negated if these precautions are not taken. For instance, if bleed water is worked back into the surface of a floor, the water:cement ratio is increased, with resultant decrease in strength and other desirable properties. It should, however, be noted that much damage to floors, particularly from hard-tyred traffic such as fork-lifts, occurs at cracks in the floor and these cracks may be the result of deficiencies in structural design rather than in surface finish.

Although test results do not satisfactorily describe the actual behaviour

of concrete in a floor, where abrasion and impact are often combined
with flexural and shearing loading, some indication can be obtained of
the comparative influence of various concrete properties and procedures.
The effect of curing has been shown by Fentress[4.68, 4.69] to be especially
significant, and particularly the need for curing to be started early, as
shown in Figs 4.12 and 4.13. These results have been confirmed by more
recent work by Senbetta and Malchow[4.70] and by Kettle and
Sadegzadeh.[4.71] In both these investigations, tests using different forms
of abrasion machines showed that, when the best curing compound had
been used, the loss due to wear was only 50 per cent of the loss when
the concrete had been air-cured. The best curing compound also gave
better results than curing under plastic sheeting or wet burlap. The
finishing procedure (wood float or steel trowel) and the time at which
the finishing was carried out was less significant than the time at which
curing was started (see Figs 4.14 and 4.15) but power floating led to
a marked improvement in wear resistance, particularly for drier concrete
mixes and when repetitions of power floating had been used.[4.71] In his
investigation, Fentress[4.69] deliberately introduced malpractices leading
to dusting, crazing and scaling in the surfaces. These surfaces were no
less resistant to the wear test for one cycle but further wear tests after
some ageing led to the complete removal of the surface layer. This
suggested that this layer had continued to shrink and had little bond with
the underlying concrete. The common defects of scaling (loss of top

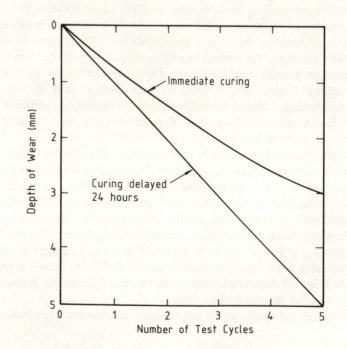

Fig. 4.12 Effect of
delay in curing on
depth of wear
(after Fentress[4.69])

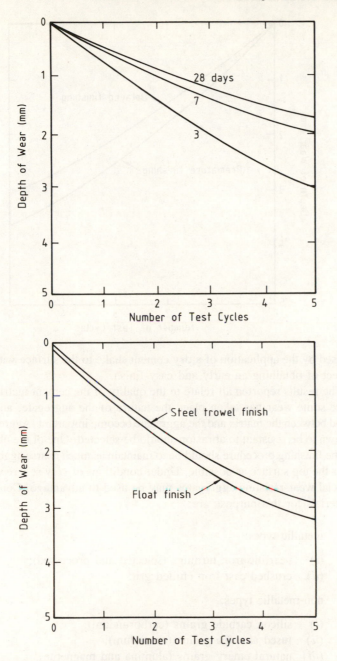

Fig. 4.13 Effect of time of curing on depth of wear (after Fentress[4.69])

Fig. 4.14 Effect of type of finish on depth of wear (after Fentress[4.69])

surface at an early age when exposed to traffic) and dusting (the presence of a loose powdery surface on hardened concrete) are believed to be primarily due to premature finishing operations undertaken while bleed water is still visible. The further common defect of crazing is frequently

Fig. 4.15 Effect of
time of finishing on
depth of wear
(after Fentress[4.69])

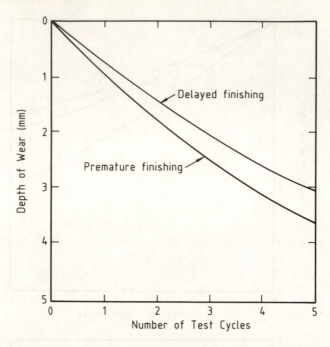

caused by the application of a dry cement shake to the surface with the object of obtaining an early and easy finish.

The results reported all relate to the quality of the cement matrix but once some wear has taken place the nature of the aggregate, and the bond between the matrix and the aggregate become important. Aggregates known to be resistant to abrasion should be selected. One of the objects of the finishing procedure should be to maintain as much coarse aggregate near the top surface as possible. Under conditions of very severe wear, special wear-resisting aggregates may be used to advantage. Common materials for this purpose are:

(*a*) metallic types:

    (*i*) pearlitic iron turnings (selected and processed);
    (*ii*) crushed cast iron chilled grit;

(*b*) non-metallic types:

    (*i*) silicon carbide grains (carborundum);
    (*ii*) fused alumina grains (corundum);
    (*iii*) natural emery grains (alumina and magnetite).

Materials from both groups if carefully used provide a considerable increase in wear resistance. Of the metallic types, the pearlitic turnings have been shown to be better than the chunky cast iron grains as they are less prone to loosening from the matrix under impact. Where slip resistance is of prime importance, or where rusting of the metals would be objectionable, the non- metallic types are preferred. Claims have been made that fibre-reinforced concrete is superior in wear resistance to conventional concrete[4.72] but it appears that the improved performance arises largely from the superior flexural resistance with a resulting reduction in cracking. Liquid hardeners have been proposed but there is not much technical support for their use except possibly where very poor and porous concrete needs some remedial action. Proper curing has been found to be more effective in improving the abrasion resistance than the application of concrete liquid hardeners. On the other hand, penetrating sealing and hardening treatments can significantly increase abrasion resistance.[4.73] Useful advice on the methods of producing satisfactory floor slabs is contained in the report of ACI Committee 302.[4.68]

Methods of Test

The performance of any floor is affected by the type of wear involved. Many test procedures have been developed over the last 60 years but many are also outmoded as the type of traffic to which industrial floors are subjected has changed. The essentials of any test method, as set out by Sawyer,[4.74] are:

(i)   the test specimen should be subjected to conditions similar to those experienced in practice;
(ii)   the test must be severe enough to cause deterioration of any concrete surface;
(iii)   the test must be sensitive to variations in surface condition;
(iv)   the test specimens should be easy to make, store and handle;
(v)   the test method should be easy to follow and the cost and time should not be excessive;
(vi)   the test results must be reproducible.

Sawyer[4.74] claimed that the German test (to DIN 51951) fulfilled all these requirements, but the requirement (i) is probably not now so well met as when the test was first developed, since industrial traffic has changed from steel wheeled dollies to hard-tyred fork-lifts. Most standard tests (e.g. ASTM C779) involve the use of quite small specimens which therefore have to be finished by hand processes. Kettle and his colleagues have built a larger machine, which although needing specimens which breach requirement (iv) has the advantage that practical site methods of power finishing can be used.[4.71]

Pavements

Pavements for road traffic and taxi-ways and run-ways for airfields require a textured surface to provide resistance to skidding and may be grooved to reduce the risk of aquaplaning. This surface must be resistant to wear from pneumatic tyred wheels and the requirements are therefore somewhat different from those of industrial floors. The ridges produced by the texturing must contain sufficient hard sand grains and must have a cement matrix of adequate strength and hence low water:cement ratio. In the USA state highway departments which receive Federal funding are required to incorporate fly ash in pavement concrete. Carrasquillo[4.75] has shown that in resistance to abrasion fly ash concrete is equal to or better than concrete with no fly ash, when tested by ASTM C944 'Rotating cutter method'. It is important to note, however, that in her tests, Carrasquillo provided wet-curing for 28 days, which is a condition which will assist fly ash concrete and is unlikely to occur on site, where curing compounds are the most usual method of curing. More useful are the comparative field tests reported by Dunstan and Joyce.[4.76] Two large areas of the cargo apron at Gatwick Airport (London, England) were paved using a high fly ash concrete and an ordinary Portland cement concrete, both of pavement quality. The concretes were required to have a flexural strength of 4 MPa at 28 days and an equivalent cube strength at 28 days (as determined from cores) of 35 MPa. The mix proportions of the two concretes are shown in Table 4.3. Both concretes were placed by a paving train of which the last element was a brush and cure unit. After four years of directly comparable trafficking and weathering, the Portland cement concrete had a pitted texture showing clean exposed aggregate and in some bays little remaining evidence of the brushed

Table 4.3 Pavement quality concretes at Gatwick Airport[4.76]

	High fly ash concrete	Normal Portland cement concrete
Aggregate		
40 mm	528 kg m^{-3}	673 kg m^{-3}
20 mm	317	398
10 mm	210	231
Sand	545	598
Cement	275	375
Fly ash	232	
Water	134 l m^{-3}	115 l m^{-3}
AEA	2000 ml m^{-3}	525 ml m^{-3}
Plasticizer	560 ml m^{-3}	600 ml m^{-3}

texture. The majority of the high fly ash bays still had a surface texture which was generally free from pitting and exposure of aggregate and the brushed texture was still in excellent condition.

Runways which are affected by high temperature jet engines require special attention to prevent erosion. The US Navy has found that a refractory concrete containing stainless steel fibres will provide satisfactory performance.[4.77] Other fibre-reinforced concretes have been less successful as fibres become loosened and form a source of 'foreign object damage to jet engines' and a 'potential injury hazard to personnel'.[4.78]

Hydraulic Structures

Hydraulic structures may be subject to physical erosion which arises from particles of rock, ice or debris carried in the flowing water; from collapse of vapour bubbles formed by pressure changes in high-velocity flows; and from fluctuations of water pressure in and on the concrete under conditions of unsteady flow. ACI Committe 210 reports on these causes and on the effect of chemical attack.[4.27]

Erosion of concrete by water-borne solids can be very severe, as quite large particles can be moved in water at quite a low velocity. Concrete in the invert of the diversion tunnel at Anderson Park Dam was worn away to a depth of about 80 mm while it was in use for 43 months carrying water which contained large amounts of silt, sand and gravel. If the hydraulic design does not provide for trash-racks or debris traps wear will occur. The rate of erosion depends on the quantity, shape, size and hardness of the transported material, on the velocity of the flow and on the quality of the concrete. Better resistance to wear is achieved by using high strength concrete containing the maximum amount of hard coarse aggregate. Where hard aggregates are not available, very strong concretes with strengths over 100 MPa can be used with high-range water-reducing admixtures and silica fume.[4.79] The special action of abrading ice has been studied by Hoff[4.80] who concludes that the phenomenon called 'ice abrasion' is more complex than a simple abrading of the concrete surface as the action of ice floes includes repeated crashing of ice into the structure as well as ice, often containing rock particles, sliding past. Further work is needed before a definitive approach to concrete mix design (and to structural design) can be established.

Vapour bubbles will form in running water wherever the pressure in the liquid is reduced to its vapour pressure. These bubbles flow downstream and on entering a zone of higher pressure collapse with great impact. Repeated collapse of such cavities near the surface of concrete will cause pitting. Cavitation erosion is readily recognized from the nature of the holes or pits formed, which are quite different from the smoother worn surface produced by abrasion from solids. Examples of locations in which cavitation damage may occur are shown in Fig. 4.16 from which it can be seen that boundary irregularities and changes of profile are the major sources of damage. Cavitation damage has not been found likely to occur in open channels when stream velocities are less than 12 m s^{-1}. In closed conduits where the air pressure is reduced by the water flow, damage has been found to occur at stream velocities as low as 8 m s^{-1} and at higher velocities the forces are sufficient to erode away large quantities of high quality concrete and to penetrate steel plates in a comparatively short time. Once cavitation damage has started, the roughened surface provides a source for new cavities to form and the damage can be extended far downstream. The best concrete made will

Fig. 4.16 Locations of cavitation erosion: (a) offset into flow; (b) offset away from flow; (c) abrupt curvature away from flow; (d) abrupt slope away from flow; (e) void or transverse groove; (f) roughened surface; (g) protruding joint (from ACI Committee 210[4.27])

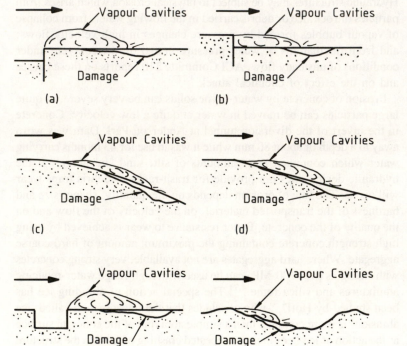

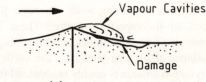

not withstand the forces of cavitation indefinitely but some things can be done to delay the onset of damage. Resistance increases with higher strength of concrete and if there is a risk of cavitation a minimum strength of 40 MPa should be demanded. In the selection of aggregates, good bond is more important than hardness as the action of cavitation is generally one of plucking out large particles. There is an advantage in limiting the maximum size of aggregate to say 20 mm. Watkins and Samarin[4.81] confirmed that the resistance to cavitational erosion depends primarily on the quality of the paste. The nature of the coarse aggregate was only found to be of significance when it became exposed after erosion of the matrix. There is a conflict of requirement relating to maximum size of aggregate when both cavitation and abrasion are factors, a situation that can often arise when cavitation removes particles of concrete which are transported downstream. The use of fibre-reinforced concrete in hydraulic structures is discussed under the heading of repair (see pp. 323−7).

In unsteady flow conditions fluctuating water pressures may cause disruption of the concrete, particularly if the concrete is permeable or cracked and internal pressures can build up. Vibration of low amplitude and high frequency has been measured in hydraulic structures. At McNary Dam an amplitude of 0.0005 mm and a frequency of 150 Hz was observed in the outlet works. Training walls in reinforced concrete have been observed to undergo vibrations at their own natural frequency and such vibration could lead to fatigue failure.

Design Errors

Most design is subject to some form of checking and approval, either within the designer's office as part of a quality assurance scheme (see Chapter 2), or by a regulatory authority or an insurer. In Germany all projects with load-bearing elements have to be passed by a licensed proof-engineer before construction and the concept of proof-engineering has spread to many other countries. Although such checking may reduce the incidence of structural collapse, it does little to prevent design errors which will in the long run lead to deterioration, malfunction and possible long-term failure. It is clearly an error of design if a concrete structure, constructed in accordance with the design, is unable to resist expected external conditions. Failure in such a case is manifested as an unacceptable difference between expected and observed performance. Failures resulting from any of the conditions discussed earlier in this chapter may therefore be justly classified as design errors. In this section we shall discuss only those errors which involve structural misbehaviour and which might therefore have been forestalled by proper proof-engineering directed as at present towards structural behaviour. Most

of these structural errors are manifested as excessive cracking which may
affect function and permit corrosion of reinforcement.

Common errors are classified under a number of headings and typical
examples are given.[4.82]

Misconceptions of Structural Action

Design procedures often include simplifying assumptions as to the way
in which the final structure will behave. If the designer does not ensure
that the structure can, in fact, behave in the assumed way cracking occurs
to the extent necessary. A common example is when a moment-free (pin-
jointed) condition is assumed and not achieved. A wall–floor joint in
a 1 ML elevated water tank is shown in Fig. 4.17. In the design of the
tank, this joint had been assumed to behave as a pin but the face marked
'bitumen paint' did not have enough separation. As a result, when the
tank wall rotated under water load the faces came into contact and the
corner behaved as a knee-joint transmitting moment. The compression
components of this moment produced a diagonal tensile resultant which
caused a substantial piece of unreinforced concrete to spall off, as shown
in Fig. 4.17(b), and exposed reinforcement then corroded.

When connected members have very different rigidities, forces may
tend to migrate from the path provided by the designer into an alternative
more rigid member. The prestressed concrete bridge superstructure
shown in Fig. 4.18 consists of precast segmental box girders connected
by an *in-situ* deck transversely prestressed. At the bearing an *in-situ*

Fig. 4.17 Wall-floor joint in an elevated water tank

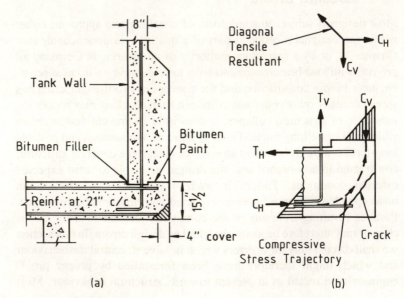

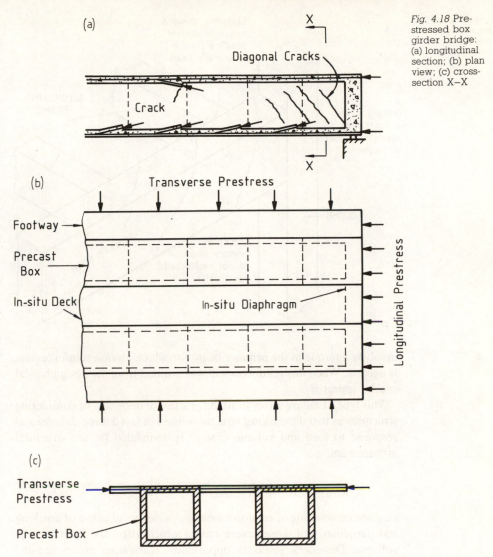

Fig. 4.18 Pre-stressed box girder bridge: (a) longitudinal section; (b) plan view; (c) cross-section X–X

diaphragm is provided. The stiffness of this diaphragm in the vertical transverse plane means that the transverse prestress intended to put compression into the portion of the deck shown hatched in the figure does not work. The flexural stiffness of the same diaphragm reduces the amount of longitudinal prestress reaching this same area and there is quite inadequate provision for the slab to carry traffic loading.

The primary beam shown in Fig. 4.19 has to transfer the negative moments from the secondary beams to the supporting columns. As the columns and the primary beams are both stiff members, this transfer

Fig. 4.19 Secondary
beams cause
torsion in a
primary beam

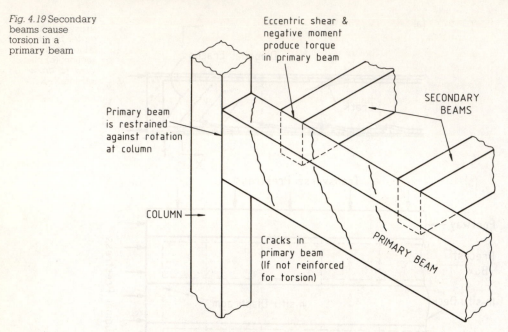

Eccentric shear &
negative moment
produce torque
in primary beam

SECONDARY
BEAMS

Primary beam
is restrained
against rotation
at column

COLUMN

Cracks in
primary beam
(If not reinforced
for torsion)

PRIMARY BEAM

involves a torque in the primary beam for which torsion reinforcement
is needed. When this reinforcement was omitted from the design helical
cracks appeared.

This type of failure arises from the practice of designers of considering
structures as two-dimensional structures when in fact a three-dimensional
response to load and volume change is demanded by the structural
arrangement.

Reinforcement Detailing

Inadequate detailing of reinforcement is a widespread cause of cracking
and particularly of those severe cracks which affect the limit state of
collapse. Designers, given the opportunity, learn from experience and
in many organizations this source of trouble is steadily reduced. Members
which appear to be particularly susceptible to severe cracking as a result
of insufficient steel or badly arranged steel are those which carry local
loads, such as corbels, supports for bridge bearings, walls supporting
column bases, prestressing anchorages and column capitals. Park and
Paulay[4.83] observe that 'to reinforce a concrete structure correctly the
designer must possess a penetrating understanding of its behaviour —
an understanding beyond the establishment of the equations of equilibrium
and strain compatibility'. The need therefore still exists for some standard
details which show successful arrangements in the situations outlined.

Conventional drawings tend to ignore the physical size of the bars and the limitations on bend shape in practical reinforcing (see Fig. 4.20). Equally important is the need to ensure that the steel is incorporated in the way it was designed.

Extensive tests on corbels by Kriz and Raths[4.84] identified six different failure mechanisms which may occur and against which reinforcing is needed (see Fig. 4.21). An arrangement of reinforcing which takes account of these potential failure modes is shown in Fig.

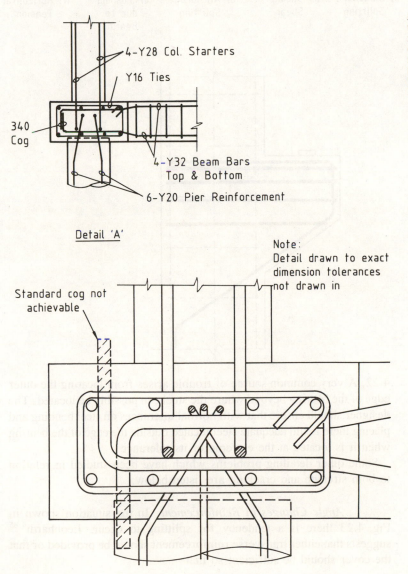

Fig. 4.20 Conventionally drawn reinforcement cannot always be fitted in (example from Gage R W, Humes ARC, NSW, Australia)

4-Y28 Col. Starters

Y16 Ties

340 Cog

4-Y32 Beam Bars Top & Bottom

6-Y20 Pier Reinforcement

Detail 'A'

Standard cog not achievable

Note:
Detail drawn to exact dimension tolerances not drawn in

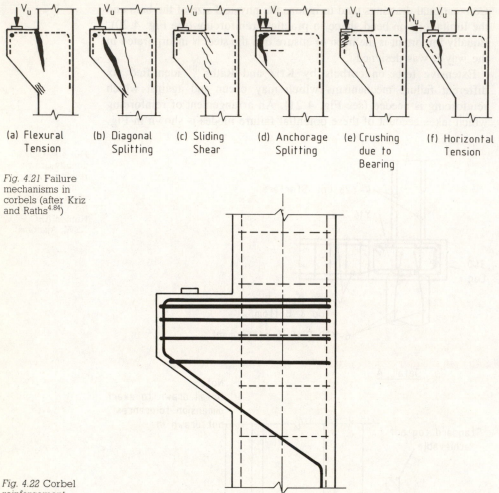

(a) Flexural Tension   (b) Diagonal Splitting   (c) Sliding Shear   (d) Anchorage Splitting   (e) Crushing due to Bearing   (f) Horizontal Tension

Fig. 4.21 Failure mechanisms in corbels (after Kriz and Raths[4.84])

Fig. 4.22 Corbel reinforcement

4.22. A very common source of trouble arises from locating the outer edge of the bearing beyond where the steel can possibly be located. The designer should ensure that with normal tolerances on steel bending and placing there is still adequate steel located outside the edge of the bearing when it is located at the extreme of its tolerance.

Some other detailing problems which have been studied in relation to both strength and cracking are listed below:

Angle Changes in Reinforcement In the situation shown in Fig. 4.23 there is a tendency for splitting to occur. Leonhardt[4.85] suggests that either transverse reinforcement should be provided or that the cover should be sufficient so that

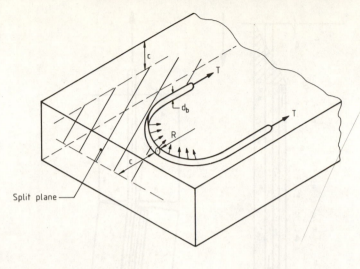

Fig. 4.23 Curved tensile bars cause transverse splitting stresses

$$R > Kd_b^2/c$$

where: K is proposed as 150 for grade 20 and 90 for grade 30 concrete.

Splices Joggled compression splices[4.86] and eccentric tension splices can both cause cracking of the sort shown in Fig. 4.24. A lapped splice in a beam can give rise to the cracking shown in Fig. 4.25 with dire results especially for the centre bars which are not adequately restrained against vertical movement.[4.83] Perhaps one of the most notorious splices ever made is that shown in Fig. 4.26 and reported by Feld[4.87] which led to a total collapse of a mess hall at Coronado, California.

Frame Corners Investigations in Sweden[4.88] and in the UK at Cement and Concrete Association[4.89–4.91] and at Nottingham University[4.92, 4.93] showed that many conventionally reinforced corner joints had strengths well below that of the members being joined, particularly for opening joints. Figure 4.27 compares results from the various investigations. The efficiency of Detail 1 is (not surprisingly) low but that for Detail 2 which has often been regarded as good practice is almost as low. As steel percentages are increased the efficiency is reduced. Details 5 and 6 showed the most promise, but only if the diagonal steel is in intimate contact with the main tension steel possibly by welding.

Some arrangements of reinforcement actually induce cracks and these should be carefully avoided. For example, severe cracking can be seen when all the top bars in a slab are terminated at the same cross section.

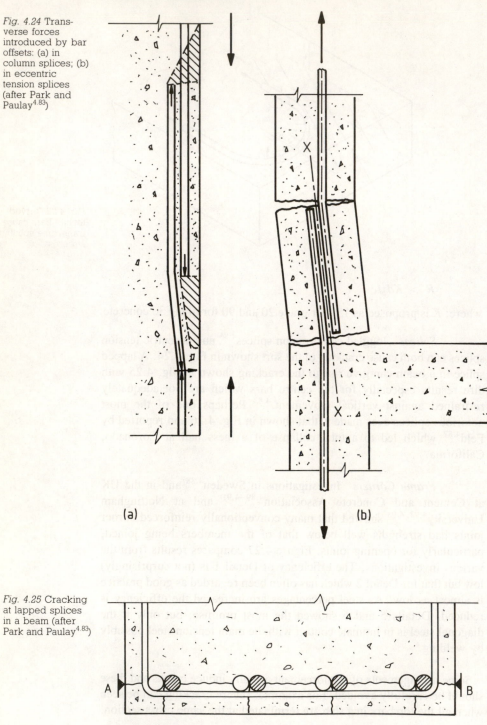

Fig. 4.24 Transverse forces introduced by bar offsets: (a) in column splices; (b) in eccentric tension splices (after Park and Paulay[4.83])

(a)   (b)

Fig. 4.25 Cracking at lapped splices in a beam (after Park and Paulay[4.83])

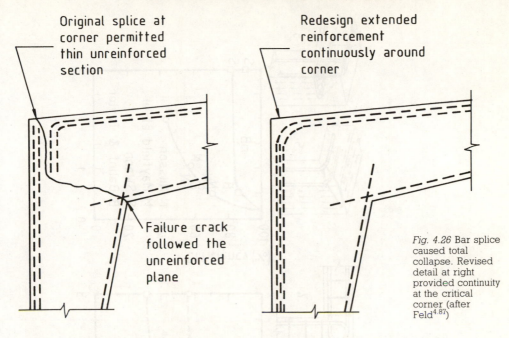

Original splice at corner permitted thin unreinforced section

Redesign extended reinforcement continuously around corner

Failure crack followed the unreinforced plane

Fig. 4.26 Bar splice caused total collapse. Revised detail at right provided continuity at the critical corner (after Feld[4.87])

Diffusion of Concentrated Loads

It is always difficult for a designer in reinforced concrete to provide adequate load paths for the diffusion into the structure of concentrated loads from column heads, prestressing anchorages and bearings. Solutions to these problems are sometimes based on the theory of elasticity applied to homogeneous masses where reinforcement is incorporated to carry the net tensile force. This process will always lead to a statically satisfactory system provided adequate anchorage of the reinforcement is present after cracks form. The result is not always the most economical one.

Restrained Shrinkage and Temperature

The reinforced concrete retaining and sea wall shown in Fig. 4.28 is a typical example of this form of cracking. The wall has a total length of 430 m, it is constructed in 9 m panels and is about 600 mm thick at the base. One or two vertical cracks completely penetrate the wall above the base and were first observed within 14 days of casting and were found subsequently to increase in width and length. As the wall contains vertical steel, it is likely that the cracks occur over this steel which is then at risk from chloride ingress. An interesting suggestion has been made by

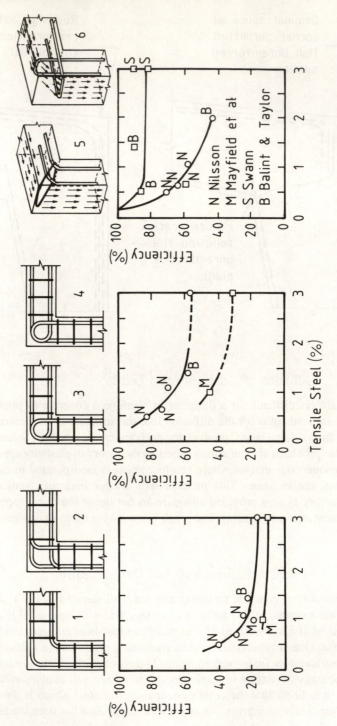

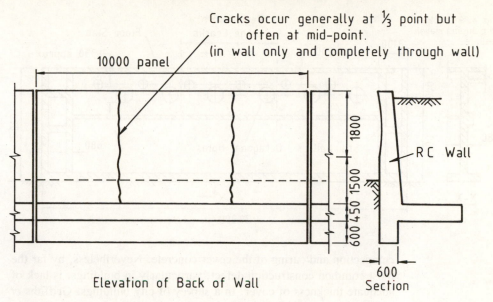

Cracks occur generally at ⅓ point but
often at mid-point.
(in wall only and completely through wall)

10000 panel

1800

1500

600 450

RC Wall

Elevation of Back of Wall

600
Section

Fig. 4.28 Thermal cracking in retaining wall

Hamada *et al.*[4.94] that rather than trying to prevent or control these cracks the designer should provide for their easy repair by preforming vertical V-shaped cracks in the first lift of concrete. This procedure should then provide cracks which are straight and easily sealed without excessive preparation.

The pumping station wall shown in Fig. 4.29 is 680 mm thick and 30 m long without a joint. The centre 10 m of the wall which includes a number of octagonal openings was found to contain major cracks up to 0.5 mm wide through which water was passing. The wall was reinforced with 0.275 per cent of steel as 20 mm bars, but this amount did not prove adequate to control the cracks. Methods of design to control cracks in water-retaining structures have been studied closely by Hughes and his colleagues at Birmingham University.[4.95–4.97]

Construction Errors

Construction and supervision deficiencies were found in an Australian survey to be a major cause of defects leading to cracking. Thirty-six per cent of the defects were due to these causes.[4.82] A well-known expert on structural failures said that he never found a failure *caused* by poor concrete but he had never investigated one that did not *contain* poor or inferior concrete.[4.98] This comment related to collapses of structures but when the definition of 'failure' is extended as we are doing here the quality of concrete does become much more important. For example, the protection afforded to steel is greatly dependent on the

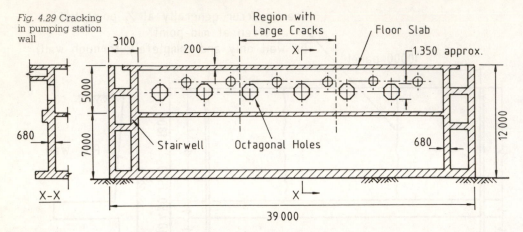

Fig. 4.29 Cracking in pumping station wall

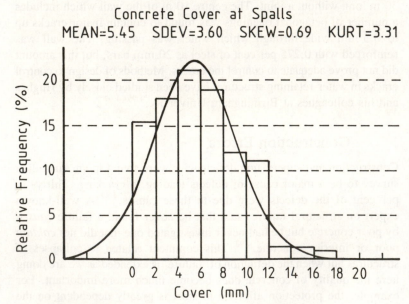

compaction and curing of the cover concrete. Nevertheless, by far the most common construction defect, particularly in buildings, is lack of adequate thickness of cover. In a survey of city buildings, Griffiths *et al*.[4.99] found that in 95 buildings the mean cover to reinforcement at locations which displayed cracking, spalling or corrosion was only 5.45 mm with the distribution of cover as shown in Fig. 4.30. The same team then examined 16 buildings under construction and summarized their views of the causes of the lack of cover found in these buildings. Table 4.4 is the result of their investigation and these findings deserve

Fig. 4.30 Distribution of cover at 227 spalls (after Griffiths *et al*.[4.99])

careful attention by all those involved in building.[4.100] Civil engineering construction may have better supervision and possibly more competent contractors but the problems are by no means eliminated in that field.

The Australian survey revealed that construction defects could be grouped into four classes. Some examples are given for each class.

A   Deficiencies in the control of concrete materials, batching and mixing.
          Use of salt water as mixing water.
          Excess fines in the aggregates.
B   Inadequate preparation before concreting.
          Salt water contamination of reinforcement.
          Lack of cover to reinforcement.
C   Inadequacies of placing and subsequent treatment.
          Plastic cracking and settlement cracking.
          Lack of curing.
D   Faults of construction planning and procedure.
          Overloading of members by construction loads.
          Loading of partially constructed members.
          Differential shrinkage between sections of construction.
          Omission of designed movement joints.
          Unexpected behaviour and restraint during prestressing.

Most of these faults in classes A, B and C can easily be avoided by proper supervision by contractors and owners. The group in class D may be more difficult to recognize and some examples are given.

Cracking occurred in the end diaphragm of a prestressed concrete box girder bridge. The diaphragm over the piers was very thick (2100 mm) and contained no reinforcement at the top under the deck as it was to be transversely prestressed. A very wide crack appeared at an early stage of construction and had to be repaired to ensure that the diaphragm operated to transfer shear loads to the bearings. Six litres of epoxy was injected under pressure before the deck slab was cast. The crack was

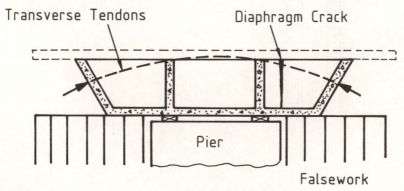

Transverse Tendons   Diaphragm Crack

Pier

Falsework

Fig. 4.31 Bridge diaphragm cracked during construction

Table 4.4 Problems relating to placing and condition of reinforcement[4.100]

Category	Problem	Cause	Frequency	Consequence
Reinforcement, chairs and ties	Reinforcement incorrectly shaped or sized	Inadequate engineering design and documentation	Frequent	Major
		Incorrect scheduling	Infrequent	Major
		Incorrect fabrication on and off site	Infrequent	Major
		Damage during handling and after placement	Infrequent	Minor
	Reinforcement in incorrect position	Deformed bar chair	Infrequent	Major
		Inadequate reference lines	Frequent	Major
		Bar chairs missing or out of place	Frequent	Major
		Inappropriate bar chairs — shape, size, material	Frequent	Major
		Clashing reinforcement	Frequent	Major
		Reinforcement cage too heavy to adjust	Infrequent	Major
		Inaccessible location	Frequent	Major
		Negligent placement and fixing	Frequent	Major
		Ties missing and loose	Frequent	Major
		Reinforcement position altered after placement due to heavy treatment of other trades	Infrequent	Minor

			Frequency	Severity
Conduits and inclusions clashing with reinforcement	Bar chairs too close to edge	Placed over critical areas such as drip drains	Frequent	Major
		Displaced due to inappropriate bar chair	Frequent	Minor
		Displaced due to rough treatment	Infrequent	Minor
	Ties too close to edge	Ties bent out towards edge	Frequent	Major
	Due to off-site problems	Lack of co-ordination of services and structure	Frequent	Major
		Position not documented	Frequent	Major
		Lack of communication between consultants	Frequent	Major
	Due to on-site problems	Inadequate indication on site of correct position	Frequent	Major
		Careless placement of conduits	Frequent	Major
		Inadequate fixing of conduits in correct position	Frequent	Major
Formwork	Formwork incorrectly positioned	Incorrect setting out	Infrequent	Major
		Negligent positioning	Infrequent	Major
		Incorrect or inadequate drawings	Infrequent	Major
		Inadequate tolerances	Infrequent	Major
	Contamination	Inadequate cleaning out	Frequent	Major
	Movement during pouring	Inadequate form thickness	Infrequent	Major
		Inadequate bracing	Infrequent	Major

almost certainly caused by slight settlement of the falsework relative to the pier, resulting in a rigid-body rotation of the wing of the diaphragm. Neither the designer nor the builder had appreciated that this was a highly probable construction eventuality.

A series of prestressed concrete railway bridges, constructed in Queensland, Australia, during 1961−62 developed cracks up to 0.25 mm in width. It was feared that these cracks might affect the long-term durability of the superstructure. The cracks occurred in thin cantilevers which extended from the sides of the box girders and cracked through their full depth to the side of the box. The design did not make provision for differential shrinkage between components of the structure and this was believed to be the cause of the observed cracking.

A two-span continuous box-girder bridge, 132 m long and 8.5 m between kerbs, developed cracks in the two abutment diaphragms as shown in Fig. 4.32. The cracks were noticed about two weeks after the girders had been post-tensioned. It was suggested by the designer that the cracks had arisen from a combination of causes which had resulted in an unexpected stress pattern in the diaphragms. Suggested causes were:

(*i*)   stresses resulting from differential creep after post-tensioning;
(*ii*)   arching of the diaphragm to the two outer bearings;
(*iii*) differential temperature effects between the boxes and the diaphragm.

The two previous cases are examples of the need for designers to consider and possibly specify appropriate construction sequences. The sequence of prestressing and the interaction between prestressing and fixing of bearings and other restraints needs careful consideration by the designer and attention by the construction engineer. The intention of the designer is not always clearly conveyed to the constructor. In the case of the splice shown in Fig. 4.26, Feld remarks that 'the designer had located a splice in order to reduce cost of shipping fabricated reinforcement. The designer stated that his intention had been that the top reinforcing bars bending around the corner be intermeshed with vertical bars. Unfortunately this intention was not conveyed to the fabricator'.

Overloading during construction, particularly when very fast construction is adopted, can arise from stacking of bricks and other building materials on structures which are only a few days old. The special problem of overloading in multi-storey buildings from propping between floors has been examined by Grundy and Kabaila and other more recent authors.[4.101−4.103] They have shown that the maximum construction load can exceed the final service load by a substantial margin and quote a case when the construction load is 50 per cent greater than the service load.

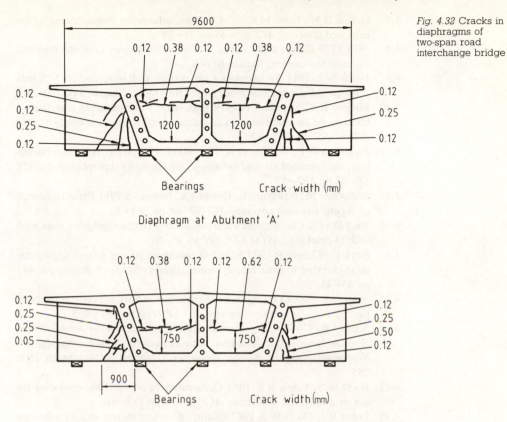

Fig. 4.32 Cracks in diaphragms of two-span road interchange bridge

References

4.1   Wright J R, Frohnsdorff G 1985 Durability of building materials: Durability research in the United States and the influence of RILEM on durability research. *Materials and Structures, Research and Testing* (RILEM, Paris) **18**(105): 205–14

4.2   Anon 1988 Concrete durability — a multibillion-dollar opportunity. *Concrete International Design & Construction* **10**(1): 33–5

4.3   Kaetzel L J, Clifton J R 1987 Maintenance and implementation of an expert system for durable concrete *ACI SP–106* pp 75–86

4.4   Clifton J R, Kaetzel L J 1988 Expert systems for concrete construction. *Concrete International Design & Construction* **10**(11): 19–24

4.5   Roper H, Lovell G, Trueman M 1989 *Auscon, an Expert System for Durable Concrete* School of Civil and Mining Engineering, University of Sydney

4.6   Philleo R E 1987 Frost susceptibility of high-strength concrete *ACI SP–100* pp 819–42

4.7   Hollon G W, Prior M E 1974 Factors influencing proportioning of air entrained concrete *ACI SP−46* pp 11−24

4.8   PCA 1979 *Design and Control of Concrete Mixtures* 12th edn Portland Cement Association, Skokie Illinois

4.9   Dolch W L 1984 Air entraining admixtures. In Ramachandran V S (ed) *Concrete Admixtures Handbook* Noyes New Jersey Chapter 5 pp 269−302

4.10 Mielenz R C, Sprouse J H 1979 High-range water-reducing admixtures: effect on the air-void system in air-entrained and non-air-entrained concrete *ACI SP−62* pp 167−92

4.11 Perenchio W F, Whiting D A, Kantro D L 1979 Water reduction, slump loss, and entrained air-void systems as influenced by superplasticizers *ACI SP−62* pp 137−55

4.12 Kobayashi M, Nakakuro E, Kodama K, Negami S 1981 Frost resistance of superplasticized concrete *ACI SP−68* pp 269−82

4.13 Stark D 1976 Characteristics and utilisation of coarse aggregates associated with D-cracking. *ASTM STP 597* pp 45−58

4.14 Stark D, Klieger P 1973 Effect of maximum size of coarse aggregates on D-cracking in concrete payments. *Highway Research Record No 441* pp 33−43

4.15 Sawan J S 1987 Cracking due to frost action in Portland cement concrete pavements — a literature survey *ACI SP−100* pp 781−803

4.16 Schiessl P 1976 *Zur Frage der zulassigen Rissbreite und der erforderlichen Betondeckung im Stahlbetonbau unter besonderer Berucksichtigung der Karbonatisierung des Betons*. Deutscher Ausschuss für Stahlbeton, Heft 255

4.17 Ho D W S, Lewis R K 1983 Carbonation of concrete incorporating fly ash or a chemical admixture *ACI SP−79* pp 333−46

4.18 Potter R J, Ho D W S 1987 Quality of cover concrete and its influence on durability *ACI SP−100* pp 423−45

4.19 Roper H, Baweja D, Kirkby G A 1985 Unpublished report, School of Civil and Mining Engineering, The University of Sydney February 1985

4.20 Wood M 1988 Stormy debate over acid rain. *The 1988 World Book Year Book* World Book Inc pp 50−65

4.21 Kong H L, Orbison J G 1987 Concrete deterioration due to acid precipitation. *ACI Materials Journal* **84**(2): 110−16

4.22 Gauri K L 1980 Deterioration of architectural structures and monuments. *Polluted Rain* Environmental Research Series **17** Plenum Press, New York pp 125−45

4.23 Webster R P, Kukacka L E 1984 *Effects of Acid Deposition on the Properties of Portland Cement Concrete: State of Knowledge* Pub No BNL 35730 Brookhaven National Laboratory 37 pp

4.24 Fisher T 1983 When the rain comes. *Progressive Architecture* **64**(7): 99−105

4.25 Figg J, Bravery A, Harrison W 1987 Covenham reservoir wave wall — a full-scale experiment on the weathering of concrete *ACI SP−100* pp 469−92

4.26 Thistlethwayte D K B 1972 *Control of Sulphides in Sewerage Systems* Butterworths

4.27 ACI Committee 210 Erosion of concrete in hydraulic structures *ACI 210R—87* American Concrete Institute 1987 22 pp

4.28 Bares R A 1978 Furane resin concrete and its application to large diameter sewer pipes *ACI SP—58* pp 41—74

4.29 American Society of Civil Engineers 1983 *Existing Sewer Evaluation and Rehabilitation* ASCE Manual & Reports on Engineering Practice No 62 116 pp

4.30 Wu G W 1986 Heat-resistant concrete pavements. *The Military Engineer* **78**(509): 487—9

4.31 Abrams M S 1971 Compressive strength of concrete at temperatures to 1600F *ACI SP—25* pp 33—58

4.32 Campbell-Allen D, Desai P M 1967 The influence of aggregate on the behaviour of concrete at elevated temperatures. *Nuclear Engineering and Design* **6**(1): 65—77

4.33 Lankard D R, Birkimer D L, Fondriest F F, Snyder M J 1971 Effects of moisture content on the structural properties of Portland cement concrete exposed to temperatures up to 500F *ACI SP—25* pp 59—102

4.34 Joint Committee of Institution of Structural Engineers and Concrete Society 1978 *Design and Detailing of Concrete Structures for Fire Resistance Interim Guidance* London

4.35 Campbell-Allen D, Low E W E, Roper H 1965 An investigation of the effect of elevated temperatures on concrete for reactor vessels. *Nuclear Structural Engineering* **2**(4): 382—8

4.36 Fisher R 1970 Uber das Verhalten von Zementmortel und Beton bei hoheren Temperaturen. *Deutscher Ausschuss fur Stahlbeton*, Heft 214 pp 61—128

4.37 Bannister J L 1968 Steel reinforcement and tendons for structural concrete. *Concrete* (London) **2**(7): 295—306

4.38 Kordina K 1976 Fire rating in buildings. *Symposium on Prestressed Concrete in Building* FIP-CIA Sydney Sept 1976 22 pp

4.39 Bardhan-Roy B K 1983 Fire-resistance — design and detailing. In Kong et al (eds) *Handbook of Structural Concrete* Pitman, London Chapter 14 pp 14—1 to 46

4.40 Lefter J 1987 Fire safety of concrete slabs. *Concrete International Design & Construction* **9**(8): 23—8

4.41 Salse E, Lin T D 1976 Structural fire resistance of concrete. *ASCE Journal of the Structural Division* **102**(ST1): 51—63

4.42 Joint Committee of Institution of Structural Engineers and Concrete Society, J Bobrowski, Chairman 1975 *Fire Resistance of Concrete Structures* London

4.43 Meyer-Ottens C 1974 Verhalten von Betonbauteilen im Brandfall. *Betong* **24**(4): 133—6 and (5): 175—8

4.44 Smith P 1963 New concrete is not for burning: investigation and repair of damage to concrete caused by formwork and falsework fire. *ACI Journal Proceedings* **60**(11): 1535—66

4.45 Green J K 1976 Some aids to the assessment of fire damage. *Concrete* (London) **10**(1): 14—17

4.46 Bessey G E 1950 *Investigations on Building Fires, Part 2. The Visible*

Concrete Structures

Changes in Concrete or Mortar Exposed to High Temperatures National Building Studies Technical Paper 4 HMSO, London

4.47   Harmathy T Z 1968 Determining the temperatures of concrete constructions following fire exposure. *ACI Journal* Proceedings **65**(11): 959–64

4.48   Elices M, Rostasy F S, Faas W M, Wiedemann G 1982 *Cryogenic Behaviour of Materials for Prestressed Concrete* FIP State of the Art Report Slough 84 pp

4.49   Rostasy F S, Wiedemann G 1981 Strength, deformation and thermal strains of concrete at cryogenic conditions. *1st International Conference on Cryogenic Concrete* The Concrete Society pp 212–23

4.50   Planas J, Corres H, Elices M, Chueca R 1984 Thermal deformation of loaded concrete during thermal cycles from 20 °C to − 165 °C. *Cement & Concrete Research* **14**(5): 639–44

4.51   Elices M, Planas J, Corres H 1986 Thermal deformation of loaded concrete at low temperatures, 2: Transverse deformation. *Cement & Concrete Research* **16**(5): 741–8

4.52   Turner F H *et al* 1982 *Guide to Good Practice — Preliminary recommendations for the design of prestressed concrete containment structures for the storage of refrigerated liquefied gases (RLG)* FIP, Slough 55 pp

4.53   Van der Veen C, Reinhardt H W 1987 Limestone as an aggregate for concrete under cryogenic conditions (in Dutch). *Cement ('s-Hertogenbosch)* **39**(3): 22–5

4.54   Popovics S 1987 Chemical resistance of Portland cement mortar and concrete. In Sheppard W L (ed) *Corrosion and Chemical Resistant Masonry Materials Handbook* Noyes Publications, New Jersey pp 293–339

4.55   Sheppard W L (ed) 1987 *Corrosion and Chemical Resistant Masonry Materials Handbook* Noyes Publications, New Jersey

4.56   Onabolu O A 1989 Some properties of crude oil-soaked concrete — 1. Exposure at ambient temperature. *ACI Materials Journal* **86**(2): 150–8

4.57   Biczok I 1967 *Concrete Corrosion and Concrete Protection* Chemical Publishing Company, New York

4.58   Kleinlogel A 1950 *Influences on Concrete* Frederick Ungar Publishing, New York

4.59   Kuhl H 1961 *Die Erhartung und die Verarbleitung der hydraulischen Bindmittel* VEB Verlag Technick, Berlin, Zement-Chemie, Band III

4.60   Lea F M 1970 *The Chemistry of Cement and Concrete* (revised edition of Lea and Desch) 3rd edn Edward Arnold 366 pp

4.61   Fulton F S 1969 *Concrete Technology — A South African Handbook* Portland Cement Institute, Johannesburg

4.62   Langelier W F 1936 The analytical control of anti-corrosion water. *Journal of the American Water Works Association* **28** Oct

4.63   Mather K 1978 Tests and evaluation of portland and blended cements for resistance to sulfate attack. *ASTM STP 663 Significance of Tests and Properties of Concrete and Concrete Making Materials* pp 74–86

4.64   Mehta P K 1988 Sulfate resistance of blended cements. In Ryan, W G (ed) *Papers, Concrete 88 Workshop* Concrete Institute of Australia July 1988 pp 337–51

4.65 Mehta P K 1980 Durability of concrete in marine environment — a review *ACI SP—65* pp 1—20

4.66 Mehta P K 1988 Durability of concrete exposed to marine environment — a fresh look *ACI SP—109* pp 1—29

4.67 Regourd M 1980 Physico-chemical studies of cement pastes, mortars and concretes exposed to sea water *ACI SP—65* pp 63—82

4.68 ACI Committee 302 Guide for concrete floor and slab construction *ACI 302.1R—80* American Concrete Institute 1980 46 pp

4.69 Fentress B 1973 Slab construction practices compared by wear tests. *ACI Journal* Proceedings **70**(7): 486—91

4.70 Senbetta E, Malchow G 1987 Studies on control of durability of concrete through proper curing *ACI SP—100* pp 73—87

4.71 Kettle R J, Sadegzadeh M 1987 The influence of construction procedures on abrasion resistance *ACI SP—100* pp 1385—1410

4.72 Neville A M (ed) 1975 *Fibre Reinforced Cement and Concrete* RILEM Symposium The Construction Press

4.73 Sadegzadeh M, Kettle R J 1988 Abrasion resistance of surface treated concrete. *Cement, Concrete & Aggregates* **10**(1): 20—8

4.74 Sawyer J L 1957 Wear tests on concrete using the German standard method of test and machine. *Proceedings of ASTM* **57**: 1145—53

4.75 Carrasquillo P M 1987 Durability of concrete containing fly ash for use in highway applications *ACI SP—100* pp 843—61

4.76 Dunstan M R H, Joyce R E 1987 High fly ash content concrete — a review and a case history *ACI SP—100* pp 1411—43

4.77 Wu G Y 1987 Steel fiber reinforced heat resistant pavement *ACI SP—105* pp 323—50

4.78 Wu G Y, Jones M P 1987 Navy experience with steel fiber reinforced concrete airfield pavement *ACI SP—105* pp 403—18

4.79 Holland T C 1983 *Abrasion-Erosion Evaluation of Concrete Mixtures for Stilling Basin Repairs, Kinzua Dam, Pennsylvania* Miscellaneous Paper No SL—83—16 US Army Engineers Waterways Experiment Station, Vicksburg

4.80 Hoff G C 1988 Resistance of concrete to ice abrasion — a review *ACI SP—109* pp 427—55

4.81 Watkins R D, Samarin A 1975 Cavitation erodability of concrete. *Proceedings of the Conference on Serviceability of Concrete* Institution of Engineers Australia, Melbourne Aug 1975

4.82 Campbell-Allen D 1979 *The Reduction of Cracking in Concrete* University of Sydney/Cement and Concrete Association of Australia 165 pp

4.83 Park R, Paulay T 1975 *Reinforced Concrete Structures* Chapter 13 The art of detailing, Wiley

4.84 Kriz L B, Raths C H 1965 Connections in precast concrete structures — strength of corbels. *Journal of the Prestressed Concrete Institute* **10**(1): 16—61

4.85 Leonhardt F 1965 Uber die Kunst des Bewehrens von Stahlbeton-tragwerken. *Beton- und Stahlbetonbau* **60**(8): 181—92; (9): 212—20

4.86 Somerville G, Taylor H P J 1972 Influence of reinforcement detailing on the strength of concrete structures. *The Structural Engineer* **50**: 7—19

4.87 Feld J 1964 *Lessons from Failures of Concrete Structures* Monograph No

1 American Concrete Institute Detroit 179 pp

4.88 Nilsson I H E, Losberg A 1976 Reinforced concrete corners and joints subjected to bending moment. *Proceedings of the ASCE* **102**(ST6): 1229–54

4.89 Swann R A 1969 *Flexural Strength of Corners of R.C. Portal Frames* Cement & Concrete Association, London TRA 433

4.90 Balint P, Taylor H P J 1972 *Reinforcement Detailing of Frame Corner Joints with Particular Reference to Opening Corners* Cement & Concrete Association, London TRA 462 Feb

4.91 Taylor H P J 1974 *The Behaviour of In-Situ Concrete Beam-Column Joints* Cement & Concrete Association, London TRA 492 May

4.92 Mayfield B, Kong F K, Bennison A, Davies J C D T 1971 Corner joint details in structural lightweight concrete. *ACI Journal* Proceedings **68**(5): 366–72

4.93 Mayfield B, Kong F K, Bennison A 1972 Strength and stiffness of lightweight concrete corners, *ACI Journal* Proceedings **69**(6): 420–7

4.94 Hamada S, Oshiro T, Hino S-I 1987 Vertical cracking of concrete tank walls. *Concrete International Design & Construction* **9**(10): 50–5

4.95 Hughes B P, Mahmood A T 1988 An investigation of early thermal cracking in concrete and the recommendations in BS 8007. *The Structural Engineer* **66**(4): 61–9

4.96 Hughes B P, Mahmood A T 1988 Laboratory investigation of early thermal cracking in concrete. *ACI Materials Journal* **85**(3): 164–71

4.97 Hughes B P, Videla Cifuentes C 1988 Comparison of early-age crack width formulas for reinforced concrete. *ACI Structural Journal* **85**(2): 158–66

4.98 Gordon C *et al.* 1975 Avoiding gross errors in concrete construction. *ACI Journal* Proceedings **75**(11): 638–46

4.99 Griffiths D, Marosszeky M, Sade D 1987 Site study of factors leading to a reduction in durability of reinforced concrete *ACI SP–100* pp 1703–26

4.100 Marosszeky M 1987 *Concrete Durability — Final Report* BRC Publication 1/87, Building Research Centre, University of New South Wales 83 pp

4.101 Grundy P, Kabaila A 1963 Construction loads on slabs with shored formwork in multistory buildings. *ACI Journal* Proceedings **60**(12): 1729–38

4.102 Agarwal R K, Gardner N J 1974 Form and shore requirements for multistory flat slab type buildings. *ACI Journal* Proceedings **71**(11): 559–69

4.103 Liu X-L, Chen W-F, Bowman M D 1985 Construction loads on supporting floors. *Concrete International Design & Construction* **7**(12): 21–6

Further suggested reading, p. 352

5 Reinforcement in Concrete

Background to Durability as Influenced by Reinforcement

At the outset, it must be said that, in a properly designed, constructed and used structure, there should be no problem of steel corrosion within the concrete during the design life of that structure. Unfortunately, this highly desirable requirement is not always achieved in practice. In a book such as this, one must therefore dwell to a great extent on the unusual, or negative, factors which promote corrosion, rather than discuss the many structures, which throughout their life-span show no problems of this sort. It should be clear, however, that concrete provides a high degree of protection against corrosion because of its alkalinity, and well-made concrete, having low permeability, minimizes the penetration of oxygen, chloride ions and carbon dioxide, all of which influence corrosion activity. The two major factors in allowing corrosion to proceed to an unacceptable degree are carbonation of the products of cement hydration (Fig. 5.1) and the depassivation of the steel by chloride ions, which may either have been present in the original concrete constituents or are introduced during service life.

Theory of Reinforcement Corrosion

It is of advantage for any engineer working in the field of concrete durability to understand some aspects of corrosion on the materials science level, and a RILEM Report[5.1] covers the topic, but excludes the special problems of hydrogen embrittlement and stress-corrosion cracking. The approach adopted by engineers is to use their understanding of the physico-chemical phenomena to formulate models which portray the actions of corrosion on structural members in service. Rigorous mathematical models such as that of Bazant,[5.2, 5.3] have not been generally applied, because localized factors perturb results significantly, and modelling these, as opposed to the global properties of the concrete, is extremely difficult. For that reason simpler models such as those proposed by Tuutti,[5.4] Beeby[5.5] and Browne[5.6, 5.7] have been proposed. Most of the models so far developed are based on diffusion theory, but

Fig. 5.1 Diagrammatic presentation of the carbonation of cement hydration products. The phenolphthalein line represents the plane along which this indicator displays a change in colour, but the pH varies across a zone rather than changing abruptly

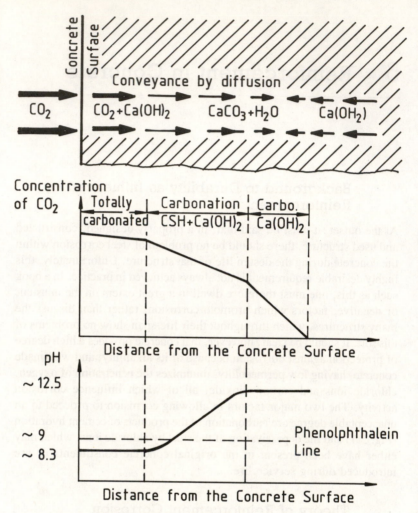

Browne notes that it is not yet established that a simple diffusion law can be used for chloride ingress with time. In many models the lifetime is divided into two periods, one during which initiation of corrosion processes occurs, and secondly the propagation stage (Fig. 5.2). According to Treadaway in[5.1] a depassivation period extends from the time of passivity destruction to the point when actions remote from the immediate vicinity of the corrosion site are required to develop (propagate) the corrosion products. The propagation period covers that time from which corrosion products form to when they generate sufficient stress to crack or spall the cover, or so degrade the steel that it cannot safely carry its design load.

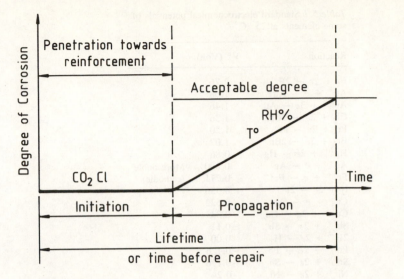

Fig. 5.2 Model of building life as influenced by corrosion, the initiation being caused by Cl^- ingress and carbonation, and the rate of corrosion propagation being influenced by the ambient temperature and internal relative humidity

For corrosion of any metal to occur the following conditions must be fulfilled:

(*i*)   an electron sink area or anode must be present at which the de-electronation reaction (i.e. metal dissolution, oxidation) occurs;

(*ii*) an electronic conductor (in most cases the metal itself) is required to carry the electrons to the electron source area (cathode) where an electronation (reduction) reaction occurs;

(*iii*) an ionic conductor must be present to keep the ion current flowing, and to function as a medium for the electrodic reaction.

In reinforced concrete the anode and cathodes develop on a single bar or on bar groups, and distances between them may be less than one millimetre or several metres. The electronic conductor is the bar itself. The ionic conductor or electrolyte is as a rule the pore solution of the concrete, but may be solutions of somewhat different composition within crevices or pits in the bar surface, as demonstrated by Roper.[5.8] Corrosion occurs due to electron-transfer reactions across the electrolyte-steel interface. Electrode potentials are measured relative to a standard half-cell electrode (an inert platinum electrode measured in an acid solution of unit hydrogen ion concentration [1 molar H^+], at 25 °C with hydrogen gas at 1 atm. pressure bubbled over its surface), which is assigned a half-cell potential of zero. Every half-cell reaction can be assigned a standard potential with respect to the standard hydrogen electrode as in Table 5.1. This table not only allows the calculation of energy associated with an electro-chemical reaction, but allows one to

Table 5.1 Standard electrochemical potentials of
some elements at 25 °C

Reaction	$V°$ (Volt)	
$F_2 + 2e \rightarrow 2F$	2.76	
$Au^+ + e \rightarrow Au$	1.68	
$Au^{3+} + 3e \rightarrow Au$	1.46	
$Cl_2 + 2e \rightarrow 2Cl$	1.36	
$Pt^2 + 2e \rightarrow Pt$	1.20	
$Br_2 + 2e \rightarrow 2Br$	1.07	
$Hg^{2+} + 2e \rightarrow Hg$	0.86	
$Ag^+ + e \rightarrow Ag$	0.800	More noble
$Fe^{3+} + e \rightarrow Fe^{2+}$	0.771	(cathodic)
$I_2 + 2e \rightarrow 2I$	0.536	
$Cu^+ + e \rightarrow Cu$	0.522	
$Cu^{2+} + 2e \rightarrow Cu$	0.341	
$Sb^{3+} + 3e \rightarrow Sb$	0.11	
$2H^+ + 2e \rightarrow H_2$	0.00	
$Pb^{2+} + 2e \rightarrow Pb$	−0.126	
$Sn^{2+} + 2e \rightarrow Sn$	−0.136	
$Ni^{2+} + 2e \rightarrow Ni$	−0.24	
$Co^{2+} + 2e \rightarrow Co$	−0.29	
$Cd^{2+} + 2e \rightarrow Cd$	−0.4	
$Fe^{2+} + 2e \rightarrow Fe$	−0.42	
$Cr^{2+} + 2e \rightarrow Cr$	−0.53	
$Cr^{3+} + 3e \rightarrow Cr$	−0.68	
$Zn^{2+} + 2e \rightarrow Zn$	−0.762	
$Ti^{2+} + 2e \rightarrow Ti$	−1.63	
$Al^{3+} + 3e \rightarrow Al$	−1.67	More active
$Mg^{2+} + 2e \rightarrow Mg$	−2.36	(anodic)
$Na^+ + e \rightarrow Na$	−2.71	
$Ca^{2+} + 2e \rightarrow Ca$	−2.71	
$K^+ + e \rightarrow K$	−2.92	
$Rb^+ + e \rightarrow Rb$	−2.96	
$Cs^+ + e \rightarrow Cs$	−2.97	
$Li^+ + e \rightarrow Li$	−3.04	

see why a zinc coating will protect an iron rod even when the coat is discontinuous, whereas a nickel coating will, if damaged, promote corrosion of the iron. Too great a dependence cannot be placed on the electromotive series in that it only considers those equilibria involving metals and their simple cations. In the corrosion process protective films may form, or hydrogen or hydroxyl ions may enter into the reactions, which themselves may or may not be potential dependent. Pourbaix diagrams are therefore more valuable to the understanding of corrosion processes. Furthermore when corrosion occurs, the observed cell voltage is less than the difference between the equilibrium electrode potentials. The non-equilibrium effects at the electrodes result from two factors:

(*i*)   changes in electrolyte concentration around an electrode (concentration polarization);

(*ii*)   restricted reaction rates at an electrode (activation polarization).

Both types of polarization are manifested by a change in the individual electrode potentials, and this change, called the overpotential, is the actual electrode potential minus the equilibrium value. The influence of the concentration overpotential (η conc) and the activation overpotential (η act) is referred to on pp. 189−93 in explaining stress-corrosion cracking. Even if the electrode reactions of a cell proceed readily and without significant overpotentials, a voltage drop (IR drop) occurs across the electrolyte when current flows. This is because the electrical resistance of the electrolyte leads to energy dissipation in the form of heat.

Some reactions during corrosion are dependent on purely chemical equilibria, while others depend on exchanges of charge. In order to present data of this nature, Pourbaix[5.9] plotted two parameters E_H (standard hydrogen potential) and pH for various equilibria on normal cartesian coordinates. In these diagrams an initial assumption is made that there exists an electrode potential for the metal concerned, at which equilibrium activity of the metal is so small that the rate of metal dissolution to maintain this activity is negligible. It has become the practice to take an ionic activity of 10^{-6} gramme ions per litre as sufficiently small for this purpose. Two Pourbaix diagrams are particularly important to the concrete engineer. They are for pure zinc (Fig. 5.3), and for iron (Fig. 5.4). Each of the lines denoted by Roman numerals in Fig. 5.3 represents a particular reaction which may occur; lines (a) and (b) represent the generation of hydrogen and oxygen respectively. The field of this diagram is divided into four segments representing areas of immunity, passivity and two of corrosion. The simple ion Zn^{++} and the complex zincate ion $ZnO_2^{=}$ are forms of zinc in solution, and hence corrosion of the zinc will occur in these segments within the bounding conditions of potential and pH. Zinc hydroxide is a compound only sparingly soluble in water and will therefore represent a solid precipitate, which may or may not form a protective film on the metal surface. This domain is representative of 'filming', and is designated by the term passivity because passivation by such a film is thermodynamically possible. The diagram for iron (Fig. 5.4) is more complex, but shows similar features. Pourbaix[5.9] notes that the data presented is for the pure metal, not for alloys, and not for solutions which form soluble complexes.

In order both to extend the scope of the Pourbaix diagram to cover experimental conditions of immunity, general corrosion, pitting and passivation, Pourbaix[5.10] used potentiokinetic tests or Evans diagrams to develop them. An Evans diagram is a plot of the change in measured current flow as a function of a change in potential applied to the metal. As the mass of metal passing into solution is a function of the current flow, the amount and rate of corrosion occurring can be obtained from such a diagram. Furthermore, certain actions such as pitting or film formation, can be inferred from the shape of the curve (Fig. 5.5) and

Fig. 5.3 Potential vs pH diagram for zinc in water at 25 °C. Lines (a) and (b) represent the generation of hydrogen and oxygen respectively, and line segments labelled with Roman numerals represent individual reactions involving Zn ions

Fig. 5.4 Potential vs pH diagram for iron in water at 25 °C. The line segments labelled with Roman numerals represent individual reactions involving Fe ions

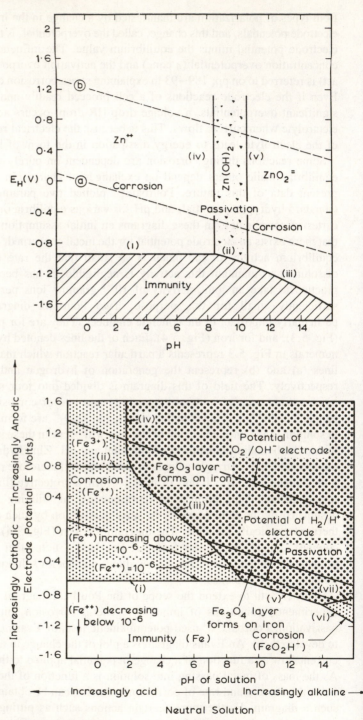

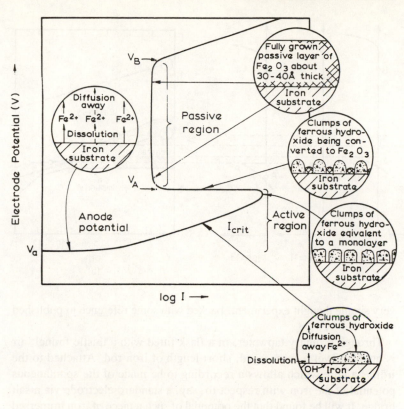

Electrode Potential (V)

log I ⟶

Diffusion away
Fe²⁺ Fe²⁺ Fe²⁺
Dissolution
Iron substrate

Passive region

Fully grown passive layer of Fe₂O₃ about 30-40Å thick
Iron substrate

Clumps of ferrous hydroxide being converted to Fe₂O₃
Iron substrate

Anode potential

Active region

I_{crit}

Clumps of ferrous hydroxide eqivalent to a monolayer
Iron substrate

V_B

V_A

V_a

Clumps of ferrous hydroxide
Diffusion away Fe²⁺
Dissolution
OH⁻ Iron substrate

Fig. 5.5 Evans diagram for corrosion of a steel showing corrosion, passivation (between points V_A and V_B) and transpassive behaviour (passivation of an electrode by electrical means. When the electrode potential reaches a critical value, V_A, the current abruptly decreases as a result of the formation of a fully grown passive film, which is disrupted only at a much higher voltage, V_B).

the evidence of the actions can be confirmed later by other techniques, such as microscopic examination of the specimen surface. Two experimental Pourbaix diagrams for conditions in chloride free and chloride contaminated solutions are given in Figs 5.6 and 5.7. In the latter, areas of pitting and imperfect passivity are observed. Evans diagrams may also be used to study reduced corrosion rates under increased resistance (Fig 5.8), and a relative decrease in the cathode area (Fig 5.9). The effect of a reduced cathodic diffusion rate has the same effect as a reduction of cathode area. It can also be concluded from Evans diagrams as to whether the corrosion process is cathodically or anodically controlled. If the cathodic reaction is proceeding slowest, then the process is under cathodic control. This condition is met in concrete, the rate of oxygen diffusion determining the rate of corrosion.

A Hypothetical Corrosion Experiment

To demonstrate how the principles considered earlier in this Chapter may be applied to the corrosion of steel in concrete, let us first consider a

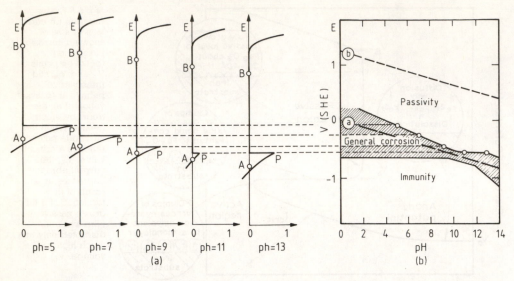

E
B

E
B

E
B

E
B

E
B

P
A

P
A

P
A

P
A

P
A

0   1
ph=5

0   1
ph=7

0   1
ph=9

0   1
ph=11

0   1
ph=13
(a)

Fig. 5.6 Experimental Pourbaix diagram for iron developed from a series of Evans diagrams run at different pH levels (after Pourbaix[5.9])

very simple thought experiment, backed with some reference to published work.

Through ordinary tapwater, in a flask fitted with a thistle funnel, air is bubbled over a submerged, short length of iron rod. Attached to the iron is a wire which allows a recording to be made of the spontaneous potential of the iron with respect to, say, a standard electrode via a salt bridge. It will be found that the potential of such a piece of iron immersed in tap water having a pH of about 7 is, provided that no bubbles accumulate on the bar, about -450 mV. These conditions can be checked on the Pourbaix diagram given in Fig. 5.4. The values correspond to a position within the 'corrosion' region of the Pourbaix diagram, and the bar will continue to corrode.

Into the tap water in a second flask, a small amount, say 10 g, of unhydrated cement is placed. It is rapidly stirred for a period of a few minutes and the potential of the rod is read when the air flow is started. At this stage too, the pH of the water is recorded. The principal soluble product of the hydrating cement is calcium hydroxide, $Ca(OH)_2$, and the initial alkalinity of the water is at least that of saturated lime water (pH = 12.4 at room temperature). In addition, the presence of relatively small amounts of sodium and potassium oxides may increase the initial alkalinity to between 12.5 and 13.2.[5.11–5.13] If the potential value is about -450 mV when the iron is first immersed, it may rapidly increase to a value of about -350 mV, indicating passivation. A thermodynamically stable compound Fe_3O_4 forms, via several intermediates beginning with $Fe(OH)_2$, as a layer on the steel, and further corrosion ceases completely for all practical purposes. This

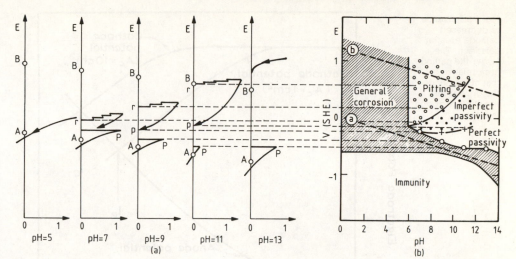

pH=5   pH=7   pH=9   pH=11   pH=13
(a)

(b)

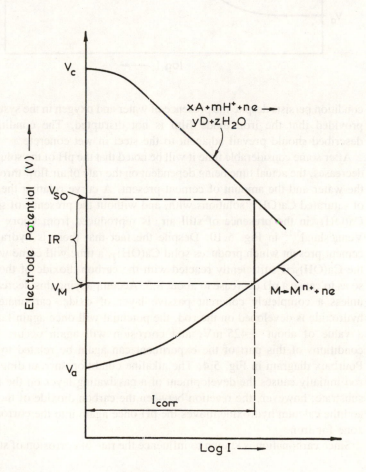

Fig. 5.7 Experimental Pourbaix diagram for iron in chloride contaminated solutions. The bounds of zones representing general corrosion, pitting and imperfect passivation are derived from Evans diagram information (after Pourbaix[5.9])

Fig. 5.8 IR drop influenced Evans diagram, showing that when a potential drop in the electrolyte is observed, the anodic and cathodic curves do not intersect at the corrosion current value

Fig. 5.9 Influence
of decreased
cathode area on
corrosion. The
larger the cathode
area relative to
that of the anode
the greater the
resulting corrosion
current

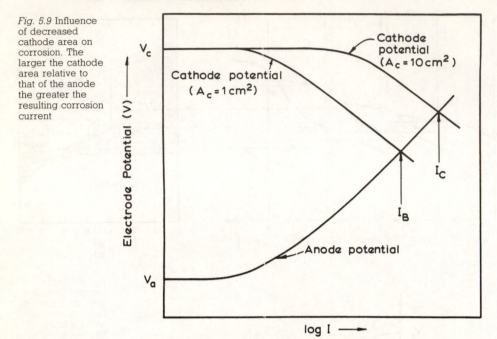

condition persists despite the presence of water and oxygen in the system,
provided that the iron oxide film is not disrupted. The conditions
described should prevail adjacent to the steel in wet concrete.

After some considerable time it will be noted that the pH of the solution
decreases, the actual time being dependent on the rate of air flow through
the water and the amount of cement present. A curve relating the pH
of saturated $Ca(OH)_2$ solutions with, and without the presence of solid
$Ca(OH)_2$ in the presence of still air, is reproduced from Gjorv and
Vennesland[5.13] in Fig. 5.10. Despite the fact that there is hydrating
cement present which produces solid $Ca(OH)_2$, a time will come when
the $Ca(OH)_2$ is sufficiently reacted with the carbon dioxide of the air
so as to cause a pH decrease to about 8.5. Accompanying this decrease,
unless a completely coherent passive layer of oxide, carbonate or
hydroxide is developed on the rod, the potential will once again fall to
a value of about -425 mV, and corrosion will again occur. The
conditions of this part of the experiment can again be related to the
Pourbaix diagram in Fig. 5.4. The alkaline condition surrounding the
rod initially causes the development of a passivating layer on the iron
substrate; however, the reaction between the carbon dioxide of the air
and the calcium hydroxide moves the pH once again into the corrosive
zone for iron.

Since carbonation is known to influence the rate of corrosion of steel,

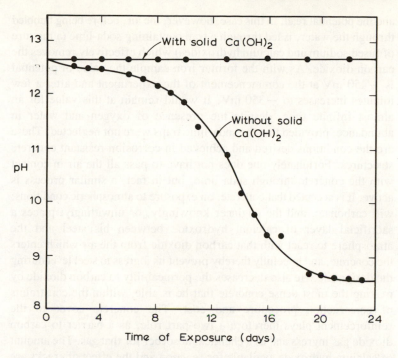

Fig. 5.10 pH of saturated calcium hydroxide solutions with and without solid calcium hyroxide present when exposed to air (after Gjorv et al.[5.13])

it is important to have some background knowledge of this phenomenon. Considerable amounts of data available from early papers[5.14–16] showed that the carbonation rate was at a maximum at about 50 per cent relative humidity (RH), and was considerably reduced for both wet, and very dry, concrete. This is a way in which the RH may influence corrosion, in that the corrosion rate of an embedded bar may be greater if concrete is kept at about 50 per cent RH, than if it remained wet continuously, an outcome different to that expected for steel exposed to the atmosphere. According to the Netherlands Committee for Concrete Research[5.17] both carbonation and corrosion are at a maximum at 60 per cent RH. Bakker[5.1] indicates that the equation which relates the carbonation depth of concrete (i.e. the level at which the calcium hydroxide is converted to calcium carbonate) with time should be a function of 15 parameters. The usual approach has been simply to consider the process as one of diffusion and to relate the depth of carbonation to a constant times the square-root of time. Bakker suggests that rather than use a constant, empirically derived correction factors must be applied to the root time formula to obtain a satisfactory model for carbonation of uncracked concrete. The influence of a number of factors on rates and depths of carbonation is discussed on pp. 106–9.

Returning to the thought experiment, consider a third flask in which a similar amount of cement is placed in the tap water; it is again stirred and the potential read. In this case, however, the air, before being bubbled through the water, is led through a trap containing soda-lime (a mixture of fused sodium and calcium hydroxides) which effectively removes the carbon dioxide. As with the former iron sample the value of potential is -450 mV at the commencement of the experiment and after a few minutes increases to -350 mV. It would remain at this value for an almost infinite time, despite the presence of oxygen and water in abundance, provided that the soda-lime traps were not neglected. These are the conditions desired and achieved in corrosion-resistant concrete structures. Fortunately one does not have to pass all the air in contact with the concrete through soda lime, but in fact, a similar process is active. It is accepted that concrete, on exposure to atmospheric conditions, will carbonate, and the engineer knowingly, or unwittingly, places a sacrificial layer of calcium hydroxide between his steel and the atmosphere to react with that carbon dioxide from the air which enters the concrete, and hopefully thereby prevent its ingress to steel level during the design life. He also decreases the permeability to carbon dioxide by making the most dense concrete that he is able, within the constraints of construction procedure and cost. The concrete cover to the reinforcement plays therefore a two-part role, as a barrier to carbon dioxide gas ingress and as a chemical absorber for that gas. The amount of calcium hydroxide available for reaction and the effect of cracks are both important in this regard and will be discussed in further detail later.

Finally, in the thought experiment, consider what will happen if, into the tapwater of a fourth flask, to which again a soda-lime trap is fitted, 10 g of cement and 0.4 g of calcium chloride is added. On repeating the measurement of current it will be found that, despite the fact that no CO_2 is present, the corrosion potential eventually becomes large, indicating that corrosion is taking place. In this case, the chloride ions are acting as depassivators, and so continued corrosion occurs. What is the pH effect of adding the salt? According to the data (Fig. 5.11) of Gjorv and Vennesland,[5.13] concentrations of sea salt up to 1 per cent in saturated solutions of $Ca(OH)_2$, and up to 4 per cent in extracts of hydrated Portland cements, do not lower the pH to less than about 12. For the same solutions with reserve basicity present in the form of solid material, concentrations of sea salt up to 10 per cent were tolerated without reducing the pH to less than that value. The effect of chlorides on pH is dependent on the reserve basicity. Well-hydrated Portland cements have contents of $Ca(OH)_2$ varying from 15 to 30 per cent by weight of unhydrated cement.[5.18] In contrast to normal Portland cements, however, other types of cement such as high-alumina cements, Portland blast furnace cements and Portland-pozzolana cements all have various amounts of $Ca(OH)_2$ bound to reactive siliceous materials, and hence the reserve basicity is correspondingly lower. That is not to say

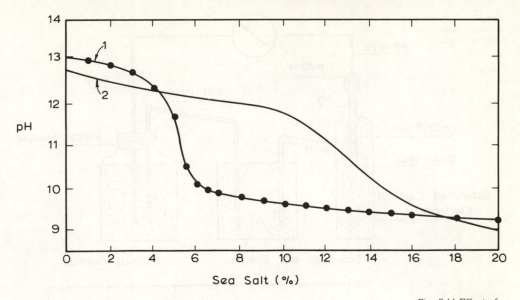

Fig. 5.11 Effect of
sea-salt on pH of
cement extracts
with (Line 2) and
without (Line 1)
solid paste present
(after Gjorv
et al.[5.13])

that other beneficial factors of such cements, such as decreased diffusion properties, will not compensate for the lower reserve basicity in reducing corrosion rates. At this point it is worthy of note that individual organizations, for example the Department of the Environment in the UK have long banned the use of calcium chloride and concrete additives containing calcium chloride in the construction of all their buildings and structures. Elsewhere the debate as to their usefulness still goes on.[5.19]

Confirmatory Experimental Work

The theoretical experiment dealt with in the former section has demonstrated the passivating action of cement paste even in the presence of abundant moisture and oxygen under certain circumstances, and the depassivating action of Cl^- ions even in the presence of cement hydrate products. It is of interest briefly to consider experimental results of some workers which substantiate the conclusions. Consider firstly the simple experimental set-up of Hamada,[5.14] (Fig. 5.12) and his findings of the change in potential with time (Fig. 5.13). These demonstrate the importance of pH in the passivation process. The results of anodic polarization of steel in concrete, as obtained by Gouda and Monfore,[5.20] demonstrate the development of the passive layer on steel, its post-passive behaviour and the breakdown or non-development of this layer in the presence of different concentrations of calcium chloride (Fig. 5.14 a,b,c).

As early recognized by Lewis,[5.21] when sodium chloride is added to cement, the formation of insoluble complexes such as

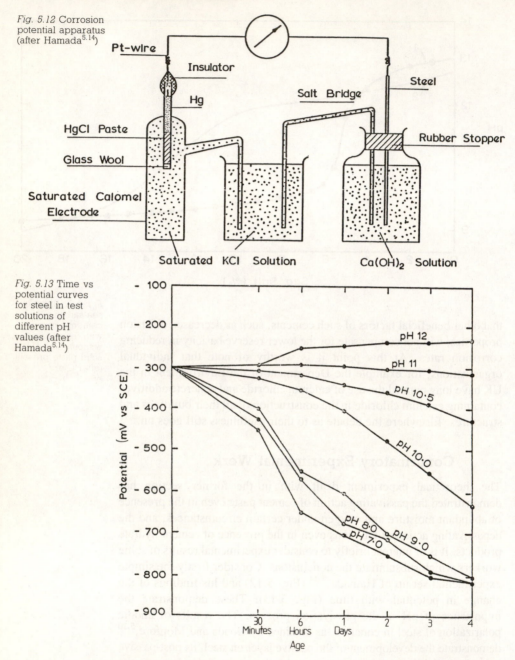

Fig. 5.12 Corrosion potential apparatus (after Hamada[5.14])

Fig. 5.13 Time vs potential curves for steel in test solutions of different pH values (after Hamada[5.14])

$3CaO.Al_2O_3.CaCl_2.10H_2O$ will have the effect of reducing the effective concentration of chloride ion. The extent to which the effective chloride concentration is reduced by the formation of the complexes was determined by analysis of available and total chloride on hydrated cement

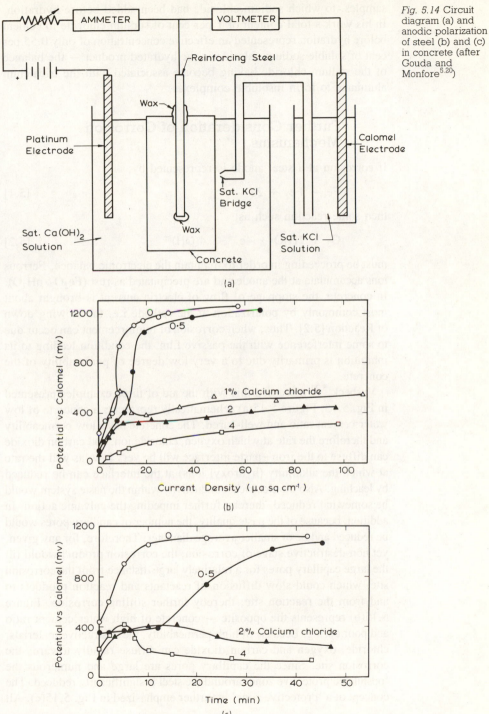

Fig. 5.14 Circuit diagram (a) and anodic polarization of steel (b) and (c) in concrete (after Gouda and Monfore[5.20])

samples, to which sodium chloride had been added before hydration. In his work a total content of 1 per cent of sodium chloride in cement before hydration represented an effective concentration of only 0.55 per cent of soluble sodium chloride in the hydrated product — the balance of the sodium chloride having become associated with the tricalcium aluminate to form insoluble complexes.

Further Consideration of Corrosion Mechanisms

If corrosion at a steel anode is represented by:

$$Fe \rightarrow Fe_2 + 2e^-$$ [5.1]

then some reaction such as:

$$O_2 + 2H_2O + 4e^- \rightarrow 4(OH)^-$$ [5.2]

must be proceeding in order to maintain the electronic balance. Ferrous ions accumulate at the anode and are precipitated as rust ($Fe_2O_3 \cdot nH_2O$). In concrete, the stoppage of flow of electric current is brought about most commonly by polarization at the cathode i.e. the slowing down of Reaction [5.2]. Thus, when corrosion of reinforcement can occur due to some interference with the passive film, the condition leading to its inhibition is primarily due to a very low degree of permeability of the concrete.

Verbeck,[5.22] discussed this with the aid of three examples presented in Fig. 5.15. Figure 5.15(a) schematically represents a concrete of low water:cement ratio and well-cured. The concrete has a low permeability and therefore the rate at which oxygen, chloride ion, and carbon dioxide can diffuse to the iron—paste interface will be very low, as will the rate at which the alkalinity (hydroxyl ions) at the interface can be reduced by leaching. Also the electrical conductivity within the paste system would be somewhat reduced, thereby further impeding the galvanic action. In addition, because of the paste quality, the number of capillary pores would be reduced and be of smaller average diameter. Therefore, for any given, yet non-destructive stage of corrosion, the corrosion product would fill the large capillary pores for a relatively large distance from the corrosion site, which could slow diffusion of reactants and reaction products to and from the reaction site, thereby further stifling corrosion. Figure 5.15(b) represents the opposite — concrete of high water:cement ratio and poor curing. Because of high permeability, the destructive materials, chloride, oxygen and carbon dioxide can diffuse rapidly towards the corrosion site. Since the capillary pores are large and numerous the 'postulated protective zone' around the steel is significantly reduced. The concept of a 'protective zone' is further emphasized in Fig. 5.15(c). All

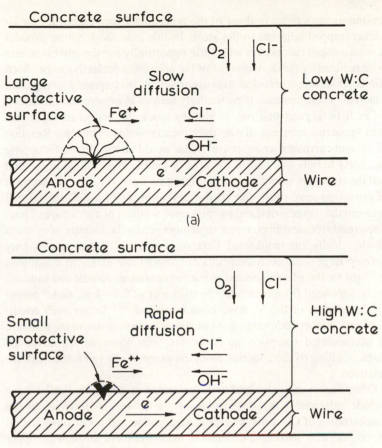

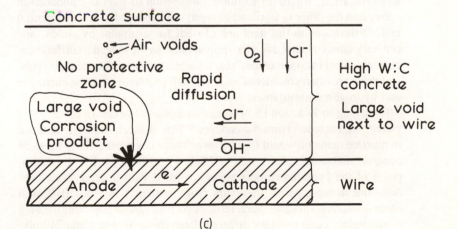

Fig. 5.15 Schematic
representation of
corrosion in
concrete: (a) low
W:C ratio, well
cured concrete;
(b) high W:C ratio;
(c) high W:C
concrete with
voids (after
Verbeck[5.22])

conditions are similar to those in the preceding figure except for an air pocket trapped adjacent to the steel. In this case the reaction product is released into this pocket with little opportunity for the corrosion rate to be reduced by the development of the postulated protective zone. Such points of severe corrosion are observed in inadequately compacted concrete. Their presence is particularly serious in prestressed members.

Let it be argued that one is able to block the access of O_2 to the steel—concrete interface all over the structure — this would stop Reaction [5.2], and corrosion would cease. What would be the effect of stopping its access to only one half of the area of the bar? One might argue that half the corrosion would occur. This is not so, however, for enhancement of corrosion may result due to the development of a differential cell between the oxygen-rich and oxygen-poor sections of a structure. Thus, since reinforcement in concrete structures generally consists of systems of electrically interconnected bars or wires, the entire system may develop large-scale corrosion cells of considerable extent, if conditions are right for their formation and continued operation. Anodic and cathodic areas, separated from each other by distances of 0.5—4 m, under potential differences of 0.5 V, have been measured.[5.23] Under such conditions serious corrosion leading to cracking of concrete due to the pressure of accumulated corrosion products has been observed. In advanced cases, spalling of the concrete from the plane of the reinforcement has occurred.

Other factors may also lead to the existence of differential cells; these include differences in alkalinity from site to site, differences in the concentration of Cl^- ions; bleeding, segregation and poor consolidation of concrete may also cause differences in environment between the upper and lower surfaces of reinforcing steel. Temperature differences within the concrete may also create differences in the electrochemical potential. Repaired areas, where for example one section of steel is embedded in epoxy and the other in Portland cement, may also produce a differential cell. Differences in the steel are caused, for example, by welds, and ordinary bars connected to zinc dipped bars may also lead to differential currents. Steel in concrete may also be subject to stray electrical currents, and these are often considered together with problems of stress corrosion and hydrogen embrittlement.

Returning to Reaction [5.2] one must ask what would occur if water were to be excluded from the concrete? This is much harder to achieve in practice than one would first believe. There is a great affinity between concrete and water, and, even well below 100 per cent RH the micropores of the hydrated cement paste are filled with hydrous solutions. Solutions in these pores (they vary from perhaps 1 to 5×10^{-9} m in mean diameter) will have characteristics such as vapour pressure, freezing characteristics and mobility different from those of free water in bulk.

Because of the large size of the capillary spaces, proportionately less of the capillary water is strongly adsorbed on the pore walls, and it is therefore more volatile than is gel water; nevertheless, the water in the capillary spaces is almost completely evaporable only at humidities below 40 per cent. Above 40 per cent relative humidity 'capillary condensation' occurs, that is, the capillaries become filled with water by precipitation from the vapour phase. Thus to ensure that relatively freely mobile water is absent from the vicinity of steel one would have to hold the humidity to below 40 per cent.

It is true that the electrical resistance of concretes is increased by air drying and that, considering the influence of an IR drop on corrosion rate, the drying of concrete is advantageous. Monfore[5.24] reports data for a series of dry, 4-inch cubes, with internal electrodes, which were given two applications of various coatings, and after the coatings had hardened, the cubes were immersed in water. Resistances measured after several periods of immersion are shown in Fig. 5.16. The epoxy coating was most effective, but after 28 days of immersion in water, the resistance was only moderately greater than that of the uncoated specimen. Some materials actually decreased in resistance after treatment. Monfore notes that these effects are in contrast to coatings on steel anchor nuts.

The Nature of Concrete Cover and Modifications Due to Cracks in Normally Reinforced Concrete

The functions of concrete cover have already been outlined. The permeability of well-compacted concrete is held, in the absence of

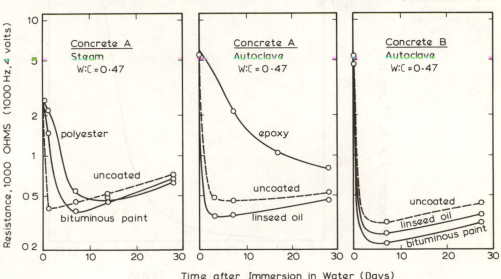

Fig. 5.16 Effect of coatings on concrete resistance (after Monfore[5.24])

cracking, to be mainly determined by the permeability of the cement paste. However, it has been shown that many rocks have greater permeabilities than mature pastes,[5.25] and at least one major corrosion problem on an Australian building facade was in part attributed to the use of very porous sandstone aggregates, chosen as aggregate for their aesthetic appeal. The curve relating water:cement ratio to permeability of paste is given in Fig. 5.17. Penetration of calcium chloride in solution into concrete and paste was measured by Ost and Monfore.[5.26] It was found that in both cases the penetration was strongly influenced by water: cement ratio and further that the ratio of penetration into concrete was greater than into the paste (Fig. 5.18). Verbeck[5.22] recalculated some of these data to obtain an almost linear relationship between permeability of concrete to chloride ions and water:cement ratio (Fig. 5.19). Whiting[5.27] developed a rapid method for the determination of chloride permeability of concrete using a driving potential of 60 V for six hours to measure the electrical charge passed through a 100 mm diameter by 50 mm thick specimen. This test known as the AASHTO T277 Method has been used with success to compare chloride penetration of concretes prior to their use on projects. It is of interest to note here that diffusivity rates of chloride ions through pastes and concrete may be significantly influenced by the associated cation.

Probably the most important recent advance in reducing chloride permeability into concretes has been the incorporation of silica fume into the mix. The presence of this material increases the electrical resistivity

Fig. 5.17 Effect of water:cement ratio on the permeability of hydrated paste to water (after Powers *et al.*[5.25])

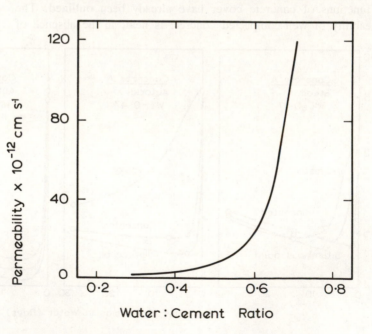

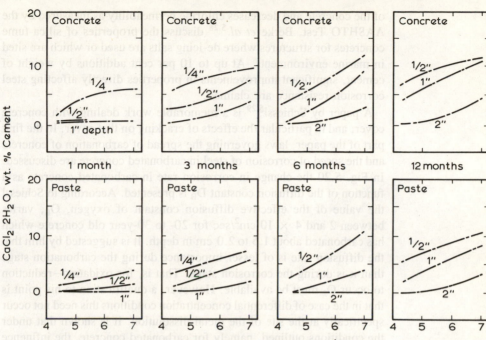

Fig. 5.18 Penetration of calcium chloride into concrete and paste from 2 per cent calcium chloride solution (after Ost and Monfore[5.26])

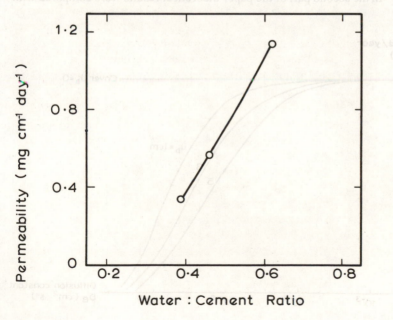

Fig. 5.19 Effect of water:cement ratio on permeability of hydrated paste to chloride ions (after Verbeck[5.22])

of the concrete and decreases chloride permeability as measured by the AASHTO Test. Berke *et al.*[5.28] discuss the properties of silica-fume concretes for structures where de-icing salts are used or which are sited in marine environments. At up to 10 per cent additions by weight of cement, significant improvements in properties directly affecting steel corrosion resistance are claimed.

A paper by Schiessl[5.29] is a memorable work dealing with concrete cover, and in particular the effects of cracking on that cover. In the first part of the paper, laws governing the spread of carbonation of concrete and the speed of corrosion of steel in carbonated concrete are discussed. In Fig. 5.20 the change in corrosion rate in carbonated concrete as a function of the diffusion constant D_B is presented. According to Schiessl the value of the effective diffusion constant of oxygen, O_E, varies between 2 and 4 $\times$ 10 cm^2/sec for 20- to 30-year old concrete which has carbonated about 1.5 to 2.0 cm in depth. It is suggested by him that the diffusion rate is of lesser importance during the carbonation stage than it is during the corrosion stage. That is, for oxidation–reduction to occur O_2 must be available. However, a qualification to his point is that in the case of differential concentration conditions this need not occur specifically at the site of the metal dissolution. It is shown that under the conditions outlined, namely for carbonated concrete, the influence of the thickness of cover is not very significant, but that there is a relationship between concrete strength (as reflecting cement content) and corrosion rate (Fig. 5.21).

In the second part of the paper theoretical results were compared with

Fig. 5.20 Change in corrosion rate in carbonated concrete as influenced by the diffusion constant (D_B) (after Schiessl[5.29])

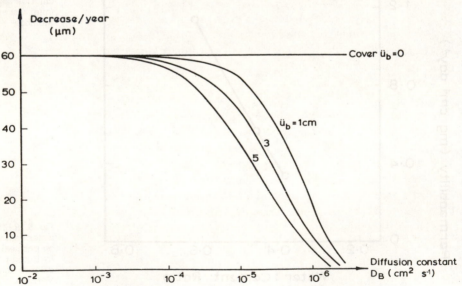

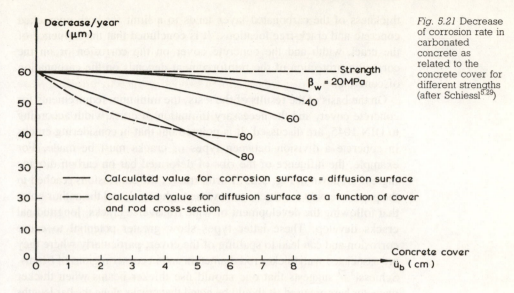

Fig. 5.21 Decrease of corrosion rate in carbonated concrete as related to the concrete cover for different strengths (after Schiessl[5.29])

experimental results. The test specimens (reinforced concrete beams with constantly open cracks and non-reinforced concrete specimens) were exposed to the air under different climatic conditions. The comparison shows good agreement, and data are presented which relate the increase in carbonation of a crack depth with crack thickness (Fig. 5.22). The

Fig. 5.22 Increase of carbonation of the crack depth with time as influenced by the crack width (w) (after Schiessl[5.29])

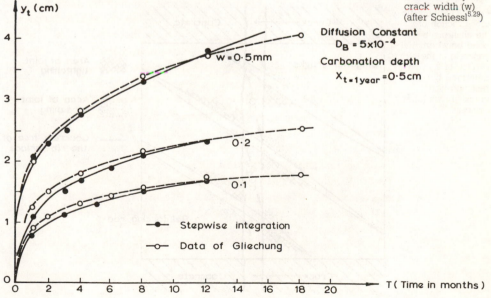

thickness of the carbonated layer tends to a limit for both uncracked concrete and crack-free locations. It is concluded that the influence of the crack width and the concrete cover on the corrosion or on the corrosion protection of the reinforcement depends on the carbonation of concrete.

On the basis of the results of the tests, the minimum requirements for concrete cover, and the necessary limitation of crack width according to DIN 1045, are discussed. It is pointed out that in considering cracks in concrete a division between types of cracks must be made. For example, the influence of the ribs of deformed bar on carbon dioxide ingress is shown in Fig. 5.23. Furthermore, the conclusions reached in the paper are only for cracks transverse to the steel, and the author notes that following the development of wide transverse cracks, longitudinal cracks develop. These latter types show greater potential to cause corrosion and can lead to spalling of the cover, particularly where they are associated with thicker rod diameters and in corner regions of beams. Schiessl[5.29] suggests that one should use thicker beams when thicker diameter bars are used. It should be noted that cracks along the bar lengths may be caused by adverse conditions such as high plastic shrinkage and settlement cracking, factors not mentioned by the author.

Taking as an example a flexural member with transverse cracks, continually open, and having an average width of 0.30 mm, and concrete cover of 25 mm he derives a probability of about 55 per cent for a corrosion layer of thickness $t_m = 0.01$ mm, to form. Were the average

Fig. 5.23 Schematic cross section of the changes in bond between concrete at the surface of a deformed bar resulting from crack growth (after Schiessl[5.29])

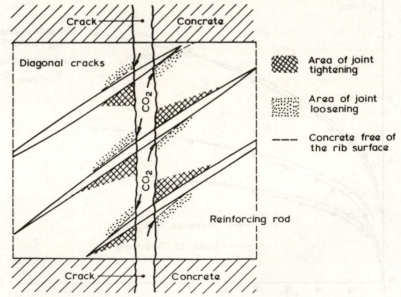

crack widths to be limited to 0.15 mm, then the probability of the development of the corrosion layer of thickness $t_m = 0.01$ mm decreases to about 35 per cent. Now, however, the number of cracks compared to the first case is nearly doubled, so the decrease in crack width does not greatly help in practice. He suggests that an advantageous step may be taken by using thinner reinforcing rod diameters, but admits that thin rods suffer greater cross sectional decrease than thicker rods for the same corrosion layer depth. It is indicated that some division of reinforcement between that required for corrosion protection and that necessary for load carrying capacity may be advisable.

In his conclusions he states that for structural members of categories of concrete strength 15 and 25 MPa, which are in the open air but protected against rain, one should increase cover by 5 mm over the then current DIN Standard, 1045, Jan. 1972, requirements. A limitation of the crack width as required by the same code did not, according to the investigation under consideration, lead to improvement of the corrosion of the reinforcement. Naturally, after consideration of the data, it will be important to see whether after long exposure periods more thin rods in a cross-section always ensure a smaller likelihood of failure than a few thicker bars. However, importance should always be given to a sensible distribution of reinforcement regardless of bar size. The width of transverse cracks must, in any case, be limited so that no longitudinal cracks appear.

Tremper[5.30] commenting earlier on the crack size debate stated that in 1947 he had found that cracks of considerable width in blocks exposed for ten years in a moist but substantially salt-free environment did not cause significant corrosion of embedded steel. Hausmann's experience[5.31] was that minor cracks penetrating to the steel are normally not damaging and frequently heal in a moist environment by alkaline material from the adjacent concrete. Gewertz[5.32] in discussing the conditions of the San Mateo-Hayward Bridge in California stated that the deterioration taking place was observed to be independent of surface cracking which normally might be considered to afford easy ingress for the corrosive effects of a marine environment. He considered that as a practical matter, therefore, it did not appear that cracks in concrete are of decisive effect with respect to the development of corrosion. Commenting on the same bridge, Mehta and Gerwick[5.33] observed that the deterioration was confined to the underside and windward faces of precast steam-cured beams; naturally-cured cast-in-place beams being unaffected. They noted that in massive steam-cured products, with heavy reinforcement consisting of closely bundled bars, differential cooling rates in different parts of the member could lead to excessive micro-cracking, which became continuous due to severe weathering conditions and led to the corrosion.

Schiessl and Bakker[5.1] note that under normal conditions, local depassivation cannot be avoided in ordinary reinforced concrete structures in the region of cracks crossing the reinforcement. On the other hand they consider that this will not lead to significant damage if the quality of the concrete cover is adequate, and the crack widths do not exceed values in the range of 0.3 to 0.4 mm. Where prestressing steel is involved any such depassivation must be avoided during service life. Mehta and Gerwick[5.33] explain the initiation of corrosion at cracks by anode formation, but note that these cracks do not play any role in increasing the cathodic area, which is essential for the progress of the corrosive attack. The effect of the absence of a large cathode on corrosion at a crack has been experimentally demonstrated by Wilkins and Lawrence,[5.34] who concluded that corrosion occurs with quite narrow longitudinal cracks, whereas steel intersected by wider transverse cracks remained passive.

It would appear from a general survey of literature that, apart from lean concrete mixes and under very severe corrosion conditions, the influence of transverse flexural cracks in normally-reinforced concrete does not play a major role in corrosion. In many cases, however, the cracks associated with corroding bars are not of this type. Many cracks examined in the field tend to run parallel with bars. They are due either to the expansive, and hence disruptive, forces caused by the volume increase of the steel as it oxidizes, or to shrinkage, differential thermal movements, or problems with concrete in the plastic state. In flexural members they may be related to stirrups or other ligatures. Elaborate calculations of crack widths due to flexural loads would appear to be valueless for corrosion protection if other types of cracks are ignored, particularly if such cracks are collinear with the steel.

Under very special conditions the effects of cover can be somewhat unexpected. Because certain types of members forming marine structures are subject to cyclic loadings at the same time as corrosion of steel is occurring, some attention has been given to problems of corrosion-fatigue of reinforcement.[5.35–5.38] It has been observed that under high-cycle fatigue conditions the fatigue life of reinforcing steel in concrete tested in air is greater than when the concrete is immersed in sea water . When load cycles were matched to expected wave frequencies, a build-up of reaction products from cement paste and sea-water blocked the cracks in the cover zone. This led to crack closure and a reduction of stress range in the bar for the same load range applied to the specimen, and enhanced endurance of the beams. Paterson[5.37] nevertheless noted continued corrosion at these crack locations despite the cyclic stiffening effects brought about by changes in the cover concrete.

Prestressed Reinforcement

As has been said of ordinary reinforcement, prestressed tendons adequately encased in Portland cement grout should show no tendency to corrode. Nevertheless, increased use of prestressed concrete utilizing higher strength tendons, longer strands, decreased spread between working and ultimate stress, the service life of older structures (between 30 and 40 years) and the critical type of structures, such as nuclear reactors and bridges in which they have been used, are all reasons for some concern amongst civil engineers as to the service life of tendons. In recent years, too, a small percentage of failures have been reported in prestressing tendons used. A survey of 242 failures in the period 1950 to 1979 was presented by Nurnberger.[5.39]

In 1985 a 32-year old post-tensioned bridge collapsed in the UK.[5.40] The simple span bridge formed by nine I-beams, each consisting of eight precast segments post-tensioned together, collapsed under self-weight as a consequence of corrosion of the post-tensioning steel at segment joints. In Germany the Southern outer roof of the Berlin Congress Hall, which was constructed in 1957, collapsed without warning on the 21st May 1980. A paper by Isecke[5.41] presents details of the failure analysis, which led to the conclusion that corrosion of the tendons at joints was responsible for the collapse. Generally failures have been traced to factors which have not permitted the passivating action of the products of hydration of Portland cement paste to act along the total length of the tendon. Data collected by Schupack and Suarez[5.42, 5.43] in the USA led them to conclude that atypical practices, including poor design details or execution, were responsible for corrosion incidents.

Possible Problems

Pitting

In both ordinary reinforcement and prestressed tendons, a thin layer of general corrosion product is acceptable, but development of very local cells, leading to pitting, and hence reduction of cross-sectional area, is particularly worrying in prestressed tendons. The development of pits may be due to bad storage conditions of tendons prior to installation, as described by Fountain[5.44] in the case of a reactor vessel construction, or to the corrosive conditions active in service as described by Phillips[5.45] for wires from the Geehi aqueduct. In both cases the mechanism can be described as in Fig. 5.24. It should be noted that at the pit, the site of corrosion, the area is depleted in oxygen and the process is auto-catalytic in nature. From Fountain *et al.*[5.44] it is apparent that,

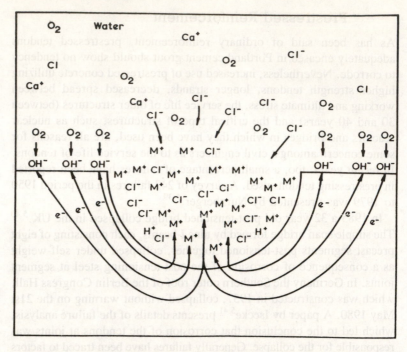

although corrosion inhibitors may be useful for protection of tendons prior to installation in certain cases, the particular one chosen led in part to the problem, and the authors suggest that strongly hydrophilic inhibitors, particularly sulphonates, should not be used for protection of tendons prior to installation. In both the first and second cases the influence of the structure of the steel is shown to be important. In the case of the Wylfa pits, growth is to a limiting pit depth, after which continued growth is in a direction parallel to the axis of the wire. In Phillips[5.45] it is stated that slight corrosion-pitting, to a maximum depth of 0.05 mm was observed, as were cracks which originated at the inner surface of the core winding where rust deposits were apparent. The cracks were at right angles to the wire surface.

Pitting may also result from stray current corrosion processes, but there seems to be very little information on this aspect of the technology. Rothman and Price[5.46] state that not much is known about the effects of stray currents on discontinuous metallic elements, such as prestressing wires wrapped around a concrete core of a prestressed concrete cylinder pipe (PCCP). They note, however, examples of interference between PCCP water mains and crossing cathodically protected gas lines, leading to failure of the unprotected prestressed pipe. Other examples of bursts relating to dc currents entering PCCP lines from traction systems and exiting at points of low resistance are quoted. With major prestressed concrete structures located close to such traction systems, or even carrying them, this is a field worthy of more detailed investigation than it has attracted so far.

A paper by Vernon[5.47] describes design procedures to control stray current in new metro construction. These methods relate to continuous bonding of reinforcing and other steel, or the use of high resistance support systems for rails. Griess and Naus[5.48] review the corrosion of steel tendons used in prestressed concrete pressure vessels, where stray current flow may be only one of several adverse conditions to be expected. Methods of laboratory testing and monitoring of stray current corrosion of prestressed concrete in sea-water are dealt with by Cornet et al.[5.49]

Stress Corrosion

Brittle fracture of prestressing steel by either stress corrosion or hydrogen embrittlement is especially dangerous from the safety standpoint because such fractures can occur with no prior macroscopic deformation of the steel and hence no advance warning. As the term implies, stress-corrosion cracking (SCC) is the result of a combined stress and corrosion state. On consideration of the stresses involved, it has been found that the residual stresses may be of equal consequence to the applied stresses. The amount of corrosive attack which occurs prior to cracking failure is often not measurable. It is significant, too, that a specific ion species in the corrosive environment is required before failure by SCC occurs (i.e. NO_2, H^+, NH_3^+, H_2S).

Many mechanisms of stress corrosion have been proposed, but at present there is no adequate theory that will explain all the observed phenomena completely. The presence of a threshold stress has been demonstrated for most alloys which have been found susceptible to SCC. A suggested theory for the mechanism is that a stress-formed crack becomes a locus of an increased dissolution activity. The electron source area is large, and hence the metal dissolution current density at the crack is very high (Fig. 5.9). Due to the stressed nature of, say, a tendon with a crack, the oxide free and kinked state of the surface of the crack allows continued progression of dissolution and corrosion with the possibility of rapid crack formation (Fig. 5.25).

A more complex model for the process within the crack follows. The total overpotential (η) (defined as the actual electrode potential minus the equilibrium value) for a particular reaction is the sum of the activation overpotential (η_{act}) and concentration overpotential (η_{conc}). For several metals the activation overpotential decreases as the strain rate increases, and since the strain rate is a function of the rate of penetration of the crack, η_{act} can be expressed as a function of the rate of penetration, or its equivalent in electric current (Fig. 5.26a). The concentration overpotential for the crack sides will be much less than for the tip (Fig. 5.26b), since the supply and the removal of products is easier. The total

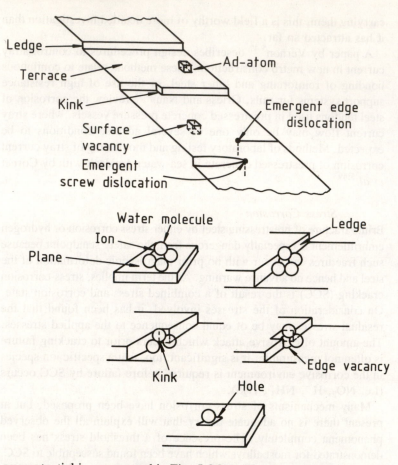

overpotential is represented in Fig. 5.26(c). The potential at the mouth
of the crack E_M will be the same as that for the surface of the metal,
and this, combined with the curve representing the IR drop down the
crack, will decide the form of the corrosion attack on the metal. Figure
5.27 shows possibilities for stress corrosion and for E too low and too
high for stress corrosion. Figure 5.28 illustrates the limiting crack depth
for stress corrosion due to the change in activation overpotential. This
theory does explain the uncertainty associated with the occurrence of
SCC. A proposed phenomenon akin to stress corrosion which has been
described is 'adsorption embrittlement'. This phenomenon is believed
to cause a reduction in the material's cohesive strength at the crack tip
due to the presence of the electrolyte.

The importance of stress corrosion in prestressing steels is not yet
finally decided. Moore *et al.*[5.50] state that their tests showed that neither
hydrogen-embrittlement cracking nor stress corrosion cracking occur
when the prestressing steel is in a chloride-bearing concrete environment.

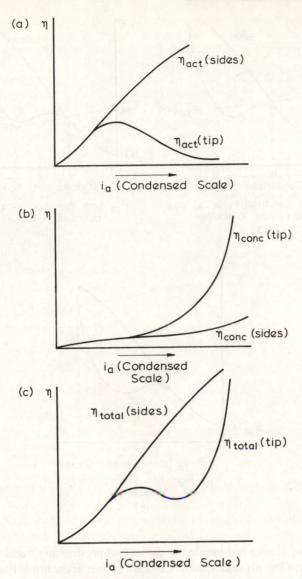

Fig. 5.26 Over-potential characteristics of a crack

However, H_2S cracking can occur and is considered to be a potentially ·serious problem if concrete cover is inadequate or if the pH of the concrete is lowered by carbonation. Work published at a Symposium on Stress Corrosion at Delft in 1971 tends to suggest that some unexplained failures may be due to this phenomenon. Alekseev[5.51] showed that under certain conditions, especially in the presence of chlorides, the tensile stresses intensify the development of local damage in the form of pits which resulted in corrosion brittleness of steel. Rehm

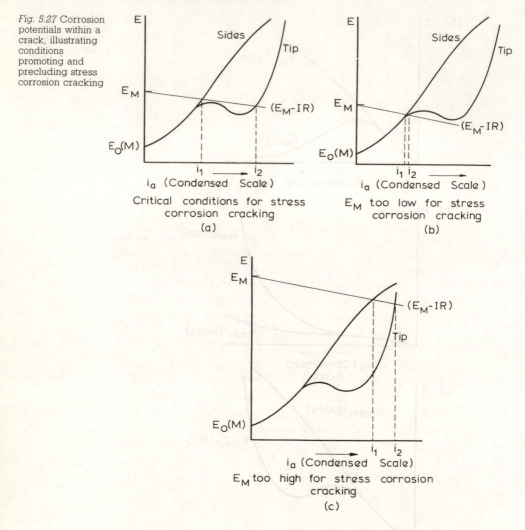

Fig. 5.27 Corrosion potentials within a crack, illustrating conditions promoting and precluding stress corrosion cracking

and Rieche[5.52] note that beside the damage of prestressing steels under conditions of low pH values, some cases are known where brittle fractures occurred in uncarbonated concrete in the absence of Cl^-, but they suggest that this phenomenon may be induced by the presence of sulphides. Brachet[5.53] has shown that stress-corrosion cracking can occur in wire maintained in tension in an environment of aerated distilled water, but suggests the mechanism to be akin to hydrogen embrittlement; an alternative possibility is the action of adsorption embrittlement.

Uncertainty attached to the importance of stress corrosion led to a symposium sponsored by the Australian Road Research Board in 1971. At that symposium the presence of stress corrosion cracks in the

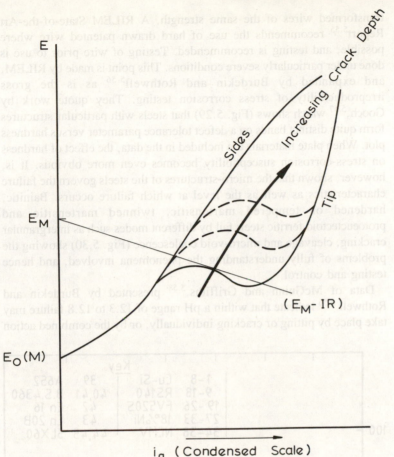

Fig. 5.28 Conditions limiting depth of stress corrosion cracking

prestressing tendons of concrete pipe in Australia and the problem of the Narrows Bridge were mentioned. Phillips[5.45] noted that in the case of the Geehi Aqueduct failure, 93 prestressed wire ends showed that they had been reduced in section by flaking corrosion and had failed in a ductile manner by overload, probably during the burst. However, multiple stress-corrosion cracks were observed adjacent to several of the ends which exhibited stress-corrosion characteristics. Isecke[5.41] noted that on portions of the fracture surfaces of the tendons from the Berlin Congress Hall failure both macro- and micro-cracks were present. He concluded that the fracture mode was typical of hydrogen-induced stress-corrosion cracking for the type of steel. Schupack and Suarez[5.54] have reported on corrosion embrittlement failures in the USA.

In laboratory tests it has generally been observed that hard drawn patented wire is susceptible to environmental cracking in a narrower range of conditions than are either quenched and tempered wires or isothermally transformed wires of the same strength. A RILEM State-of-the-Art Report[5.55] recommends the use of hard drawn patented wire where possible, and testing is recommended. Testing of wire prior to use is done under particularly severe conditions. This point is made by RILEM, and expanded by Burdekin and Rothwell[5.56] as is the gross irreproducibility of stress corrosion testing. They quote work by Gooch,[5.57] which shows (Fig. 5.29) that steels with particular structures form quite distinct bands on a defect tolerance parameter versus hardness plot. When plate materials are included in the data, the effect of hardness on stress-corrosion susceptibility becomes even more obvious. It is, however, shown that the micro-structures of the steels govern the failure characteristics as well as the level at which failure occurs. Bainitic, hardened or tempered martensitic, twinned martensitic and protoeuctectoid ferritic steels fail by different modes such as intergranular cracking, cleavage and micro-void coalescence (Fig. 5.30) showing the problems of fully understanding the phenomena involved, and hence testing and control.

Data of McGuinn and Griffiths,[5.58] presented by Burdekin and Rothwell,[5.56] indicate that within a pH range of 12.3 to 12.8 failure may take place by pitting or cracking individually, or by the combined action

Fig. 5.29 Defect tolerance of high strength steels. Closed symbols denote twinned martensite; open symbols denote other micro-structures; dashed numbers indicate intergranular failure (after Gooch[5.57])

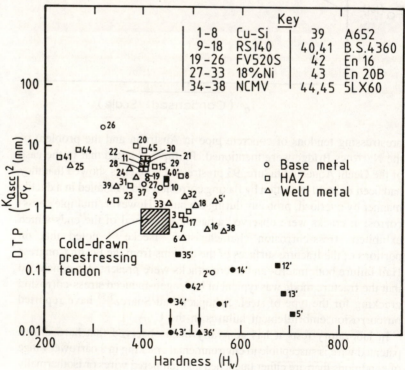

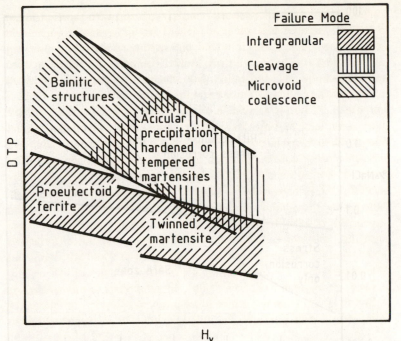

Fig. 5.30 Failure mechanisms of high strength steels (after Gooch[5.57])

of pitting and cracking if chlorides are present (Fig. 5.31). However Griffiths suggests that a zone exists within this pH range, wherein tendons might be expected to be immune from both cracking and pitting under air saturated, freely corroding conditions. For this to occur Cl^- concentrations at pH 12.4 must be as low as 60 ppm. Rather more serious for long-term durability of prestressed reinforcement are the findings of Alonso and Andrade,[5.59] who studied the active/passive potential regions for prestressing steel immersed in sodium bicarbonate solutions, and found that SCC was possible in carbonating concrete. They defined concentrations of 0.1 and 0.01 M of sodium bicarbonate and sodium carbonate respectively as the limits between activity and passivity for rebars immersed in them. Such concentrations exist in pore solutions of concretes in the final stages of carbonation. Whether this condition will pose a significant threat to tendons in ageing concretes and grouts is as yet unknown.

It would appear that stress-corrosion cracking is a rare phenomenon in prestressed concrete; nevertheless under adverse conditions it is held by several research workers to have occurred in practice. Safeguards must be to ensure that the 'adverse conditions' blamed for such problems never occur as a result of either ignorance or bad practice. Such adverse conditions particularly relate to the presence of specific ions mentioned previously.

Fig. 5.31 Failure
mechanisms of
prestressing
strand (after
McGuinn and
Griffiths[5.58])

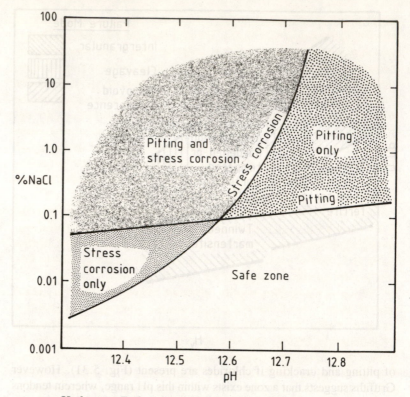

Hydrogen Embrittlement

The embrittlement of steel by hydrogen is characterized by the sensitivity of the embrittling phenomenon to slow strain rates and to elevated temperatures. This type of sensitivity suggests that the embrittlement is controlled by the diffusion of hydrogen. A steel which has been embrittled by hydrogen shows the following characteristics:

(*i*)   the notch tensile strength may be less than normal and directly reflects the loss of ductility due to H_2;

(*ii*)   delayed failure may occur over a wide range of applied stress;

(*iii*) there is only a slight dependence of the time-to-failure on the applied stress;

(*iv*)   there is a minimum critical value of stress below which the failure does not occur.

It is remotely possible that due either to a chemical reaction or even to incorrect corrosion protection procedures hydrogen could be evolved near to tendons in concrete. The question is what becomes of hydrogen atoms adsorbed on the metal surface? They may be desorbed in either a chemical or electrodic reaction as hydrogen molecules which diffuse out into the solution, or they may diffuse into the metal in the form of adsorbed hydrogen (Fig 5.32a,b). The diffusion coefficients for hydrogen into iron are of the same order as these for ions in aqueous solution. Furthermore the diffusion coefficient is the same for polycrystalline and single crystal iron, thus the hydrogen must diffuse through the lattice (interstitial diffusion). The higher the lattice strain or distortion of the iron the larger the accumulated concentration of hydrogen at that site (Fig. 5.32c). At some stage, the adsorbed hydrogen atoms reach imperfection sites and here become adsorbed hydrogen atoms which combine to form hydrogen molecules. Pressures now become enormous (10^5 to 10^7 atm) and the surrounding metal is stressed beyond its elastic limit leading to cavities and blisters (Fig. 5.32d,e).

If within a crack such abnormal stress concentration arises, the material at the crack apex is locally stressed into the plastic deformation region. Anodic dissolution may now dissolve away the kinky surface, further plastic deformation creates fresh kinky surface inside the crack and thus yielding aids metal dissolution (Fig. 5.32d).

Cahill[5.60] discusses two failures which were reported to have occurred after the tendons had been stressed. In both these cases galvanized strand was used. In the first, failure occurred between four and 48 hours after the strand was loaded, and the strand failed at a notch caused by the grips. It was determined that the wire itself had a higher hydrogen content than usual. The second failure occurred after several weeks and was said to be due to water in the duct which had caused extensive corrosion of the zinc. It was concluded in this paper that corrosion occurring when prestressing wire is in contact with zinc results in delayed failure, although the hydrogen content is not significantly greater than that of uncoated specimens corroded for similar times, and that hydrogen induced into a prestressing wire by a cathodic current causes delayed failure. On the effects of galvanized strand Moore et al.[5.50] note that galvanized coatings lowered steel strength and increased relaxation but they state that hydrogen embrittlement by cathodic charging of the steel did not occur for stressed specimens, with breaks in the galvanized layer, that were tested for long times in a chloride environment. Burdekin and Rothwell[5.56] note that based on FIP findings, galvanized tendons can with caution be used in practice. They instance the use of such tendons on the Forth Road Bridge without hydrogen embrittlement problems.

It was concluded by the Snowy Mountains Hydro Electricity Authority that the chief cause of wire fracture in the Geehi Aqueduct failure was hydrogen-embrittlement cracking irrespective of whether or not stress corrosion occurred by anodic metal dissolution.[5.45] From the literature it appears that under certain conditions hydrogen embrittlement can occur; the frequency is fortunately low. Prediction of the precise conditions under which it will occur in practice does not yet seem possible.

Fig. 5.32 Hydrogen
embrittlement
processes: (a)
adsorption at
crack tip; (b)
decohesion by
hydrogen influx to
dilated lattice; (c)
lattice distortion by
hydrogen; (d)
hydrogen absorp-
tion; and (e)
accumulation of
hydrogen species
at defects

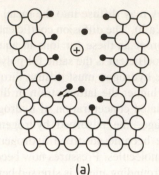

(a)

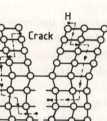

(b)

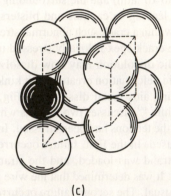

(c)

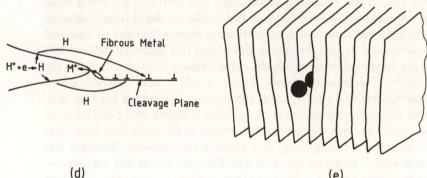

(d)

(e)

Cover, Cracks and Grouting or other Methods of Protection

Moore *et al.*[5.50] found that for pre-tensioned beams exposed to 3.5 per cent NaCl solutions, covers of greater than 20 mm are required to provide protection for the steel, even in short-time testing. They also found that open transverse cracks only 0.1 mm wide at the concrete surface permitted rapid corrosion to occur. It is understandable that the permissible crack size for corrosion of prestressing tendons will not be the same as for normal reinforcing. The RILEM Report[5.55] suggests that the thickness and permeability of concrete cover should, ideally, be specified and recommends that means should be developed for confirming, by non-destructive tests, that the values specified are actually being achieved. It suggests that cover thickness might be examined by gamma radiography where geometry is relatively simple. With respect to cracking it states that in partially pre-tensioned concrete some cracking within controlled limits is permissible and in post-tensioned systems some relaxation might be possible, in so far as cracks might be permitted to approach supplementary reinforcement or even ducting, as closely as they are allowed to approach reinforcement in reinforced concrete. It recommends that this should be made the subject of a special study and appropriate codes of practice formulated. In pre-tensioned systems the concrete requirements should be similar to those for high standard reinforced concrete. Water:cement ratio should be low and permeability kept to a minimum. RILEM suggests that for these designs one should ensure the absence of cracks in concrete at full load.

Post-tensioned systems present a different aspect of cement technology, *viz* grouts and grouting. Portland cement grouts have been found to be extremely efficient in preventing corrosion provided that there are no air pockets left within the duct. Once again the data of Moore *et al.*[5.50] have proved valuable. They showed that corrosion of steel can occur in the cavities of an improperly filled duct and state that it is important that ducts be completely filled. Such cavities or voids were further studied, and it was found that:

(*i*)   voids form more readily at higher flow velocities;
(*ii*)   more voids form when steel fills a high proportion of duct area;
(*iii*) voids in the grout tend to disappear when grouting pressures are maintained constant after the grouting is completed;
(*iv*)   voids can be caused by the presence of bleed water in pockets, since this bleed water is reabsorbed after the grout hardens, thus leaving a void.

In the case of critical structures, full-scale grouting tests have been performed, such as those described by Schupack[5.61] for a 54 m vertical 54-strand tendon to be used for a nuclear power station containment structure.

The use of special admixtures for grout improvement must be considered for new installations and repair materials. The incorporation of silica fume as a concrete grout constituent is growing. Silica fume together with a high-range water reducer enhances the pumpability and reduces the segregation and bleeding of the grout. This is brought about by the spherical shape of the particles and their fineness. They are about 100 times finer than cement and fill with solid material those spaces which would otherwise be occupied by the liquid phase of the grout. Apart from increasing the solids volume of a grout, silica fume displays excellent pozzolanic properties. The low achievable water:cement ratios, ranging from 0.20 to 0.40, give such grouts their low electrical conductivity and low permeability characteristics. A general description of silica fume use in high-strength concrete is given in a report by ACI Committee 226,[5.62] whereas the more specific use of it in grouts is considered by Domone and Tank.[5.63] The technology of silica fume incorporation in grouts has not yet been developed to a stage where no features require further investigation. For example experienced workers such as Perenchio et al.[5.64] note ineffective protection to bare strand by encasement in a silica fume-modified grout. They suggest that unsatisfactory proportioning may have been responsible for this finding, which appears to be contrary to most published results. Hope and Ip[5.65] describe grout formulations based on silica fume and a superplasticizer. Into these grouts they incorporate aluminium powders to react with alkalis derived from the cement, thereby offsetting volume changes caused by the settlement of cementitious materials and plastic shrinkage of the grout. It is of interest to note that potential hydrogen embrittlement is not seen as precluding the use of such grouts.

Wire-wound or wrapped systems have had a somewhat chequered history. Shotcrete coverings to wires are praised by some engineers and damned by others. Nurnberger[5.39] discusses damage to 36 wrapped tanks, and finds that about 55 per cent of these cases were attributable to too small a cover, too high a porosity of the enveloping mortar or concrete, or lack of adherence of shotcrete to the concrete element. In the application of shotcrete considerable skill is necessary to avoid problems of development of air pockets. Study of the rate of carbonation of shotcrete would be of interest. Crowley[5.66] suggests that wires be covered with at least 25 mm of shotcrete, and suggests the use of screed-wires on tank walls to ensure this thickness is maintained. The development of including chopped steel strand into the shotcrete to prevent the development of wide cracks is significant in this field, and is dealt with in Chapter 7, pp. 261–6.

Non-grouted systems rely on protection other than that of cement paste hydration products and this calls for some significant departures from the technology applied in bonded systems, certain of which are dealt with by ACI-ASCE Committee Report ACI 423.3R.[5.67] One procedure for post-tensioned systems is to cast into the concrete already greased monostrand tendons sheathed in a plastic tube, and, at some given time stress them and seal the anchor assembly with a mortar plug. The grease used is usually a proprietary brand which contains an inhibitor. Sheathing, grease and other element specifications are covered by a publication released by the Post-Tensioning Institute.[5.68] The advantages claimed for such a system are the significant savings in manpower and time for installation. Such systems have a fairly long history of use in the USA.

According to Nurnberger[5.39] failures in six cases of unbonded tendons were due to inadequate corrosion protection caused principally by damage to paper wrapping. The method of paper wrapping formerly used on the tendons has since been discarded. Tendons are now protected within plastic tubing. In the newer systems the end anchorage remains vulnerable, as its metal component remains exposed and must be effectively protected. Schupack and Suarez[5.54] describe failures of tendons where end anchorage pockets have been improperly filled. Perenchio et al.[5.64] suggest that further work could profitably be done on: (i) improving end fixings by epoxy coatings, (ii) the use of polyethylene duct or epoxy coated steel duct, and (iii) the use of epoxy-coated prestressing strand.

The most important field for the development of large-scale non-bonded tendons has been in nuclear reactors, and a paper by Hildebrand[5.69] discusses some of the procedures and the rare problems which have arisen in using corrosion-inhibiting compounds for the protection of prestressing systems.

Special Modifications to Concretes and Reinforcement

Where conditions of exposure are particularly adverse, methods may be adopted for use in normally reinforced, and prestressed members, which may reduce the corrosion effects. An approach which has been of particular interest to bridge engineers has been the use of overlays, membranes and sealers for deck slabs. Aspects of these are dealt with in Chapter 7, pp. 272–82, and are considered in some detail in *ACI 222R-85*.[5.70] Babaei and Hawkins[5.71] give a detailed evaluation of bridge-deck protective strategies, and indicate that the use of coated bar and concrete sealers is increasing, whereas increased covers and interlayer membranes are less often applied. It should be noted that Hay and Virmani[5.72] suggest that none of the overlays, membranes or sealers tested by them stop corrosion in underlying salty concrete, and conclude that such repair procedures are not permanent, although they may be cost-effective in reducing deterioration rates. Another approach has been to use mineral admixtures such as silica fume, slag and fly ash to reduce concrete permeability, and thereby improve durability. In this section of the book only treatments which in some way modify the materials themselves, and substitutions for such materials as are normally used, will be considered.

Corrosion Inhibitors

A corrosion inhibitor is an admixture that is used in concrete to prevent the metal embedded in concrete from corroding. There exist various types of inhibitors.

(*i*) Anodic (alkalis, phosphates, chromates, nitrites, benzoates). Anodic inhibitors function by decreasing the reaction at the anode. They may react with the existent corrosion product to form an extremely insoluble adherent coating on the metal surface. Organic inhibitors replace water at sites on the Inner Helmholtz plane, thus decreasing corrosion.

(*ii*) Cathodic (calcium carbonate, aluminium oxide and magnesium oxide). Cathodic inhibitors act to stifle the cathodic reaction. They are generally less effective since they do not form films on the anode.

(*iii*) Mixed. A mixed inhibitor may affect both anodic and cathodic processes.

(*iv*) Dangerous and safe. A safe inhibitor is defined as one which reduces the total corrosion without increasing the intensity of attack on unprotected areas, while dangerous inhibitors produce increased rates of attack on such areas. Such increased rates can be due to lack of sufficient inhibitor to prevent complete protection, or the presence of crevices into which the inhibitor does not rapidly diffuse. Anodic inhibitors are generally dangerous, except sodium benzoate. Cathodic inhibitors are generally safe, but zinc sulphate is an exception.

Dean et al.[5.73] have presented a somewhat different classification based on actions such as barrier layer formation, neutralization and scavaging. These represent processes by way of which the passivation is achieved. It is of interest to note that barrier layer formation is probably best achieved by simply completely coating steel with a well-cured low-water cement paste, which needs no extra admixtures at all. A rapid method for studying the action of inhibitors on corrosion of steel embedded in concrete containing calcium chloride was given by Gouda and Monfore.[5.20] From this paper their curves for three anodic inhibitors are given in Fig. 5.33. The amounts of the salt used are as percentage by weight of cement.

In ACI 212.3R,[5.74] released in 1989, it is noted that numerous

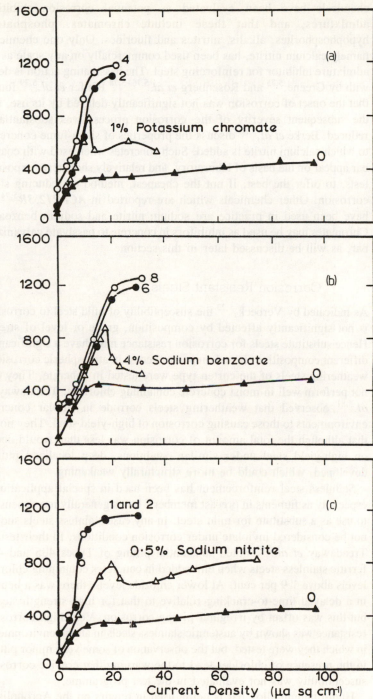

Fig. 5.33 Current
density vs potential
diagrams for
anodic inhibitor
actions (after
Gouda[5.20]). Effect
of (a) potassium
chromate, (b)
sodium benzoate,
and (c) sodium
nitrite on the
anodic polarization
of steel in
concrete
containing 2%
calcium chloride,
1.6% sodium
chloride, and 2%
calcium chloride
respectively

chemicals have been evaluated as potential corrosion-inhibiting admixtures, and that these include chromates, phosphates, hypophosphorites, alkalis, nitrites and fluorides. Only one chemical, namely calcium nitrite, has been used commercially on any scale as an admixture inhibitor for reinforcing steel. The passivating action is dealt with by Greene,[5.75] and Rosenberg et al.[5.76–5.78] Pfeifer et al.[5.79] found that the onset of corrosion was not significantly delayed by its use, but the subsequent severity of the corrosion process was substantially reduced. Berke et al,[5.80] discuss the properties of silica-fume concretes to which calcium nitrite is added. Such concretes when used with coated bar appear on the basis of laboratory, and relatively short-term exposure tests, to offer the best, if not the cheapest, method of reducing steel corrosion. Other chemicals which are reported in *ACI 212.3R*[5.74] to have been used in practice are sodium nitrite and sodium benzoate. Chromates may be used as inhibitors in concrete to passivate galvanized bar, as will be discussed later in this section.

Corrosion Resistant Steels

As indicated by Verbeck,[5.22] the susceptibility of mild steel to corrosion is not significantly affected by composition, grade or level of stress. Hence substitute steels for corrosion resistance must have a significantly different composition. Based on some success in atmospheric corrosion, weathering steels of the corten type were tested in concrete. They did not perform well in moist concrete containing chlorides. Treadaway *et al.*[5.81] observed that weathering steels corrode in similar concrete environments to those causing corrosion of high-yield steel. They noted that although the total amount of corrosion was less than would occur on high-yield steel under similar conditions, deep localized pitting developed, which could be more structurally weakening.

Stainless steel reinforcement has been used in special applications, especially as fitments in precast members, but is generally too expensive to use as a substitute for mild steel. In any case stainless steels should not be considered inviolate under corrosion conditions. In their testing Treadaway *et al.*[5.81] noted substantial pitting of Types 405 and 430 ferritic stainless steels when embedded in concretes containing chloride levels above 1.9 per cent. At lower chloride levels, there was a benefit in a delayed time-to-cracking relative to that for high-strength steels, but this was offset by irregular pitting corrosion. Very high corrosion resistance was shown by austenitic stainless steels in all the environments in which they were tested, but the observation of some very minor pitting in the presence of chlorides lead to the warning that crevice corrosion susceptibility was not evaluated in the test programme.

High-titanium alloy bar is being used in repairs on the Acropolis in

Athens. This bar is grouted into holes drilled into the marble slabs, and
the grouts are based either on Portland cement or epoxy.

Coatings for Steel

Coatings are sometimes considered for steel that is to be embedded in
concrete exposed to adverse corrosive conditions. There are both benefits
and disadvantages to their use and any benefit can only be optimized
by carefully considering the specific job. The more obvious of these
considerations are:

(*i*)   do the expected service life and structure exposure warrant coating
of the steel?

(*ii*)   if coating is desirable, is a field job required or may the coating
be applied prior to fabrication of the reinforcing for the structure?

(*iii*) do transportation and subsequent fabrication pose a significant
danger to the coating?

(*iv*)   in view of the exposure conditions, is the choice of coating dictated
by these conditions rather than adoption of other measures?

Two groups of coatings are advocated for use, the non-metallic or
organic and secondly the metallic.

Organic Coatings

Materials originally tested as organic coatings included coal-tar enamel,
epoxy, asphalt, chlorinated rubber, vinyl, phenolic, neoprene and
urethane. In considering the literature most of these were seen to have
significant disadvantages, but the epoxy group appeared to have the best
potential for use. Despite the fact that epoxy coatings provided excellent
corrosion protection of prestressing steel in tests by Moore *et al.*[5.50] the
authors stated that they believed them to be lacking in wear resistance
and ease of application. Furthermore, they were also relatively high in
cost. Work was subsequently undertaken by Clifton *et al.*[5.82] to
ascertain the feasibility of using organic coatings, especially epoxies,
to protect the steel reinforcing bars embedded in concrete of bridge-decks
from rapid corrosion. This corrosion is caused by the chloride ions from
the most commonly applied de-icing salts, sodium chloride and calcium
chloride. Altogether, 47 different coating materials were evaluated to
some extent, consisting of 21 liquid and 15 powder epoxies; five polyvinyl
chlorides; three polyurethanes; one polypropylene; one phenolic nitrile;
and one zinc-rich coating. The chemical and physical durabilities,
chloride permeabilities and protective qualities of coatings were assessed.
The bond strengths between coated and uncoated bars and concrete were
measured by both pullout and creep tests, and the results compared.

Results indicated that both epoxy and polyvinyl chloride coatings, if properly applied, could be expected to adequately protect steel reinforcing bars from corrosion. However, only the epoxy-coated bars had acceptable bond and creep characteristics when embedded in concrete. The powder epoxy coatings overall performed better than the liquid epoxies, and four powder epoxy coatings were identified as promising materials to be used on reinforcing bars embedded in concrete decks of experimental bridges. The findings led to a large input of finance for research on the bar by the US Department of Transportation and epoxy-coated bar became the usual reinforcement for bridge-decks subject to freezing conditions in the USA and Canada. There is a great deal of information on the properties of this bar and its use. It is summarized in the Final Report of the Federal Highways Administration on Coated Reinforcing Steel for Bridge Decks,[5.83] and the ASTM Standard for the Bar and the ACI 222R-85 Report.[5.70]

The epoxy coat acts to isolate the steel from contact with oxygen, moisture and chloride. However, at damaged points on the coat, corrosion may commence. Where such damage exists on the bar coupled to uncoated steel, the performance of such bar is still considered to be satisfactory, but not as good as when all bar is coated. There is yet some question of the possible flow of epoxy from highly stressed areas of the bar under fatigue conditions, as noted by Roper,[5.38] and further uncertainty has been noted with respect to long-term bond and possible split development. Johnston and Zia[5.84] tested slab specimens to compare strength, crack width and crack spacing, and beam-end specimens were tested under both static and fatigue conditions. The slab specimens showed little difference in crack width and spacing, deflections or ultimate strengths for coated and uncoated bar. The slabs containing epoxy-coated bar generally failed in flexure rather than in bond at approximately 4 per cent lower loads than those with uncoated bar.

The beam-end specimens were flexural-type specimens in which load was applied to the reinforcing bar. Splitting occurred along the reinforcing bars, but failure was primarily by either pullout or yield of the reinforcing steel. Based only on tests that ended in pullout failure, the uncoated bar was found to have developed 17 per cent more bond strength than the epoxy-coated bar. To allow for this reduction in bond of the epoxy bar, it was recommended that development lengths be increased by 15 per cent when epoxy bar is used.

Treece and Jirsa[5.85] tested beams with lap splices in a constant moment region, and bond strengths of epoxy-coated bars were compared to that of uncoated bar. In this study all failures were caused by splitting of the cover in the splice region, due to a reduction of adhesion between steel and concrete, resulting in almost complete loss of any frictional capacity between the materials. Treece and Jirsa warn that where splitting failures occurred, a reduction of about 35 per cent in bond strength was found, and that the 15 per cent increase in bond length should only be considered applicable to bars with large cover or wide spacing. They suggest that development lengths of 1.5 should be used with epoxy bar with small cover or close spacing, where splitting is likely. They note that the influence of transverse reinforcement in preventing splitting may modify this recommendation. Despite some continuing uncertainties, most of which are expressed in an ACI Workshop Report,[5.86] the bar is being used at an increasing rate in parking garages, marine structures and other reinforced concrete that may be exposed to salt water.

Perenchio et al.[5.64] have considered the use of epoxy coated strands for prestressed concrete. The Report ACI 439.4R[5.87] presents details on available epoxy-coated strand in the USA. Wedge anchorages are available that develop the full strength and ductility of the 2070 MPa ultimate strength, 1860 MPa yield strength strand. The higher available strength simplifies replacement of corroded tendons with this material, and it is claimed that since the strand is flexible, damage to the coating during handling is reduced.

For special purposes, organic coatings other than epoxy have occasionally been used. In Germany PVC has been used on welded wire fabric; however little data on long-term durability is available on such materials.

Metallic Coatings

Metallic coatings of rebar are capable of providing protection to the black steel in one of two ways. Metals with a more negative corrosion potential (less noble) than steel, such as zinc and cadmium, provide sacrificial protection to the steel embedded in concrete, although the development of passivating products on the coating is of significance in the long term. Metals and alloys with a less negative corrosion potential (more noble) than the bar steel, such as nickel and stainless steel, protect the reinforcement only as long as the coating is unbroken, since the bar steel is anodic to the coating. The steel is protected by such metals simply by encapsulation.

Readily available metallic coated reinforcing is limited to galvanized material. This is despite the fact that Bird and Strauss[5.88] noted that cadmium would be a better choice of coat under marine exposure conditions, as in the presence of Cl^- ions, zinc coating does not always provide increased protection. Cadmiun suffers from a cost disadvantage when compared to zinc, and its derivatives are highly toxic. Nickel-clad bars have only been produced in the USA. The external protective surface is in this case produced by rolling a steel billet clad with a nickel surround to form the bar. Baker et al.[5.89] showed this type of bar to have excellent resistance to chloride attack when used in concrete. Stainless steel-clad bar manufacured in much the same way as the nickel-clad bar, but with a much thicker cladding (up to 10 mm) over the black steel core, has been successfully used in the UK in the repair of historic structures such as churches. Only where cost is of secondary importance can bar of the last two types mentioned be considered for use.

Probably the most detailed account of the properties of galvanized reinforcement in concrete is given by Andrade and Macias.[5.90] Their work explains to a great extent the controversies as to the benefit of using galvanized bar. They nominate three factors which are critical in influencing the corrosion behaviour of galvanized steel in concrete, although they believe them to be irrelevant in the corrosion of black steel bar under the same conditions. These factors are:

(i) the alkali content of the cement, which results in different pH values in the aqueous phases of the concrete pores;
(ii) the type of metallurgical structure of the galvanized coating, which is controlled by the type of base steel, the bath temperature and time of immersion in the bath;
(iii) the amount of moisture contained in the concrete pores.

The influences of these three conditions on the corrosion characteristics of galvanized bar are as follows:

(i) The more alkaline the cement liquor the less protective it is to the galvanized bar.
(ii) The galvanized steel should have a coat thickness of not less than 60 μm, and should be greater the more aggressive the environment; furthermore it should have an external pure zinc layer, as it is then more resistant to attack than when alloyed layers are exposed. This pure zinc layer should not be less than 10 μm thick.
(iii) When galvanized rebars are used, changes in concrete humidity may greatly affect the development of passivation, so that curing at high humidity is of importance. Furthermore, cyclical changes in humidity content of the concrete pores may sometimes affect corrosion rates dramatically, but these effects are not easily predictable. Water should not be allowed to drain through the concrete.

Andrade and Macias[5.90] believe that galvanized steel may be resistant to chloride attack, but only if the protective layer of calcium hydrozincate which forms on the bar is compact and continuous, and the remaining unreacted coating is thick enough to resist pitting if there is a tendency for it to occur. In carbonated concrete, even in the presence of chlorides galvanized bar always gives a better performance than black steel.

Chromating treatment (50 to 70 ppm of CrO_3 in the mixing water, or previous chromate dipping of the bars), which prevents hydrogen evolution in the presence of hydroxyl ions, is necessary to avoid bond loss as a consequence of bubble entrapment adjacent to the reinforcement. The actions of hydrogen evolution and suppression were considered by Bird.[5.91]

Cook[5.92] discussed the use of high-strength galvanized wire in prestressed concrete. Galvanized prestressing strand is available in the USA (ACI 439.4R[5.87]), and such strand was used on the first prestressed concrete bridge built in that country. Much work on galvanized strand has been undertaken in France, and a report by Creton and Raharinaivo[5.93] deals with the field application of hot-dipped galvanized prestressing steels. Other workers, particularly in Germany, express concern with regard to risks of hydrogen embrittlement of such high-strength coated steel.

Reports on the degree of success in mitigating the effects of corrosion by use of galvanized bar vary significantly. It is generally agreed that when galvanized bar is used, a longer period of service elapses before the effects of corrosion become noticeable. The argument is whether such a delay in the onset of corrosion problems has a cost-benefit. Certainly in the case of carbonating concrete in the absence of significant levels of chloride ions, its use can generally be justified for use in thin sections having limited cover. In the presence of significant amounts of chloride ions, the case for its use is less well-supported. The problem which confronts any worker in this field is that in earlier descriptions of the performance of structures, the criteria now known to be of significance were often never assessed or even recorded. Reassessment of a full range of structures on a regional basis may provide the most valuable guidance to the engineer proposing to use galvanized bar.

Cathodic Protection

If the Pourbaix diagram for iron (Fig. 5.4) is examined, a domain is found to exist wherein no corrosion occurs. The domain of immunity of iron or steel exists only when the metal has a more negative potential than it achieves under any naturally occurring condition. The concrete technologist can make use of this immune state by providing energy to the steel so as to develop this negative potential, and thereby eliminate any tendency for the steel to corrode. He does this by the installation of a cathodic protection system, either to a newly built structure, or to one which already shows signs of corrosion. In the latter case the corroding anode areas, which are current discharging ones can be converted to non-corroding cathodic areas or current recieving ones by the application of an adequate direct current into the electrolyte surrounding the reinforcing steel. The electrolyte is the pore liquor of the concrete.

An elementary corrosion cell comprising one anode and one cathode can be represented electrically by the circuit shown in Fig. 5.34(a). From this, using Ohm's law, it follows that

$$I_{corr} = \frac{E_c - E_a}{R_c + R_a}$$

When cathodic protection is applied to such a system it requires the provision of a new anode as shown in Fig. 5.34(b). Conventional positive current is made to flow from this anode, through the electrolyte, to the structure to be protected. An equivalent circuit for elementary corrosion cell plus cathodic protection is shown in Fig. 5.34(c). This circuit could be made more complex by the inclusion of metallic resistance, electrode capacitance, polarization and other terms, but these can be conveniently omitted in a discussion of principles. They should not, however, be neglected in a real system.

If: I_a = conventional positive current from the anode
I_c = conventional positive current to the cathode
I_{cp} = conventional positive cathodic protection current

then, from Kirchhoff's laws applied to the analogue circuit:

at X $\quad I_a + I_{cp} - I_c = 0$ $\qquad$ [1st law]

or $\quad I_a + I_{cp} = I_c$

round loop AXC $\quad E_c - E_a = I_aR_a + I_cR_c$ $\qquad$ [2nd law]

Substituting $I_a + I_{cp}$ for I_c in the second of these:

$$E_c - E_a = I_aR_a + (I_a + I_{cp})R_c$$
$$= I_a(R_a + R_c) + I_{cp}R_c$$

thus $\quad I_a = \dfrac{(E_c - E_a) - I_{cp}R_c}{R_a + R_c}$

This means that I_a, the corrosion current, can be made small by raising $I_{cp}R_c$ to the value of $(E_c - E_a)$, and of course for no corrosion the aim is to make I_a equal to zero.

It is of importance that if R_c is made large then I_{cp} can be kept small and still achieve full cathodic protection. For this the adherent paste surrounding the bar may be considered as a coating in the same way as wrapping is used on steel pipelines to reduce current demand.

In establishing a circuit for a CP system (Fig. 5.34) the source of emf, E_{cp}, may be an external dc power source (battery, dc generator,

Fig. 5.34 (a) elementary corrosion cell; (b) cathodic protection circuit; (c) equivalent circuit for cathodic protection system

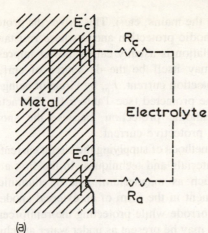

(a)

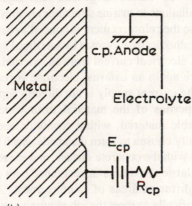

(b)

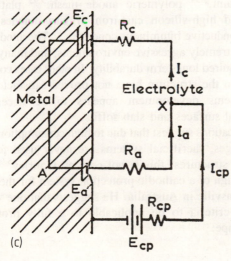

(c)

transformer/rectifier fed by the mains, etc.). This drives the protective current I_{cp} through the cathodic protection anode whose resistance to earth is R_{cp}. Such an installation is usually called 'power impressed'. Alternatively, the anode may itself be the dc source of emf, E_{cp}, providing the necessary protection current, I_{cp}, by virtue of being more active than the subject to be protected (see Table 5.1). This action is called sacrificial anode cathodic protection, because the anode is consumed in providing the protective current.

Consider further the two methods of supplying the circuit current from the viewpoint of anode materials and techniques. In the first a more negative metal or alloy such as magnesium, zinc or aluminium is connected to the reinforcement in the form of an external anode, and is allowed to sacrificially corrode while protecting the reinforcement. As a rule this type of anode may be present as under water attachments to marine structures, or may be buried in ground-beds adjacent to pipelines. Many concrete installations in marine structures are unwittingly cathodically protected because the reinforcement is bonded to bare steel sections which are purposely cathodically protected by sacrificial systems. In the second type of system, electrical current from an external source is applied to the circuit. Once again an external electrode is required. However, since in this case the current supply is not directly dependent on the electro-chemical properties of the material from which it is manufactured, the most durable material, within economic and other functional constraints, is usually chosen to form the anode. Such anodes need to be in electrical contact with the concrete surface, and are required to supply current in spite of large variations in resistance between the surface of the concrete and different areas of reinforcement. Anodes continue to be developed specifically to meet the job requirements. They include conductive paint,[5.94] polymeric anode mesh,[5.95] platinized titanium mesh,[5.96] and high-silicon cast-iron primary anodes with secondary anodes of conductive bituminous concrete.[5.97] The anodes are subject in service to extremely agressive environments, and many have failed to provide the required long-term durability. In other cases coatings required to fix them to the structure have not proved satisfactory in service. Further systems development appears to be necessary, particularly for vertical surfaces and slab soffits.

Although most publications suggest that due to high consumption rates and low driving voltages, sacrificial systems are never used for the protection of concrete structures, this is not correct. Gourley[5.98] was responsible for the design of a cathodic protection system for the Ross River Pipeline at Townsville in Australia. He suggests that one would expect the following criteria to be applicable to a fully protected prestressed concrete pipe:

(*i*)   an immediate change of -300 mV when the current is switched on;

(*ii*)   a final potential of between -450 and -1100 mV after the system has stabilized.

The higher figure in (*ii*) is interesting in the light of the hydrogen generation limit possibly being exceeded. The protection is achieved by a sacrificial method using banks of zinc anodes, the number in any bank being dependent on the soil resistivity. The pipeline was made electrically continuous by means of copper connectors linking up longitudinal strips embedded in the surface of each pipe core. The latter were linked to the tails of the circumferential windings at anchor blocks.

Without doubt the chief impetus to the study of cathodic protection of concrete structures by cathodic protection has been bridge-deck corrosion problems in the USA and Canada. Although some attempts have been made using sacrificial zinc sheet anodes, all the known installations on decks are impressed current systems. ACI Committee 222 reports[5.70] that successful applications to concrete stuctures were accomplished as early as 1946, but its application to brige-decks occurred in 1974. The cause of the delay in implementing the technology lay in the need to develop highly conductive overlays and closely spaced anode sets to minimize resistance and to provide uniform distribution of current. Three further reasons are given by Schell and Manning[5.94] for the relatively few cathodic protection systems at present being installed on bridges as compared with the number receiving other treatments. These are that the optimum time to apply cathodic protection is when the bridge is being constructed or in the early stages of corrosion, before physical distress develops; a satisfactory air-void system must be present in concrete to be left *in-situ*; and, many bridges are prestressed in part. If the cathodic protection system installation is delayed, then repair becomes very costly. In the absence of a satisfactory air-void system, future damage due to freeze—thaw conditions will not be halted by such an installation. In the case of prestressing there is the fear that evolution of hydrogen gas will occur with the accompanying problems of hydrogen embrittlement. Furthermore, for cathodic protection to be effective, the steel in the concrete must be electrically continuous, so that a uniform potential can be established. There exists the problem that in post-tensioned structures the tendons may be shielded from current supply by the metallic ducts which surround them. Problems related to hydrogen evolution and tendon shielding were noted in a 1985 study reported on by Schell and Manning.[5.94] Work is underway in Canada to assess their significance in practice.

Hay and Virmani,[5.72] in discussing cathodic protection of bridge-decks, state that the expense of the systems placed in California using duron iron anodes and coke breeze overlays, has stimulated searches for alternative systems. They describe the use of conductive polymer mortar used in slotted and strip/overlay systems. In the slotted system, slots (19 mm deep, 199 mm wide, at 305 mm centres) contain 0.8 mm platinum−niobium copper core wires or carbon strands, which are surrounded by conductive polymer grout. In the strip/overlay system, conductive polymer grout strip mounds (25 mm wide by 12.5 mm high at 305 mm centres) in conjunction with carbon strand are placed over the prepared bridge deck surface. The anode strip mounds are overlaid with Portland cement concrete or latex modified concretes to provide the new riding surface as described in FHWA Reports.[5.99, 5.100] It is reported that the systems were functioning satisfactorily; however, the fear was expressed that as chlorine gas was liberated at the face of the slot, it would combine with any moisture present to form hydrochloric acid which would attack the Portland cement paste. Systems are being sought to disperse the chlorine gas or minimize its impact. If chloride ions are continually being added to the system, and it is installed where chlorine gas is not readily dispersed, the impact not only on the materials, but on human beings must be considered. One of the authors inspected a prefabricated mesh anode protection system in a high-rise Asian city apartment block. The continued source of chlorides was from the use of sea-water for sewage and washing purposes. A strong aroma of chlorine gas was present throughout the building. Lankes[5.101] has highlighted a further problem where plant piping, tanks, bulkheads etc. must be cathodically protected, and the reinforcing bars of other structures are electrically interconnected with the metal to be protected. In this case cathodic protection demand must include the current demand of the interconnected reinforcing bars. Such current demands may be high.

Robinson[5.102] discusses the use of cathodic protection to prevent steel corrosion of reinforced and prestressed concrete structures that have been damaged, subjected to stray current interference or installed in severely corrosive environments. He states that at the time of writing there existed no widely accepted cathodic protection criteria for steel in concrete. Structures were sometimes protected as if they were bare or organically coated steel structures. The assumption may be made that the steel is protected only if sufficient current is applied to maintain a steel potential of −850 mV to a copper sulphate electrode (CSE). For concrete that is submerged in water or is buried underground the −850 mV CSE criterion is easily attained at reasonable current densities, say 25 μA/sq ft of bar area. In the bridge-deck case, where oxygen is abundant and pore-blocking by water is not active, it may not be possible to fulfil this criterion requirement even at currents approaching 3 mA/sq ft of deck surface. In this case a shift of 400 mV for all half-cell potentials has been recommended by ACI 222.[5.70] Alternatively E-log I curves must be used to determine the required current inputs. Generally a combination of criteria is used such as a minimum negative potential shift of 300 mV when measured with the CP current flowing; or a minimum negative polarization potential shift of a given value varying between 100 and 300 mV with current interrupted, as determined by measuring polarization decay.

Uncorroded steel in concrete normally has a potential ranging from 0 to −300 mV to a CSE, which is 300 to 500 mV more noble than bare steel in soil. It has been demonstrated by Hausmann[5.103, 5.104] that the corrosion of steel in concrete exposed to a high chloride environment can be prevented if sufficient cathodic protection current is applied to shift the steel polarization potential to −500 mV to a CSE. For a structure subject to stray current discharge Hausmann[5.31] suggests that the polarization potential should be shifted to −400 mV to a CSE, which is in agreement with the ACI figure. It has been pointed out[5.101, 5.102] that the initial current density requirements decrease with time, presumably due to oxygen depletion, so that from values of say 1.6 mA/sq ft of concrete area, at the commencement of protection they fall to about 0.07 mA/sq ft. This is a value considerably lower than Hausmann's[5.31] 0.25 mA/sq ft. More recently NACE members have been concentrating on the establishment of a more precise criterion for protection based on the decay from the instant off potential, obtained in the same way as for cathodically protected steel structures.[5.105]

Apart from the danger of hydrogen evolution due to over-protection of a structure leading to prestressing steel embrittlement, another real or perceived problem exists if current input is higher than the minimum required for corrosion protection, or indeed if current is applied at all. This problem is the effect of direct current on the bond strength of steel. Rosa in 1913,[5.106] applied currents between 100 and 1000 mA/sq ft to reinforced concrete, and showed that the bond strength between steel and concrete decreased dramatically. The concrete softened at the bar surface due to a build-up of sodium and potassium ions transported to that interface by the current flow. Locke[5.107] quotes the results of other studies done at levels varying between 3 and 180 mA/sq ft. In one of three studies no decrease in bond strength was detected. In two others bond strength did drop, in one case after about four years at the low applied current level of 3 mA/sq ft. Lin[5.108] went one step further in studying the bond reduction effects of current flow. He loaded beams in flexure to develop specific crack widths, and maintained loadings to simulate service conditions while subjecting the steel to current flow. He found that for a given flexural crack width, the greater the current flow, the sooner longitudinal cracking occurs. Corrosion rates were not related to different flexural crack widths for cracks which remained open during current flow. He warns that flexural bond strength loss at lapped splices may be of particular concern under severe corrosion conditions. The authors of this book believe that the nature of the cement, its maturity and the reactivity of the aggregate may all significantly influence the degree of impact such mass transport effects will have in practice. It is the opinion of most workers in the field that at the levels of current input of well-designed systems, the effects of bond strength loss will prove to be unimportant over the design life of the structure.

Costing of a cathodic protection system must necessarily be divided into two parts: initial capital cost and annual running and maintenance cost. The initial capital cost depends on the system chosen and the area of concrete to be protected. Installation costs will vary from country to country. For the most part the rectifiers, transformers, meters, alarms and cabling are standard items, and hence readily costed. The costs of special anodes depend on the type chosen. Based on the costs of coke breeze, asphalt and bituminous concrete system costed in Sydney for a 2 500 sq m installation, initial capital cost for such a system would be between $A125 000 and $A175 000 at 1987 values. Annual running and maintenance costs when broken down into individual items suggest that inspection costs will be greater than the cost of anode replacement, or power consumed, which is the least cost of all.

It appears that the application of cathodic systems for protection of concrete structures offers some real hope to the concrete technologist, but the field remains open for the introduction of innovative methods to overcome problems of both technique and cost. For example a possible method of attaining protection using electro-chemical means, without the high costs of providing permanent installations, may be the removal of chlorides from concrete as demonstrated by Slater and his co-workers.[5.109] In this a somewhat higher current is passed between a temporary anode and the bar until the chloride concentration of the concrete is sufficently lowered by outward diffusion to reduce the corrosion rate to an acceptable level. The method is not without problems such as heating of the concrete and increasing its permeability. The first is overcome by contol of input current, and the second by impregnation or coating of the concrete after chloride removal. Naturally the best solution would be to construct so as to preclude steel corrosion during the design lives of concrete structures.

Some Aspects of Repair

Almost throughout this book aspects of repair of concrete structures suffering in part from reinforcement corrosion are considered, and they will not be repeated here. Basically two conditions prevail when an engineer is confronted by repair of corrosion induced durability problems. In the first case the damage is superficial and, once this is confirmed, repairs may be conducted in the most economical way. As a rule if patching, crack filling and coating procedures are adopted, the problems simply reappear later, hopefully at other locations. In doing patch repairs the removal of insufficient concrete to ensure that chloride contamination is adequately reduced is the greatest cause of corrosion reactivation. The creation of local cathodic areas by the very act of applying a patch is also important in promoting further corrosion. Providing the damage is not endangering the strucural stability, temporary repairs of this sort can be continued for a prolonged period. If such repair is repeated too often or the structure is difficult to work on, the costs of adopting such a repetitive procedure may become burdensome. Furthermore, if continued corrosion occurs, the steel will eventually not perform its intended function. Outlines for the adoption of such procedures appear in publications by Pullar-Strecker,[5.110] a Concrete Society Technical Report,[5.111] a CIRIA Technical Note[5.112] and innumerable other reports and individual papers on specific types of structures and materials.

In the second case, structural problems result or may result from the degradation, and some of the more complex and costly procedures outlined in this chapter and elsewhere in this book have to be adopted. In the future the most troublesome problems confronting engineers in the area of structural repair are expected to be inspection, assessment and rectification of prestressed concrete structures. A Concrete Society Technical Report,[5.113] outlines procedures for inspection and assessment of data, stressing the importance of considering the consequences and possible modes of failure. Two examples given in this publication are situations wherein:

(i)   if corrosion occurs, it will be detected visually prior to significant deterioration and the structure has a secondary load distribution system which will provide an alternative route for loading;

(ii) if corrosion occurs, it will not be detected visually and the structure does not have a secondary distribution system.

Conditions in (i) allow the structure or element to be considered from the corrosion damage vewpoint in much the same way as if it were not prestressed. In case (ii) if there is evidence of a generally unsatisfactory environment, and signs of corrosion are elsewhere present, structural modification may clearly be required and it is probable that the structure may need to be taken out of service until such work is completed.

References

5.1   Schiessl P (ed) 1988 *Corrosion of Steel in Concrete* RILEM Report of the Technical Committee 60—CSC 96 pp

5.2   Bazant Z P 1979 Physical model for steel corrosion in concrete sea structures — theory. *ASCE Journal of Structural Div* **105**(ST6) Proceedings Paper 14651: 1137—53

5.3   Bazant Z P 1979 Physical model for steel corrosion in concrete sea structures — applications. *ASCE Journal of Structural Div* **105**(ST6) Proceedings Paper 14652: 1155—66

5.4   Tuutti K 1982 *Corrosion of Steel in Concrete* Research Report 4/82 Swedish Cement and Concrete Research Institute, Stockholm

5.5   Beeby A W 1978 Corrosion of reinforcing steel in concrete in relation to cracking. *The Structural Engineer* **56A**(3): 77—81

5.6   Browne R D 1982 Design prediction life for reinforced concrete in marine and other chloride environments. *Durability of Building Materials* Amsterdam **1**: 113—25

5.7   Browne R D 1986 Practical considerations in producing durable concrete. In *Improvement of Concrete Durability* Thomas Telford, London pp 97—130

5.8   Roper H 1983 Fatigue and corrosion fatigue surfaces of concrete reinforcement, corrosion, microstructure and metallography. *Microstructural Science* **12**: 37—44

5.9   Pourbaix M 1971 Potential — pH diagrams and metallic corrosion. In Ailor W H (ed) *Handbook on Corrosion Testing and Evaluation* John Wiley Chapter 26 pp 661—87

5.10 Pourbaix M 1966 *Atlas of Electrochemical Equilibria in Aqueous Solutions* Pergamon Press, New York 644 pp

5.11 Kalousek C L, Jumper C H, Tregoning J 1943 Composition and physical properties of aqueous extracts from Portland cement clinker pastes containing added materials. *National Bureau of Standards Journal of Research* **30**: 215—55

5.12 Shalon R, Raphael M 1959 Influence of sea water on corrosion of reinforcement. *ACI Journal* Proceedings **55**(12): 1251—65

5.13 Gjorv O E, Vennesland O 1976 Sea salts and alkalinity of concrete. *ACI Journal* Proceedings **73**(9) Sept 1976 pp 512—16

5.14 Hamada M 1968 Neutralization (carbonation) of concrete and corrosion of reinforcing steel. *Proceedings of the 5th International Symposium on the Chemistry of Cement* Tokyo Vol III pp 343—84

5.15 Meyer A 1968 Investigations on the carbonation of concrete. *Proceedings of the 5th International Symposium on the Chemistry of Cement* Tokyo Vol. III pp 394—401

5.16 Verbeck G 1958 Carbonation of hydrated Portland cement *ASTM STP 205* pp 17—36

5.17 Netherlands Committee for Concrete Research 1975 *Carbonation of Lightweight Concrete* (in Dutch) Rep 73, Zoeterveer 25 pp

5.18 Pressler E E, Brunauer S, Kantro D L, Weisse G H 1961 Determination of the free calcium hydroxide contents of hydrated Portland cements and calcium silicates. *Analytical Chemistry* **33**: 877—82

5.19 ACI Forum 1985 Influence of chlorides in reinforced concrete. *Concrete International Design and Construction* **7**(9): 13—19

5.20 Gouda V K, Monfore G E 1965 A rapid method for studying corrosion inhibition of steel in concrete. *Journal of the Portland Cement Association Research and Development Laboratories* **7**(3): 24–31

5.21 Lewis D A 1962 Some aspects of the corrosion of steel in concrete. *Proceedings of the 1st International Congress on Metallic Corrosion* Butterworths, London pp 547–52

5.22 Verbeck G J 1975 Mechanism of corrosion of steel in concrete *ACI SP–49* pp 21–38

5.23 Stratfull R P 1975 Half-cell potentials and the corrosion of steel in concrete *Highway Research Record No 433*

5.24 Monfore G E 1968 The electrical resistivity of concrete. *Journal of the Portland Cement Association Research and Development Laboratories* **10**(2): 35–48

5.25 Powers T C, Copeland L E, Hayes J C, Mann H M 1954 Permeability of Portland cement pastes. *ACI Journal* Proceedings **51**: 255–95

5.26 Ost B, Monfore G E 1966 *Penetration of Chloride into Concrete* Portland Cement Association Research & Development Bulletin No 192 23 pp

5.27 Whiting D 1981 *Rapid Determination of the Chloride Permeability of Concrete* Report No FHWA–RD–81–119 Federal Highway Administration, Washington DC 174 pp

5.28 Berke N S, Pfeifer D W, Weil T G 1988 Protection against chloride-induced corrosion. *Concrete International Design & Construction* **10**(12): 45–55

5.29 Schiessl P 1976 Regarding the question of admissible crack width and the required concrete cover in reinforced concrete with special consideration of the carbonation of concrete (in German). Heft No 255 *Deutscher Ausschuss fur Stahlbeton* pp 39–49

5.30 Tremper B 1966 Corrosion of reinforcing steel *ASTM STP 169–A*

5.31 Hausmann D A 1964 Electrochemical behavior of steel in concrete. *ACI Journal* Proceedings **61**(2): 171–88

5.32 Gewertz M W 1957 *Causes and Repair of Deterioration to a California Bridge Due to Corrosion of Reinforcing Steel in a Marine Environment, Part 1 Methods of Repair* Bulletin 182 Highway Research Board

5.33 Mehta P K, Gerwick B C 1982 Cracking-corrosion interaction in concrete exposed to marine environment. *Concrete International Design & Construction* **4**(10): 45–51

5.34 Wilkins N J M, Lawrence P F 1983 The corrosion of steel reinforcements in concrete immersed in seawater. In Crane A P (ed) *Corrosion of Reinforcement in Concrete Construction* Ellis Horwood, Chichester Chapter 8 pp 119–41

5.35 Bannister J L 1978 Fatigue and corrosion fatigue of torbar reinforcement. *The Structural Engineer* **56A**(3): 82

5.36 Arthur P D, Earl J C, Hodgkiess T 1979 Fatigue of reinforced concrete in seawater *Concrete* (London) **13**(5): 26

5.37 Paterson W S 1980 Fatigue of reinforced concrete in seawater *ACI SP–65* pp 419–36

5.38 Roper H 1983 Investigations of corrosion, fatigue and corrosion fatigue of concrete reinforcement. *Corrosion '83* Paper 169 NACE 12 pp

5.39 Nurnberger U 1981 Analysis and evaluation of failures in prestressed steel.

Proceedings of the 3rd Symposium on Stress Corrosion of Prestressing Steel FIP, Madrid 13 pp 220–9

5.40 Anon 1985 Welsh bridge failure starts tendon scare. *New Civil Engineer* Dec 12 1985: 12–13

5.41 Isecke B 1983 Failure analysis of the partial collapse of the Berlin Congress Hall. *Prakt Met* **20**: 118–29

5.42 Schupack M A 1978 A survey of the durability performance of post-tensioning tendons. *ACI Journal* Proceedings **75**(10): 504–5

5.43 Schupack M, Suarez M G 1982 Some recent corrosion embrittlement failures of prestressing systems in the United States. *Journal PCI* **27**(2): 38–55

5.44 Fountain M J, Blackie O, Mortimer O 1975 Corrosion protection of prestressing tendons. *International Conference on Experience in the Design, Construction and Operation of Prestressed Concrete Pressure Vessels and Containment for Nuclear Reactors* Institution of Mechanical Engineers (University of York) England

5.45 Phillips E 1975 *Survey of Corrosion of Prestressing Steel in Concrete Water-Retaining Structures* Technical Paper No 9 Australian Water Resources Council

5.46 Rothman P S, Price R E 1984 Detection and considerations of corrosion problems of prestressed concrete cylinder pipe. *ASTM STP 906 Corrosion Effect of Stray Currents and Techniques for Evaluating Corrosion of Rebars in Concrete* pp 92–107

5.47 Vernon P 1986 Stray-current corrosion control in metros. *Proceedings of the Institution of Civil Engineers* Part 1 **80**: 641–50

5.48 Griess J C, Naus D J 1980 Corrosion of steel tendons used in prestressed concrete pressure vessels. *ASTM STP 713 Corrosion of Reinforcing Steel in Concrete* pp 32–50

5.49 Cornet I, Pirtz D, Polivka M, Gau Y, Shimizu A 1980 Laboratory testing and monitoring of stray current corrosion of prestressed concrete in sea water. *ASTM STP 713 Corrosion of Reinforcing Steel in Concrete* pp 17–31

5.50 Moore D G, Klodt D T, Hensen R J 1970 *Protection of Steel in Prestressed Concrete Bridges* Report No 90 National Co-operative Highway Research Program

5.51 Alekseev S N 1971 Some aspects of stress corrosion of high strength reinforcement. *Symposium on Stress Corrosion* Delft

5.52 Rehm G, Rieche G 1971 Fractures of high-strength steels caused by corrosion-processes in alkaline solutions. *Symposium on Stress Corrosion* Delft

5.53 Brachet M 1970 Report based on some years of observations of stress corrosion phenomena in high strength steel. *Annales de l'Institute Technique du Batiment et des Travaux Publics* (267–8): 82–100

5.54 Schupack M, Suarez M G 1981 Corrosion embrittlement failures of prestressing systems in the United States. *FIP Symposium on Stress Corrosion Cracking of Prestressing Steel* Madrid Spain 15 pp

5.55 RILEM 1976 Committee CRC, Corrosion of reinforcement and prestressing tendons, a state of the art report. *Materials and Structures* **9**(51): 187–206

5.56 Burdekin F M, Rothwell G P 1981 *Survey of Corrosion and Stress Corrosion in Prestressing Components used in Concrete Structures with Particular Reference to Offshore Applications* Cement and Concrete Association, Wexham Springs, Slough 36 pp

5.57 Gooch T G 1974 Stress corrosion cracking of welded joints in high strength steels. *Welding Journal* **53**(7): 287–306

5.58 McGuinn K F, Griffiths J R 1977 Rational test for stress corrosion crack resistance of cold drawn prestressing tendons. *British Corrosion Journal* **12**: 152–7

5.59 Alonso M C, Andrade M C 1989 The electrochemical behaviour of steel reinforcements in sodium carbonate and sodium bicarbonate solutions in relation to stress corrosion cracking. *Corrosion Science* **29**(9): 1129–39

5.60 Cahill T 1971 Delayed failures in prestressing wire and strand. *Symposium on Stress Corrosion* Delft

5.61 Schupack M 1975 Grouting aid for controlling the separation of water for cement grout for grouting vertical tendons in nuclear concrete pressure vessels. *International Conference on Experience in the Design, Construction and Operation of Prestressed Concrete Pressure Vessels and Containment for Nuclear Reactors* Institution of Mechanical Engineers (University of York) England

5.62 ACI Committee 226 1987 Silica fume in concrete. *ACI Materials Journal* **84**(2): 158–66

5.63 Domone P L, Tank S B 1986 Use of condensed silica fume in Portland cement grouts *ACI SP–91* pp 1231–60

5.64 Perenchio W F, Fraczeck J, Pfeifer D W *Corrosion Protection of Prestressing Systems in Concrete Bridges* Report 313 NCHRP Transport Research Board Washington DC 25 pp

5.65 Hope B B, Ip A K C 1988 Grout for post-tensioning ducts. *ACI Materials Journal* **85**(4): 234–40

5.66 Crowley F X 1965 Shotcrete covercoats for prestressed concrete tanks *ACI SP–49* pp 11–20

5.67 ACI-ASCE Committee Report ACI 423.3R 1989 Recommendation for concrete members prestressed with unbonded tendons *ACI Structural Journal* **86**(3): 301–18

5.68 *Post-Tensioning Manual* Post-Tensioning Institute 4th edn 406 pp

5.69 Hildebrand M 1975 Evaluation of corrosion-inhibiting compound for the protection of prestressing systems. *International Conference on Experience in the Design, Construction and Operation of Prestressed Concrete Pressure Vessels and Containment for Nuclear Reactors* Instutition of Mechanical Engineers (University of York) England

5.70 ACI Committee 222 Corrosion of metals in concrete. *ACI 222R–85* American Concrete Institute 1985 30 pp

5.71 Babaei K, Hawkins N 1988 Evaluation of bridge deck strategies. *Concrete International Design & Construction* **10**(12): 56–66

5.72 Hay R E, Virmani Y P 1985 North American experience in concrete bridge deterioration and maintenance. *Concrete Bridges — Investigation, Maintenance and Repair* The Concrete Society, London 14 pp

5.73 Dean S W, Derby R, Von Dem Bussche G T 1981 Inhibitor types. *Materials Performance* **20**(12): 47–51

5.74 ACI Committee 212 Chemical admixtures for concrete. *ACI 212.3R−89* American Concrete Institute 1989 31 pp

5.75 Greene N D 1982 Mechanisms and application of oxidizing inhibitors. *Materials Performance* **21**(3): 20−22

5.76 Rosenberg A M, Gaidis J M, Kossivas T G, Previte R W 1977 A corrosion inhibitor formulated with calcium nitrite for use in reinforced concrete *ASTM STP 629* pp 89−99

5.77 Rosenberg A M, Gaidis J M 1979 The mechanism of nitrite inhibiton of chloride attack on reinforcing steel in alkaline aqueous environments. *Materials Performance* **18**(11): 47−54

5.78 Lindquist J T, Rosenberg A M, Gaidis J M 1979 Calcium nitrite as an inhibitor of rebar corrosion in chloride containing concrete. *Materials Performance* **18**(9): 36−46

5.79 Pfeifer D W, Landgren J R, Zoob A 1987 *Protective Systems for New Prestressed and Substructure Concrete* Report No FHWA−RD−86−193 Federal Highways Administration

5.80 Berke N S, Pfeifer D W, Weil T G 1988 Protection against chloride-induced corrosion. *Concrete International Design & Construction* **10**(12): 45−55

5.81 Treadaway K W J, Cox R N, Brown B L 1989 Durability of corrosion resisting steels in concrete. *Proceedings of the Institution of Civil Engineers* Part 1 **86**: 305−33

5.82 Clifton J R, Beeghly H F, Mathley R G 1975 *Non-metallic Coatings for Concrete Reinforcing Bars* Building Science Series 65, National Bureau of Standards

5.83 Federal Highways Administration 1978 *Coated Reinforcing Steel for Bridge Decks* NEEP No 16 Final Report

5.84 Johnston D W, Zia P 1982 *Bond Characteristics of Epoxy Coated Reinforcing Bars* Report No RHWA/NC/82−002 Department of Civil Engineering, North Carolina State University, Raleigh 163 pp

5.85 Treece R A, Jirsa J O 1989 Bond strength of epoxy-coated reinforcing bars. *ACI Materials Journal* **86**(2): 167−74

5.86 ACI workshop on epoxy-coated reinforcement 1988 *Concrete International Design & Construction* **10**(12): 80−4

5.87 ACI Committee 439 Steel reinforcement — physical properties and US availability. *ACI 439.4R−89* American Concrete Institute 1989 14 pp

5.88 Bird C E, Strauss P J 1967 Metallic coating for reinforcing steel. *Materials Protection* **6**(7): 48−52

5.89 Baker E A, Money K L, Sanborn C B 1977 Marine corrosion behavior of bare and metallic-coated reinforcing rods in concrete *ASTM STP 692* pp 30−50

5.90 Andrade M C, Macias A 1987 Galvanized reinforcements in concrete. In Wilson A D, Nicholson J W, Prosser H J (eds) *Surface Coatings — 2* Elsevier Applied Science Chapter 5 pp 137−82

5.91 Bird C E 1964 The influence of minor constituents in Portland cement on the behaviour of galvanized steel in concrete. *Corrosion Protection and Control* **7**: 17

5.92   Cook A R 1979 *High Strength Galvanized Wire in Prestressed Concrete*
       Paper presented at the 12th International Galvanizing Conference, Paris
       98 pp

5.93   Creton B, Raharinaivo A 1983 *Field Application of Hot-dip Galvanized
       Prestressing Steels* ILZRO Project No ZE−220 Final Report Sept 1983

5.94   Schell H G, Manning D G 1989 Research direction in cathodic protection
       for highway bridges. *Materials Performance* **28**(10): 11−15

5.95   Drachnik K J 1986 Application of a polymeric anodemesh for cathodic
       protection of a reinforced concrete structure *ASTM STP 906* pp 31−42

5.96   Martin B L 1988 Mesh-based cathodic deck protection. *Concrete
       International Design & Construction* **10**(12): 24−6

5.97   Stratfull R F 1974 Experimental cathodic protection of a bridge deck.
       Transportation Research Record No 500 Transportation Research Board
       pp 1−15

5.98   Gourley J 1976 The cathodic protection of prestressed concrete pipelines.
       Corrosion Australasia **3**(1): 4−7

5.99   Turgeon R 1984 *Cathodic Protection of Bridge Decks in Pennsylvania*
       Report No FHWA−PA−84−013 Pennsylvania Department of Trans-
       portation, Harrisburg PA Oct 1984

5.100 *Cost Effective Bridge Concrete Construction and Rehabilitation in Adverse
       Environments* FCP Annual Progress Report Project 4K Federal Highway
       Administration, Mclean VA Sept 1982

5.101 Lankes J B 1976 Cathodic protection of reinforcing bars. *ACI Journal
       Proceedings* **73**(4): 191−2

5.102 Robinson R C 1975 Cathodic protection of steel in concrete *ACI SP−49*
       pp 83−93

5.103 Hausmann D A 1967 Steel corrosion in concrete. *Materials Protection*
       6(11): 12−23

5.104 Hausmann D A 1969 Criteria for cathodic protection of steel in concrete.
       Materials Protection **8**(10): 23−5

5.105 Slater J E 1983 Corrosion of metal in association with concrete. *ASTM
       STP818* pp 1−83

5.106 Rosa E B, McCollom B, Petters O S 1913 Electrolysis of concrete.
       Materials Bureau Standards Technology Paper No 18

5.107 Locke C E 1986 Corrosion of steel in Portland cement concrete:
       fundamental studies. *ASTM STP 906 Corrosion Effect of Stray Currents
       and the Techniques for Evaluating Corrosion of Rebars in Concrete* ASTM
       Philadelphia pp 5−14

5.108 Lin C Y 1980 Bond deterioration due to corrosion of reinforcing steel
       ACI SP−65 pp 255−70

5.109 Slater J E, Lankard D R and Moreland P J 1976 Electrochemical removal
       of chlorides from concrete bridge decks. *Materials Protection* **15**(11): 21−6

5.110 Pullar-Strecker P 1987 *Corrosion Damaged Concrete — Assessment and
       Repair* Construction Industry Research and Information Association,
       London 99 pp

5.111 *Repair of Concrete Damaged by Reinforcement Corrosion* Concrete Society
       Technical Report No 26 The Concrete Society, London 1984 30 pp

5.112 *Protection of Reinforced Concrete by Surface Treatments* Construction Industry Research and Information Association Technical Note 130 CIRIA, London 1987 72 pp

5.113 *Durability of Tendons in Prestressed Concrete — Recommendations on Design, Construction, Inspection and Remedial Measures* Concrete Society Technical Report No 21 The Concrete Society, London 1982 7 pp

Further suggested reading, p. 352

6 Maintenance and Repair Strategies

Inspection and Maintenance

There is a very wide divergence of opinion and practice on the extent to which structures, and especially concrete structures, should be subjected to regular inspection. As a generalization it appears that buildings, and especially buildings originally constructed by developers, get least regular attention. At the other end of the scale, public works, and especially those related to transportation, often receive regular attention and maintenance at quite short intervals. There is however no uniformity in approach and there are no established norms. This situation follows from the lack of any rational economic basis for determining the frequency of inspection or for setting criteria for the need for repair.

The aims of maintenance work can be classified as:

(*i*)   the avoidance of accidents, which may harm people or plant;
(*ii*) the continued operation of a facility;
(*iii*) the protection of the capital investment in the asset.

There is little choice relating to class (*i*). Something has to be done. Decisions in relation to classes (*ii*) and (*iii*) result from the policy of the owners and management, but these decisions may often have little or no rational basis. Indeed decisions often result from following a management style established long in the past by the owner. Information supplied to the Warren Centre for Advanced Engineering in the University of Sydney[6.1] on the need for and the extent of maintenance on wharf structures around Australia showed that many owners of wharfs did no inspection on a regular basis, or no inspection at all, of the state of their concrete structures. A few carried out repairs when damage was noticed. Some inspected everything annually. It could not be concluded that the structures owned by those in the last group were going to last longer or be cheaper in the long run. Much of the damage reported was corrosion of reinforcing steel, most of which was related to inadequate concrete and shortage of cover, both matters which should have been forestalled in the design and construction stages. One owner in Victoria reported that 'minor repairs to concrete structural elements are required in order to prevent serious structural problems at a later date'. At Port Hedland in northern Western Australia, where there is a large tidal range and the temperature ranges from 5 °C to 45 °C, severe cracking and reinforcement corrosion is reported after 15 years and, in spite of annual inspections, major and expensive repairs are required. The owner of a coal loading facility at Gladstone, in Queensland, stated: 'The Board does not design for obsolescence of a structure at the end of a particular shiploading contract. . . . Designing for limited life usually means increased maintenance towards the end of the life, especially if the contract is extended. Doubtless this can be justified by discounted cash flow analysis. However, the people who do the analysis are rarely around to justify their figures when the maintenance is being carried out.'

Frequent and well-organized inspections are probably the most effective method of reducing maintenance costs. By such inspections it is possible to build up a data-base relating to a particular structure, or to a class of structures. The development of defects can then be traced and decisions can be made on the point in time when repair becomes economical.

Unfortunately 'How frequent?' and 'What is well-organized?' are questions to which there is no definitive answer. Wynhoven and Hunton[6.2] addressing the question of building facades, in an Australian context, have suggested the following program of inspection:

(i)   prior to completion of the defects maintenance period (one year);
(ii)   prior to completion of any guarantee period;
(iii) otherwise two to four year intervals.

For reinforced concrete not exposed to very aggressive conditions, the rate of deterioration (if any) is not likely to be so rapid that inspections more frequent than every four years are needed. In very severe conditions, such as those discussed in Chapter 4, pp. 113−19, inspections every year may well pay off. The wharfs already referred to which had annual inspections were in many cases supported on steel piles which were cathodically protected. The cathodic protection system needed inspections every year, or even more frequently, and the concrete was given a look over at the same time. Without the cathodic protection system inspection, it is unlikely that the concrete structure would have got such frequent attention, and it probably did not need it anyway.

The *FIP Guide to Good Practice* on the subject of inspection and maintenance[6.3] includes a table which is said to give a general indication of the intervals that might be applicable for 'routine' and 'extended' inspection. Routine inspection is largely visual and no special equipment or access is needed. Extended inspection requires a closer and more intensive examination of all elements of the structure and therefore needs special access, remote viewing techniques and so on. Intervals are proposed for three classes of structure and for three environmental and loading conditions.

The structure classes are:

Class 1 — where possible failure would have catastrophic consequences and/or where the serviceability of the structure is of vital importance to the community;

Class 2 — where possible failure might cost lives and/or where the serviceability of the structure is of considerable importance;

Class 3 — where it is unlikely that possible failure would lead to fatal consequences and/or where a period with the structure out of service could be tolerated.

Environmental and loading conditions are:

Very severe — the environment is aggressive and there is cyclic or fatigue loading;

Severe — the environment is aggressive, with static loading, or the environment is normal, with cyclic or fatigue loading;

Normal — the environment is normal, with static loading.

The proposed intervals (in years) are as set out in Table 6.1. These intervals should, we believe, be regarded as absolute maxima and we suggest that in most cases more frequent inspections should be planned. Inspection intervals are, in the end, a matter of engineering and economic judgement and within the proposed ranges of time, the designer should be asked to establish an inspection program to suit his structure in relation to use, siting, construction and design.

Inspections carried out at the designated intervals will only be effective if there is consistent and orderly reporting. For this to happen, forms

Table 6.1 Indication of inspection intervals[6.3]

Environmental and loading conditions	Classes of structure					
	1		2		3	
	Routine	Extended	Routine	Extended	Routine	Extended
Very severe	2*	2	6*	6	10*	10
Severe	6*	6	10*	10	10	—
Normal	10*	10	10	—	Only superficial inspections	

* Midway between extended inspections

for recording observations must be developed and levels of deterioration established so that subjective judgements are eliminated as far as possible. It is important that the inspector should not be asked to make judgements and decisions during the inspection. If, for example, the inspector wants to give views as to causes of deterioration, these views should be put in a clearly separated part of the report. The whole process should be designed so that the inspection is done with as little expenditure as possible on wages. The design of inspection forms needs care and experience so that the information collected will be reliable and useful and will form the basis for subsequent decisions on more extensive investigation or repair.

The various stages of the process are set out in Table 6.2. The most difficult part of the process is the establishment of criteria for deciding the action that should follow the observation of a defect in an inspection. Some aspects are discussed below (pp 234–42).

A good example of an inspection scheme devised for concrete grain silos is given by Johinke and Tickner.[6.4] A silo (generally one of a large number of similar structures) which had the defects shown in Fig. 6.1 can be graded according to the classification system shown in Fig. 6.2. The condition of a number of facilities can then be summarized as shown in Fig. 6.3. Classifications of this type are well-suited to data processing programs, which allow easy recording and simple access to historical data. There is clearly an economic advantage in instituting such inspection processes at the start of the use of the structure, even though it may be difficult to establish the classes of possible defects before any have appeared. If the process is not instituted at the beginning of the life of the facility, then the initial inspections to record data may be time-consuming and expensive.

The collected data can provide the basis for determining repair priorities. In the silo example, the forms shown in Fig. 6.4 were used.

Table 6.2 Stages of inspection and maintenance

A   Inspection
     Collect data at specified intervals in specified form

B   Analysis
     (1)   Add latest information to data-base which contains all earlier information
     (2)   Examine progression of defects
     (3)   Relate defects to action criteria

C   Action possibilities
     (1)   Note and wait for the next inspection
     (2)   Alter inspection frequency
     (3)   Institute repairs
     (4)   Further detailed investigations
     (5)   Put safety procedures in place

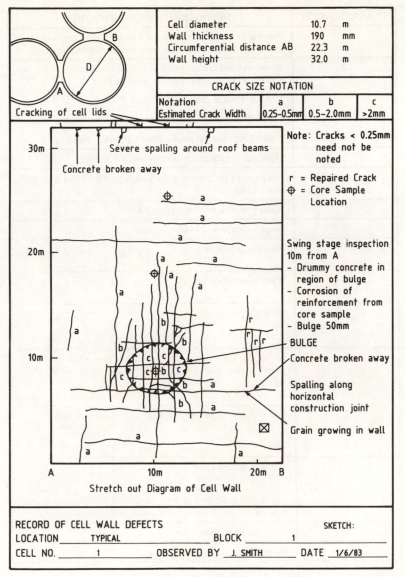

Fig. 6.1 Defects in silo cell wall (after Johinke and Tickner[6.4])

A simple scheme to provide a management information system in relation to highway bridges is reported by Andrews.[6.5] The system, which is based on and replaces the traditional bridge register, is used to develop tables indicating the need to reconstruct and the need to repair in a priority order. In this system, the priority for reconstruction is determined by calculating a number which is the product of factors based on gross load capacity; traffic congestion/delay; accidents; local pressures

DEFECT	CLASSIFICATION	DEFINITION OF CLASSIFICATION
CRACKING OF CELL WALLS	1	Bulging in wall over a height of 3 metres or greater, with associated severe vertical, horizontal and diagonal cracking
	2	Severe vertical cracking over height of 10 metres or more. In a 5 metre width of wall cracking is worse than: * 3 No vertical cracks greater than 2 mm wide * or 6 No vertical cracks greater than 0.5 mm wide * or 10 No vertical cracks greater than 0.25 mm wide
	3	Significant cracking - between classes 2 and 4.
	4	Not greater than 5 vertical cracks less than 0.25 mm wide in 5 metre width
WEATHER	1	Water penetration in more than 4 areas
PENETRATION	2	Water penetration in 4 or less areas
THROUGH WALLS	3	No water penetration
WATER PENETRATION	1	Water penetration occurs
THROUGH OVER-SILO STRUCTURE	2	No water penetration
SPALLING OR CRACKING AT * Cell lids	1	Severe cracking with visible displacement of concrete with a potential danger of falling concrete or loss of support to structure
* Over-silo beam supports * Access manholes	2	Severe cracking, but no visible danger of falling concrete
* Plant or gallery supports	3	Minor or no cracking
DAMAGE TO SUBSTRUCTURE	1	Significant settlement or joint movement leading to water penetration and reinforcement corrosion and concrete spalling
	2	Water penetration at joints
	3	No defect

Fig. 6.2 Classification of defects (after Johinke and Tickner[6.4])

and traffic management schemes. The lowest number indicates the greatest priority for reconstruction in a bridge 'league table'. A similar number is determined for the priority for maintenance of components of a bridge. The importance of a component ranges from 1.0 for main structural members, including foundations, to 1.5 for fenders and service ducts (see Fig. 6.5). The inspector gives to each component a 'condition factor', which ranges from 0.1 to 1.0, where 1.0 indicates a condition not needing any repair. The final repair priority number is the product

SITE: TYPICAL				ASSESSMENT BY: J. JONES							
DATE: 1/6/83				DATE: 1/8/83							
ELEMENT	DEFECT CLASSIFICATION										
	CELL NUMBER										
	TOWER	1	2	3	4	5	6				
STRUCTURAL DEFECTS											
* Cracking of cell walls	-	1	3	2	2	3	4				
* Weather penetration - Through walls		3	2	1	2	3	3				
- Through over-silo structure		1	2	1	1	2	2				
* Spalling at - Over-silo beams		1	3	2	2	2	3				
- Cell lids		2	2	1	3	1	3				
- Plant or gallery supports	1										
- Access manholes		3	1	3	3	3	2				
* Substructure	1										
MATERIAL DEFECTS											
* Laminated concrete	1	1	3	2	3	3	3				
* Reinforcement cor-rosion	1	2	3	2	3	3	3				
* Joint defects	1	1	2	2	2	3	3				
* Crumbling or spalling concrete	1	1	3	1	2	3	3				
TEST RESULTS											
* Chloride	1	3	-	3	-	-	-				
* pH	2	2	-	2	-	-	-				
* Density	2	2	-	1	-	-	-				

Fig. 6.3 Assessment summary (after Johinke and Tickner[6.4])

of the importance of the component, its condition, and the road classification. This particular method has been in use by a UK County Council, where it provides an information system which is both useful and useable as an engineering management tool. The simplicity of the system is very attractive and provided that careful thought is given to

Fig. 6.4 (a) and (b) classification of repairs and repair summary (after Johinke and Tickner[6.4])

MANUAL OF STANDARD ASSESSMENT PROCEDURES

(a)

TYPE	DESCRIPTION
T	Temporary
C	Cosmetic
P	Protective Maintenance
S	Structural
	EXTENT OF WORKS
1	Major repairs (> $30,000)
2	Minor repairs (< $30,000)
	IMPORTANCE OF FACILITY
1	Terminal
2	Sub terminal
3	Minor
	IMPORTANCE OF ELEMENT
	Storage or Handling Capacity
1	> 25,000 tonnes
2	> 5,000 tonnes
3	< 5,000 tonnes

REPAIR AND MAINTENANCE ASSESSMENT

(b)

SITE: TYPICAL						BLOCK 1						
INSPECTED: J. SMITH						ASSESSMENT BY: J. JONES						
IMPORTANCE OF FACILITY - 1						DATE: 1/8/83						
ELEMENT	REPAIR CLASSIFICATION											
	CELL NUMBER											
	TOWER	1	2	3	4	5	6					
ELEMENT IMPORT-ANCE	1	3	3	3	3	3	3					
STRUCTURAL DEFECTS												
* Cracking of cell walls		S1										
* Weather penetration - Through walls		*		P2								
- Through over-silo structure		P2		P2	P2							
* Spalling at - Over-silo beams		T2										
- Cell lids				T2		T2						
- Plant or gallery supports	T2											
- Access manholes			S2									
* Substructure	P2											
MATERIAL DEFECTS												
* Laminated concrete	S1	C-2										
* Reinforcement cor-rosion	S1											
* Joint defects	S1	C-2										
* Crumbling or spalling concrete	S1	C-2		C-2								
TEST RESULTS												
* Chloride	S1											
* pH												
* Density				C-2								

COUNTY COUNCIL
HIGHWAYS & TRANSPORTATION DEPT.

BRIDGE LOCATION _____
INSPECTED BY (SIG.) _____

BRIDGE INSPECTION REPORT

BRIDGE No. _____
SHEET _____ OF _____
SPAN No. _____

Fig. 6.5 Bridge
inspection report
form (after
Andrews[6.5])

CONDITION REPORT					
ITEM NO.	ITEM DESCRIPTION	(a)*	(b)**	(a)x(b)	DESCRIPTION OF CONDITION AND REMEDIAL WORK REQUIRED (REFERRED TO BY ITEM NO.)
1	Foundations	1.0			
2	Invert/Apron	1.4			
3	Fenders	1.5			
4	Piers/Columns	1.0			
5	Abutments	1.0			
6	Wing Walls	1.1			
7	Retaining Wall/ Revetment	1.2			
8	Approach Embankments	1.4			
9	Bearings	1.1			
10	Main Beams	1.0			
11	Transverse Beams	1.0			
12	Diaphrams/Bracings	1.1			
13	Concrete Slab	1.0			
14	Metal Deck Plates	1.0			
15	Jack Arches	1.0			
16	Arch Ring	1.0			
17	Spandrels	1.0			
18	Tie Rods	1.1			
19	Drainage Systems	1.3			
20	Waterproofing	1.2			
21	Surfacing	1.4			
22	Service Ducts	1.5			
23	Expansion Joints	1.2			
24	Parapets/Handrails	1.3			
]25	Access Gantry				
26	Walkways/Ladders				
27	Machinery				
LOWEST (a) x (b) x ROAD FACTOR = PRIORITY No.					

* Location Factor   ** Condition Factor (0.1 to 1.0)

the numbers chosen in setting up such a system it could provide very useful guidance in making decisions about inspection, maintenance and repair.

A much more elaborate computer-based system developed for the State of New York, USA, is described by Kamp.[6.6] This system contains a data-base for over 6000 state-owned bridges and a further 11 000 owned by counties, towns, villages and cities. The information is used to prepare bridge maintenance works programs, to ensure bridge safety and to prepare a bridge capital construction program for the state. As Kamp points out, 'bridge inventories need to be kept current, reflecting changes in bridges and the replacement of existing bridges with new. The data on file must be constantly checked for reliability.' Any authority embarking on a data file for maintenance must be fully satisfied that funds are going to continue to be available to keep the data current. If this cannot be assured the exercise will be nothing but an expensive excursion into computing.

Structural Appraisal

A series of questions have to be addressed in any investigation of the state of a deteriorated structure and its possible need for repair.

(*i*)   Does the present condition mean that safety is impaired?
(*ii*)   What is the cause, or causes, of the structure reaching its present state?
(*iii*) Is it likely that, without repair, the structure will become dangerous?
(*iv*)   Does the present condition affect the operation of the facility?
(*v*)   Is it likely that, without repair, the operation will be affected in the future?

Safety must always be the top priority. Once that has been decided, no further action can be taken sensibly until the cause of the trouble has been established. Unless this is done, there is always a risk that the treatment of a symptom may exacerbate the condition. For example, a conclusion that corrosion of prestressing anchorages in a parking station was caused by wetting from rain led to a decision to enclose the anchorages behind a protective casing. The humid condition within the casing, associated with a concentration of chlorides, accelerated the corrosion. It cannot be emphasised too often that the execution of repairs, where the cause of damage remains unknown, is just a waste of resources.

Before any of the five questions can be answered, it is often necessary for an extensive condition survey to be carried out. If the regular inspection procedures advocated on pp. 226−8 have not been put in place, then any condition survey is likely to be an expensive and lengthy operation, often made more difficult by the lack of satisfactory access. For example, the usual method of examining the facade of a high-rise building is by means of the window-cleaners' boat. Corner panels are not then accessible for proper inspection. The designer and owner can avoid this problem by making an early decision to provide building maintenance units (not just window cleaning units) so that experienced people can carry out inspections and perform periodic maintenance of the whole building facade if needed. For bridge and wharf structures, specialist equipment which allows easy access has been developed and is available for purchase by large organizations, such as transport authorities, and for hire by others (see Fig. 6.6).

ACI Committee 201 has prepared a useful check list of matters that should be examined in a condition survey.[6.7] Part of such a survey involves finding out about the materials and construction methods originally used and often this information is not available if the inspection starts at some long interval after construction has been completed. This is a further reason for maintaining an inspection procedure right from the beginning.

Of the five questions listed, question (i) must be addressed and answered before any further action is taken. Lack of safety can take many forms, of which the most dramatic constitutes the collapse of all, or part, of a structure. Loss of inherent strength, through deterioration of concrete or corrosion of reinforcing steel, may seriously reduce the capacity of a dam, a wharf, or a bridge to carry its extreme design load. Degradation of a structure may also reveal weaknesses in design. For example, the development of radial and circumferential crack patterns round the

Fig. 6.6 Bridge inspection equipment (photograph provided by Barin spa, Italy)

column heads of a flat slab building proved that the original design concept was not properly developed. Additional load put into the building could have caused a disastrous shear failure and progressive collapse of the whole building. The survey must, for this reason, include an examination of the design processes. *ACI 210.1R* does not include this aspect of a condition survey and Table 6.3 is provided here to overcome this deficiency.

Much more common than potential collapse is the possibility that pieces falling from a structure may injure passers-by (the deciduous building syndrome). This safety problem may frequently be avoided by the regular removal of potential spalls, the provision of protective awnings, or the repair of the cause, which is often, but not always, corroding steel.

A systematic approach to determining causes and evaluating further action has been proposed by Corcoran[6.8] and he gives useful guidance on what is a difficult and perplexing task. The process of establishing causes involves two distinct phases: investigation and synthesis. In each it is important that no aspects are ignored and no data are rejected. The 'off-beat' result may be the key that unlocks the mystery.

Establishing Causes

Investigation
In-situ investigation This investigation involves a physical examination of the structure. There is a tendency to make the examination superficial but it is important that everything that can be observed should be noted and recorded, generally in sketches supported by photographs. A complete photographic coverage is extremely valuable as detailed and repetitive examination becomes possible, in conditions of much greater comfort and concentration than are available on site. The amount of detail in a good photograph is more than is visible to the unaided eye and a magnifier often reveals additional information. The location for the taking of samples, and the methods to be used should be decided and recorded. Keeping accurate records of samples is necessary and becomes even more so if they are to stand up in any legal proceedings. Sloppy methods leave an investigator wide open to attack by legal counsel.

Laboratory investigation Laboratory work is frequently expensive and should only be undertaken when specific items of information are needed, such as the presence of chlorides, the cement content, the compressive strength of cores. A detailed specification of the tests required should be provided, after discussion with the laboratory people. In examining a sample it may well be that previously unnoticed

Table 6.3 Design information

A   Design of structure
  1   Standards for live-loading and wind-loading
  2   Frequency of live-loading
      2.1   Floor loading
      2.2   Impact and inertia loads
      2.3   Crane, hoist and lift loads
      2.4   Hydrostatic and/or earth pressure
      2.5   Others
  3   Design allowances for other loadings
      3.1   Construction loading, e.g. stacking of materials, propping
      3.2   Differential shrinkage
      3.3   Differential temperature, e.g. air-conditioning
      3.4   Foundation settlement
      3.5   Differential vertical movements from creep
      3.6   Blast, flood, earthquake, severe wind
      3.7   Swelling soils
  4   Details of design provision for wind-loading
  5   Has design load been exceeded, during construction or subsequently?
  6   Code of practice followed for detailed design
  7   Reinforcement grade and type, if different from that used
  8   Cover specified
  9   Calculations of crack widths
  10 Design provision for shrinkage of concrete and relative movement between concrete and other
      materials, e.g. brickwork
  11 Design life of structure
  12 Special provision for precast and prestressed members
      12.1 Lifting points for precast units
      12.2 Reinforcement at anchorages and angle changes in tendon profile
      12.3 Account taken of attached members diverting prestress

B   Variations from design
  1   Concrete
      1.1   Deliberate changes of concrete strength
      1.2   Accidental changes of concrete strength
      1.3   Use of admixtures not as initially specified
  2   Reinforcement
      2.1   Substitutions
      2.2   Inaccurate placing, displacement
      2.3   Splices not as designed
      2.4   Wrong reinforcement, omissions
  3   Construction joints
  4   Changes from specified procedures
      4.1   Transporting, placing, compacting, finishing
      4.2   Stripping times
      4.3   Falsework removal
      4.4   Reshoring
      4.5   Prestressing sequences

C   Supervision
  1   Amount of supervision
  2   Status of supervisor
  3   Person responsible for supervision, e.g. engineer, architect, quality control organization
  4   Availability of construction records

---

aspects are revealed. For this reason, the laboratory should be asked to report anything that is noticed during the preparation and testing of samples.

Desk investigation   All the available facts about the structure should be examined. Site instructions and diary notes by inspectors are particularly valuable in revealing otherwise unsuspected construction difficulties and modified construction procedures.

Interviews   Discussion with people who have been involved during construction (and during subsequent repair) may often reveal aspects of operations which would have otherwise been lost. The interviewer has to tread a careful path, especially as the person being interviewed is often retired or no longer connected with the parties to the job and is reluctant to be drawn into what he sees as possible court proceedings.

Synthesis

When a substantial amount of the material has been gathered in the investigation stage, it becomes necessary to put it all together to establish what the problem really is, what are the probable causes and what factors are influencing the deterioration. The way to proceed is by proposing a hypothesis and testing the resultant scenario against the observed facts. Many such hypotheses and scenarios may be looked at and many eliminated, until only a few remain. Further testing of the remaining hypotheses may reveal the need for further investigation to resolve unanswered questions. The essential feature of this stage is that facts be not ignored or distorted. Complete self-discipline and honesty are prerequisites. The possibility of more than one mechanism operating to cause deterioration should not be overlooked. In the process of testing hypotheses, the dictum of Sherlock Holmes should be borne in mind: 'How often have I said to you that when you have eliminated the impossible, whatever remains, *however improbable*, must be the truth?' (A. Conan Doyle *The Sign of Four*).

Once the most probable causes have been identified, the process of evaluation must begin. The phases to be considered under 'Evaluation' are three: elimination of external causes, permanent correction, and temporary correction.

Evaluation

Elimination of External Causes

Diversion   Almost the only situation under which aggressive conditions can be diverted from an existing structure is when the attack is related to fluid flow. As discussed in Chapter 4, pp. 137−9, if cavitation erosion is the cause of disruption, repairs are unlikely to be successful unless the flow pattern is modified. Attack on concrete by 'lime-hungry' water may be reduced by arranging for the water to flow over limestone beds before it reaches the attacked concrete. In the case of one-off incidents, such as a fire, it may be worth considering the introduction of more effective fire prevention and detection systems to reduce the likelihood of a repetition of the attack.

Foundation Drainage and Strengthening   The differential settlement of structures is a common cause of cracking and distress. This has sometimes been found to be due to a change in the moisture conditions in the foundation, often brought about by the change in drainage conditions resulting from works on neighbouring sites or from the planting or removal of large trees. If further settlement is to be prevented, action to maintain an adequate foundation is essential.

Restriction of Use   Damage which is caused by excessive load may be prevented from recurring by restricting the loads that are allowed to be put on a structure. In some cases, damage has been diagnosed as being due to occasional overloads such as the stacking of test weights for a crane. A restriction which prevents this sort of loading is not likely to be economically unacceptable.

'Permanent' Correction

The examination of possible forms that permanent repair may take will depend in the first instance on whether external causes have been eliminated, wholly or in part. If such elimination is possible, it may still not be economically viable and repairs to resist the continuing causes must still also be examined. Possible ways for repair are restoration, strengthening and protection. These routes are not mutually exclusive and many good solutions will involve a bit of each.

Restoration   If the structure, as originally designed but not necessarily as built, is deemed to be adequate for future use then a restoration to the as-designed condition may prove to be the best way to proceed. There are of course difficulties. The common occurrence of inadequate cover, particularly behind vertical surfaces, is very difficult to correct effectively.

Strengthening   Many ingenious ways have been devised of strengthening structures to enable them to carry additional load. Additional structural members such as casings for columns, the attachment of steel plates by gluing and the redistribution of actions by prestressing have all been successfully used. The relation of the new material to the old is important in any strengthening. Deliberate arrangements must be made for transferring existing load into new elements, as otherwise the new element can only carry loads which are additional to those already in place. The relative stiffness of the old and new elements is also of great importance. A new element of less stiffness than the adjacent old element will not carry its intended share, even though it may have strength adequate to do so. Under the heading of strengthening should also be included the increase of resistance to aggressive attack. A dock floor disrupted by sulphate ground-water can be replaced by one made from a concrete containing sulphate-resisting cement.

Protection   Coatings and sheathing may be considered as ways to effect repairs so that the structure is thereby protected from the attacking environment.

Temporary Correction

Propping   If the cause of deterioration is diagnosed as over-loading, or if the examination suggests that deterioration has advanced to such a stage that design loads can no longer be safely carried, it may be possible to continue use of a building or bridge structure by introducing props. There are two main difficulties. Adequate footings have to be found for the prop loads. The introduction of props may redistribute loads so that the existing reinforcing is in the wrong place. Some careful design is often needed to overcome these problems.

Patching   Spalls and pop-outs may be replaced by mortar or concrete patches. Unless the cause is removed the process can not be more than a temporary expedient and is often of little use.

Action

The range of options available must be set out with schematic drawings, and estimates of times and costs. The technical advantages and disadvantages of each proposal and the effects on the operation of the facility should be made clear. The choice of option will, in the end, be an economic one, based on the owner's economic assessment but this assessment can only be properly done if the necessary technical advice is made available. Documentation and operational method will follow the decision on an option.

Economic Appraisal

The owner of a structure which has deteriorated has to make a decision, on economic grounds, whether:

(*i*)   to allow the structure to continue to deteriorate and finally to demolish it;

(*ii*)   to undertake short-term repairs, in the knowledge that further work is going to be needed before the end of the useful life of the structure is reached;

(*iii*)   to institute a major refurbishment with the object of providing a structure which will continue to operate for the remainder of its life with only routine maintenance.

We are not in a position to give guidance on reaching this decision, which must be based on the specific position of the structure and its owner, in relation to such matters as cash flow, present value and predicted future costs and values. In assessing the course to be followed under either (*ii*) or (*iii*) above, the structure must be considered not only as a whole but also as an assemblage of components. Any one component may be (*1*) left alone, (*2*) given minor repairs, (*3*) totally refurbished, or (*4*) replaced. The complete economic analysis of all possible courses is difficult and extensive. In the first instance, attention should be concentrated on those components which are likely to involve the greatest expense. The principle of attending to the 'big bucks' first is a good one. One point worth making is that (*i*) above is likely to be more often the right economic solution than is at present accepted. The reason that it is not often adopted is that owners are concerned about the public image that is created by deteriorating structures and it is not easy to persuade the public or employees that a deteriorating structure is not necessarily an unsafe one.

Economics also have a useful part to play in deciding on a proper program of maintenance. Such a program can only be set up rationally when a substantial amount of data have been collected on the effect of various maintenance procedures on similar structures. So far, only concrete pavements have been sufficiently studied and even here judgement has to be used to reach a useful conclusion. An outline of a method of making use of maintenance costs and performance records in concrete pavements has been developed for application in the State of Indiana by Fwa and Sinha.[6.9] At project level, their analysis provides a tool for evaluating the effects of routine maintenance on a pavement. At network level, analysis can examine the regional variation of routine maintenance and how these effects are influenced by pavement characteristics, traffic and climate. Their methods can be used to provide background against which decisions on the economic level of maintenance can be made.

References

6.1   Harris Sir Alan, Campbell-Allen D, Bridge R Q (eds) 1983 *Marine Works for Bulk Loading* The Warren Centre for Advanced Engineering, The University of Sydney 386 pp

6.2   Wynhoven J H, Hunton D A T 1985 Maintenance and repair of building facades. Concrete Institute of Australia *12th Biennial Conference Proceedings*, Melbourne

6.3   Bilger W, Murphy W E, Reidinger J 1986 *Inspection and Maintenance of Reinforced and Prestressed Concrete Structures* FIP Guide to Good Practice Thomas Telford, London 7 pp

6.4   Johinke B L, Tickner N D 1985 *Asset Management — Application to Concrete Structures* Concrete Institute of Australia 12th Biennial Conference Proceedings, Melbourne

6.5   Andrews P 1986 BRAINS — Bridge record and inspection system. *Construction Repairs & Maintenance* 2(6): 17–21

6.6   Kamp R N 1982 Bridge inventory and inspection programs in New York State. *IABSE Reports* **38** IABSE Symposium Washington DC pp 31–9

6.7   ACI Committee 201 Guide for making a condition survey of concrete in service *ACI 201.1R–68* American Concrete Institute (Revised 1984) 1984 14 pp

6.8   Corcoran B J 1985 Strengthening and repair of concrete structures. *Engineered Repair of Concrete Structures* Lecture 5, The University of Sydney & Concrete Institute of Australia 20 August 1985

6.9   Fwa T F, Sinha K C 1986 An analysis of routine maintenance effects on rigid pavements *ACI SP–93* pp 39–56

Further suggested reading, p. 352

7 Materials and Processes for Repair

Special Concretes and Mortars

Under certain circumstances, particularly when repairs to concrete structures are being made, the designer or contractor may choose to use a special concrete or mortar. The choice is generally governed by the desire to modify a mechanical property of the concrete, change the time at which such property achieves a given value, or to improve the durability of the concrete under the anticipated exposure conditions. As examples of concretes tailored to change the mechanical properties, high-aluminate cement concretes, which shorten setting time and increase early strengths, and shrinkage-compensated concretes, which expand much more during early hydration, thus compensating for subsequent shrinkage, may be given. In virtually every case of such modified material usage, there is some penalty other than monetary cost which must be paid. In the case of the high-aluminate cement concretes, drying shrinkage and swelling in water are higher than those of equivalent strength Portland cement concretes, and durability may be adversely affected. When using shrinkage-compensated cement, the expansions of the concrete must be restrained by proper positioning of reinforcement, or the expansions may lead to severe cracking or complete disintegration of such concrete.

Special Cements for Accelerated Strength Gain

In repairs of certain structures, particularly roadways and bridges, it may be desired that early strength gain should be as rapid as possible. The engineer may, as a first approach, consider using admixtures so that ordinary types of Portland cement can be used. The chief chemical admixture now used for this purpose is superplasticizer of one type or another. Formerly high doses of calcium chloride were advocated but this procedure has been rejected on the basis of corrosion problems associated with calcium chloride use. The time of setting of Portland cement concrete and its strength gain may be shortened by the use of calcium aluminate cement. Because of problems associated with the conversion, under hot humid conditions, of the calcium aluminate hydrates from one form to another, and the resultant strength losses, other types of cements have been preferred.

Regulated set cement is a modified Portland cement which contains a substantial amount of calcium fluoro-aluminate.[7.1] The cement meal contains a substantial amount of fluorite as a substitute for limestone. The burning process is not without problems due to the release of small amounts of fluoro compounds. When prepared and ground the initial and final set of this type of cement occurs almost simultaneously and therefore the time between mixing and set is often referred to as the handling time. As a rule this varies between two and 45 minutes. The strength level is adjusted by controlling the amount of calcium fluoro-aluminate in the cement. The time of set is reduced and the compressive strength gain increased in regulated cement mortars and concrete by an increase in the cement content of the mix, reduction of the water:cement ratio, increased temperature of the mix and increase in curing temperature. The chemical reactions of this type of cement are much more energetic than those of Portland cements. For that reason retardation is necessary. Conventional retarders for Portland cement are not effective in controlling the set of regulated set cement. However, citric acid is used in the mix as a retarder. Where practical, the setting action can be effectively controlled by reducing the mix temperature. Such reductions in the temperature of the mix are also advantageous, as the heat of hydration is considerably higher than that of Portland cement concrete. For this reason all exposed surfaces of newly-placed regulated set Portland cement concrete must be protected from moisture loss. Chlorinated rubber, butadiene—styrene sealing as well as polyethylene sheets or wet burlap are recommended for this purpose.

Special cements based on chemical reactions which are completely different from those of normal Portland or similiar cements are now part of the technology. These include fast-setting magnesium phosphate and aluminium-phosphate cements, which when used for concrete patching of pavements allow traffic flow after only 45 minutes.[7.2]

Expansive Cements

One of the most important factors in ensuring satisfactory performance of mortars and concretes used for repair purposes is to reduce as much as possible the differential dimensional changes between the patch and the substrate concrete. This may, under certain circumstances, be achieved by reducing or compensating for the shrinkage properties of the patch material by the use of an expansive cement in the repair mortar or concrete. An expansive cement, when mixed with water, forms a paste that, after setting, tends to increase in volume to a significantly greater degree than Portland cement paste. This expansion may be used to compensate for the volume decrease due to shrinkage or to induce tensile stress in reinforcing (chemical post-tensioning).[7.3]

Three different types of expansive cements have been recognized and are designated in the USA as Type K, Type M and Type S. Type K is an expansive cement containing anhydrous tetracalcium alumino-sulphate which is burnt simultaneously with a Portland cement composition, or burnt separately when it is to be interground with Portland cement clinker, or blended with Portland cement, calcium sulphate and free lime. Expansive cement Type M is a mixture of Portland cement, calcium-aluminate cement and calcium sulphate. Expansive cement Type S is a Portland cement containing a large computed tricalcium-aluminate content and modified by an excess of calcium sulphate above the usual optimum content. In all cases the specific surface or fineness of an expansive cement has a major influence on its expansion characteristics. The increase in specific surface accelerates very early formation of ettringite in the plastic mix, and, as a result, with the increase in the specific surface for a given sulphate content, the amount of expansion decreases with increasing surface area. Commercially available Type K and S cements in the USA are proportioned to produce relatively low expansions in shrinkage-compensating concrete. Within the normal range of cement usage in concrete, an increase in expansion can be obtained by increasing the total cement content of the mix as shown in Fig. 7.1 for shrinkage compensated concrete made with Type K cements. In Fig. 7.2 the expansion and shrinkage versus age for restrained laboratory specimems of Type K and Type I ASTM cement concretes are illustrated. It should be noted that, on drying, the shrinkage-

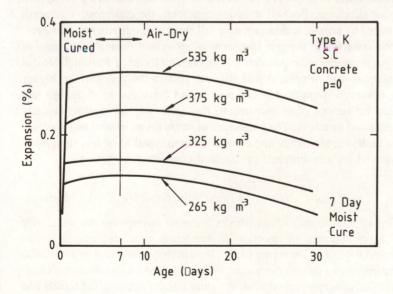

Fig. 7.1 Effect of cement content on the expansion characteristics of shrinkage-compensating concrete made with Type K cement (after ACI Committee 223[7.4])

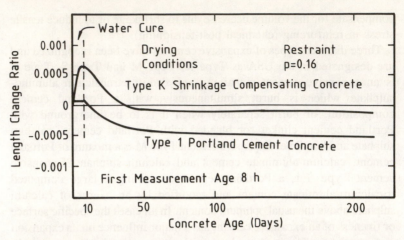

compensating concrete does not attain the original length which it had after expansion during the water-curing cycle, but that it does experience a permanent shrinkage. The Type I cement concrete shrinkage is, however, considerably greater than this. The first measurement of this series of concretes was taken at the age of eight hours.

Such factors as workability, bleeding, time of set, unit weight and yield and strength are similiar to those for normal Portland cement concretes, providing that reasonable restraint to the expansion is attained. Bond strengths have been found to be higher or equal to those of concrete using normal Portland cement and moduli of elasticity seem to have values similar to those for normal concrete. Use of shrinkage-compensating cement leads to improved resistance to cracking caused by restrained drying shrinkage. For self stressing concretes, the compressive strength as tested by normal techniques may fall short of requirements; however, if the concrete in the test specimen is restrained during its expansive stage, strengths are equivalent or superior to those of Portland cement concretes, with a general rule that the greater the restraint the higher the observed strength. Loss of prestressed force due to shrinkage and creep influences these cements in the same way as in mechanically prestressed members. The magnitude of stress losses in steel and concrete due to drying shrinkage and creep are about equal to or less than those observed for conventional prestressed concrete.

Polymer Concrete and Polymer-Modified Concrete

Polymer emulsions such as latexes are used to improve the adherence of fresh concrete or mortars to hardened concrete, and are thus particularly suited for repair work. The emulsions are used as a colloidal suspension of a polymer in water. When combined with concrete a latex-modified concrete is produced. Polymer latexes improve the tensile and flexural strengths of the repair materials, increase their durability as well as increasing early bond properties. It may be seen therefore that in seeking the best materials for repair purposes the concrete technologist may sometimes choose an organic polymer to modify and improve the materials for repair (see *ACI Standard ACI 503.4-79*). Strictly speaking, to talk of polymer concretes only when one is dealing with organic polymers, is a misnomer, since the build up of cementing materials in normal Portland cements depends on polymerization of the silicates. However, by usage, reference to polymers in concrete generally relates to organic molecules incorporated into that material or used with inorganic aggregate materials.

An organic monomer is a molecule which is capable of combining chemically with other molecules to form a high molecular weight material known as a polymer. A polymer which results from the chemical bonding between the monomer units may develop into a long chain-like structure during polymerization or consist of short randomly oriented molecules which are cross-linked. Thermoplastics have long linear chains and exhibit reversible change in their properties on heating and cooling. Thermosets are polymers which do not show such reversible changes with temperature. There is a full range of materials which exhibit characteristics from thermoplastic to thermoset types depending on the degree of cross-linking. Increasing the amount of cross-linking raises the glass transition temperature and eventually a thermoset results. In general polymers are relatively chemically inert and may be chosen to have high tensile and compressive strengths. Their disadvantages are that they have a lower modulus of elasticity and a higher creep than most Portland cement products. Furthermore they may be degraded by oxidizing agents, ultraviolet light, chemicals, certain micro-organisms and in particular heat. In the presence of organic solvents they may crack severely or even become soluble.

Polymers are used in the production of three types of polymer concrete composites: polymer-impregnated concrete (PIC), polymer concrete (PC), and polymer Portland cement concrete (PPCC). Almost full coverage of information on the use of polymers in concrete exists in six special publications by the American Concrete Institute. These report the proceedings of six symposia held since 1972 on polymers in concrete.[7.5–7.10] The development and use of polymer concrete and polymer-impregnated concrete is outlined by Dikeou in a Progress in Concrete Technology Volume of CANMET.[7.11] The report of the ACI Committee 548R-77 also contributes to this topic.[7.12]

Polymer Concrete
Polymer concrete is formed by polymerizing a monomer mixed with aggregate at ambient temperature using curing agents or a chemical catalyst. A large range of organic systems have been used, the earliest

using polyester-styrene and epoxy-resin systems. The epoxy resins are relatively high in cost and at the present time systems based on methylmethacrylate and stryene are most popular. As the presence of moisture at the aggregate monomer interface is disadvantageous, silanes are used to act as a coupling agent and displace water from the aggregate surface. Interfacial bond is increased and the strength of the composite is improved. Mix designs are based on gradings requiring a minimum amount of monomer or resin for compaction to occur. Fine grain fillers such as Portland cement, silica flour or fly ash may be used to improve workability and minimize resin use. The plastic polymer concrete can be placed and compacted by vibration in a manner similiar to that used with conventional concrete. However, solvents are required to ensure immediate cleanup of equipment.

Some mechanical properties of polymer concretes are presented in Table 7.1.[7.11] According to Dikeou conventional reinforced concrete theory does not predict conservative values for the ultimate moment capacities for flexural PC elements and strength of materials approaches together with numerical methods must be used to predict these. Short-term deflections of flexural elements may be predicted using a model for a cracked reinforced flexural element. On the other hand, long-term deflections may be significant due to the viscoelastic nature of the material. When polymer concrete is used with fibre reinforcement, as the content of fibres increases the moment capacity, ductility and resistance to crack propagation as measured by modulus of rupture increase. Ultimate shear in reinforced concrete flexural members has been found to be conservative for conventionally reinforced PC and shear reinforcement required in PC sections is about the same as in normally-

Table 7.1 Mechanical properties of polymer concretes[7.11]

Polymer	Polymer:aggregate ratio	Density (g/cm³)	Compressive	Tensile	Flexural	Modulus of elasticity 10⁴ MPa
Polyester	1:10	2.52–2.34	110–125	12–14	35–40	2.8–3.5
Polyester	1:9	2.33	69	–	17	2.8
Polyester-styrene	1:4	–	82	–	–	–
Epoxy + 40% dibutyl phthalate	1:1*	1.65	50	130	–	0.2
Epoxy + polyaminoamide	1:9	2.28	65	–	23	3.2
Epoxy-polyamide	1:9	1.9–2.1	90–99	–	30–35	–
Epoxy-furan	1:1*	1.7	60–70	7–8	0.1	–
MMA–TMPTMA	1:15	2.4	137	10	22	3.5

* Polymer Mortar

reinforced concrete. Pull-out bond strength tests to measure anchorage of reinforcement have been found to vary from 4 to 7 MPa, for smooth bars embedded in PC.

Polymer concrete has been used for surface patching and full depth patches of bridge decks, as for example the work on the Major Degan Expressway in New York City.[7.11] Manufacture of wall panels and pipes are other uses to which this material has been put. In the case of pipes special classes are manufactured for fresh water service as well as sewage and waste waters.[7.8] The material has been used as a corrosion-resistant pipe liner, and could be used as such in repair procedures. PC repairs have been carried out on Grand Coulee Dam by the Bureau of Reclamation.[7.8]

Polymer Portland Cement Concrete

Although most organic polymers and their monomers are somewhat incompatible with mixtures of Portland cement, water and aggregate, PPCC can be produced by adding to the fresh concrete either a polymer in the form of an aqueous solution or a monomer which is polymerized *in-situ*. Alternatively powdered emulsions can be added to the concrete. Latexes of rubber, acrylics and vinyl acetates have been found to be effective for use in PPCC provided that anti-foaming agents are added to control excessive air entrainment brought about by presence of the organics. Moist-curing required for the hydration of cement is allowed for one to three days and is folllowed by a dry curing period during which any increase in temperature leads to increases in early strengths.

Two-layer composite beam systems have a wide range of important applications in the repair of deteriorating infrastructure systems. The top layer may be made of a concrete which is strong both in tension and compression and has relatively negligible permeability such as polymer Portland cement concrete or polymer modified concrete. Nawy[7.8] has studied the shear transfer behaviour between such two-layer systems and concludes that they can be designed to resist the shear force at the interface without using shear reinforcement through the interface. Special polymer concretes for underwater placement are at present being developed. An acryl-type polymer which enhances the resistance of concrete to be separated in water has been described by Sakuta *et al.*[7.8] Polymer-modified mortars are used for local patch repairs of most types of deteriorating Portland cement concrete surfaces. They are chosen for their improved durabilty and better adhesion characteristics. Resistance to freeze—thaw, to abrasion and to impact is improved by use of these materials. Creep is, however, higher than in conventional concrete but this can be under certain circumstances advantageous in delaying the onset of cracking.

Polymer-Impregnated Concrete

Conventional Portland cement concrete may be dried and then saturated with a liquid monomer such as methylmethacrylate (MMA) or styrene [S]. Polymerization of the monomer may be achieved by gamma radiation or thermo-catalytic means. Liquid vinyl monomer systems such as methylmethacrylate have thermal decomposition temperatures of about 260 °C but the useful temperature range may be raised by suitable cross-linking. Dense concrete requires a lower polymer loading for full impregnation than a more porous concrete. However, impregnation is by no means easy in the case of dense concretes and pressure or vacuum techniques are needed. It may be acceptable only to partially impregnate when improved strength is not needed and only greater durability is desired. The thermal catalytic polymerization process has been used extensively for PIC and is the most practical method of polymerization. Polymerization of monomer-saturated concretes while immersed under water is possible. The strengths of PIC are high in compression, tension and flexure. Strengths can be over four times that of control specimens. The strength-producing effectiveness of monomers depends on the efficiency of impregnation, the conversion of monomer to polymer, the formation of a continuous polymer phase and the mechanical properties of the polymer. Fully impregnated PIC materials show an increase in modulus of elasticity. However, with partial impregnation their moduli of elasticity do not differ significantly from the control materials despite the fact that the stress–strain curves are more linear. Creep of polymer-impregnated concrete is considerably lower than that of the equivalent Portland cement concrete. The thermal coefficient of expansion is about 25 per cent higher than untreated concrete. PIC demonstrates significantly less water absorption and permeability than does conventional concrete. As a result the resistance of PIC to most forms of environmental attack including freeze–thaw, sulphates and acids is significantly improved. On the other hand, when exposed to fire, PIC shows a gradual reduction in strength with increased temperature and above the glass transition temperature the polymer will liquify. However, as long as the thermal degradation temperature of the polymer is not exceeded the PIC will regain its strength upon cooling. Surface burning tests on several types of PIC specimens indicated no fuel contribution to the fire and minimum flame spread rates. Smoke density in these tests was minimal.

From a structural viewpoint PIC exhibits very linear stress–strain behaviour and the strain at failure is slightly greater than that for ordinary concrete while the strength is generally three to five times greater. Elastic theory may be used to predict behaviour of PIC members right to the ultimate strength. Higher percentages of steel and/or steel with higher yield points can be used with PIC and the member can still behave as an under-reinforced beam. PIC should reduce long-term deflections and will improve anchorage of reinforcement. Dikeou discusses the behaviour of beams made from partially-impregnated concrete and suggests that numerous new commercial applications for PIC are emerging. He includes examples of highway bridge impregnation and the impregnation of concrete surfaces in dams, building structures and prestressed cylinder piles. The main drawback to the use of polymer-impregnated concrete is the high cost of both materials and impregnation procedures.

Sulphur-infiltrated Concrete

Malhotra[7.13] in a paper on sulphur concrete and sulphur-infiltrated concrete suggests that sulphur-infiltrated concrete offers job site applications such as in the repair of deteriorated structures and bridge decks. He goes on to state that it is doubtful if cast-in-place concrete can be economically sulphur-infiltrated. Furthermore he suggests that because of the low melting point of sulphur, applications of sulphur infiltrated concretes would have to be limited to structures in which concrete is not exposed to temperatures exceeding 100 °C. He states that where good chemical resistance, high mechanical properties and high impermeability are of special importance this form of concrete can play a useful role.

Factors Affecting the Choice of Special Concretes

Two factors are paramount in the choice of a special concrete. The first is the properties of the concrete itself; the second is the cost of using such a concrete, as all special concretes demand a premium cost above that of a normal concrete. The approach by means of which comparison can be made between the properties and costs of such materials is well outlined in a paper by Pfeifer and Perenchio.[7.14] Their work was undertaken to study the chloride ion intrusion into relatively impermeable special concretes. The following concretes were studied:

(*i*)   normal concrete (0.40 water:cement ratio);
(*ii*)   Iowa low-slump concrete (0.32 water:cement ratio);
(*iii*) low-slump concrete using a high-range water-reducing admixture (0.32 water:cement ratio);
(*iv*)   acrylic latex-modified concrete (0.31 water:cement ratio);
(*v*)   styrene—butadiene latex-modified concrete (0.31 water:cement ratio);
(*vi*)   epoxy-modified concrete (0.15 water:cement ratio);
(*vii*) polymer concrete based on methyl methacrylate resin.

In all cases the range of concrete strength was between 5 000 and

7 600 psi. The data in Fig. 7.3 show the best performance of these different concretes on the basis of water-absorption characteristics. Required curing periods to achieve the best performance are also shown. The MMA polymer concrete showed no weight gain on soaking. All of the modified concretes, and particularly the epoxy-modified concrete, showed reduced water uptakes when compared with a normal concrete; however, the acrylic-modified concrete showed the unique chacteristic of rather high rates of water uptake, even after extended periods of soaking. Leaving aside the acrylic latex-modified material, the authors then considered the extra costs for these concretes assuming that materials only were responsible for that extra cost. These extra costs, considered on the basis of materials purchase only, varied as follows:

$7 per volume unit for the Iowa low-slump concrete;
$11 per volume unit for the high-range water-reducing admixture concrete;
$250–$325 per volume unit for the styrene-butadiene latex-modified concrete;
$375–$600 per volume unit for the epoxy modified concrete;
$700–$1000 per volume unit for the MMA polymer concrete.

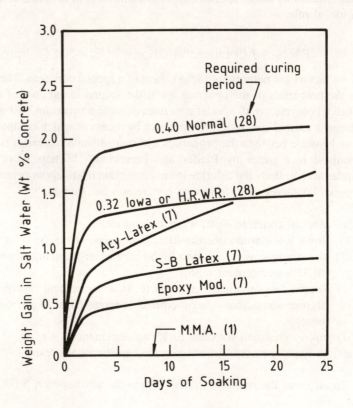

Fig. 7.3 Salt water-absorption of special concretes (after Pfeifer and Perenchio[7.14])

prestressing have all been successfully used. The relation of the new material to the old is important in any strengthening. Deliberate arrangements must be made for transferring existing load into new elements, as otherwise the new element can only carry loads which are additional to those already in place. The relative stiffness of the old and new elements is also of great importance. A new element of less stiffness than the adjacent old element will not carry its intended share, even though it may have strength adequate to do so. Under the heading of strengthening should also be included the increase of resistance to aggressive attack. A dock floor disrupted by sulphate ground-water can be replaced by one made from a concrete containing sulphate-resisting cement.

Protection   Coatings and sheathing may be considered as ways to effect repairs so that the structure is thereby protected from the attacking environment.

Temporary Correction
Propping   If the cause of deterioration is diagnosed as over-loading, or if the examination suggests that deterioration has advanced to such a stage that design loads can no longer be safely carried, it may be possible to continue use of a building or bridge structure by introducing props. There are two main difficulties. Adequate footings have to be found for the prop loads. The introduction of props may redistribute loads so that the existing reinforcing is in the wrong place. Some careful design is often needed to overcome these problems.

Patching   Spalls and pop-outs may be replaced by mortar or concrete patches. Unless the cause is removed the process can not be more than a temporary expedient and is often of little use.

Action

The range of options available must be set out with schematic drawings, and estimates of times and costs. The technical advantages and disadvantages of each proposal and the effects on the operation of the facility should be made clear. The choice of option will, in the end, be an economic one, based on the owner's economic assessment but this assessment can only be properly done if the necessary technical advice is made available. Documentation and operational method will follow the decision on an option.

is related to fluid flow. As discussed in Chapter 4, pp. 137–9, if cavitation erosion is the cause of disruption, repairs are unlikely to be successful unless the flow pattern is modified. Attack on concrete by 'lime-hungry' water may be reduced by arranging for the water to flow over limestone beds before it reaches the attacked concrete. In the case of one-off incidents, such as a fire, it may be worth considering the introduction of more effective fire prevention and detection systems to reduce the likelihood of a repetition of the attack.

Foundation Drainage and Strengthening The differential settlement of structures is a common cause of cracking and distress. This has sometimes been found to be due to a change in the moisture conditions in the foundation, often brought about by the change in drainage conditions resulting from works on neighbouring sites or from the planting or removal of large trees. If further settlement is to be prevented, action to maintain an adequate foundation is essential.

Restriction of Use Damage which is caused by excessive load may be prevented from recurring by restricting the loads that are allowed to be put on a structure. In some cases, damage has been diagnosed as being due to occasional overloads such as the stacking of test weights for a crane. A restriction which prevents this sort of loading is not likely to be economically unacceptable.

'Permanent' Correction
The examination of possible forms that permanent repair may take will depend in the first instance on whether external causes have been eliminated, wholly or in part. If such elimination is possible, it may still not be economically viable and repairs to resist the continuing causes must still also be examined. Possible ways for repair are restoration, strengthening and protection. These routes are not mutually exclusive and many good solutions will involve a bit of each.

Restoration If the structure, as originally designed but not necessarily as built, is deemed to be adequate for future use then a restoration to the as-designed condition may prove to be the best way to proceed. There are of course difficulties. The common occurrence of inadequate cover, particularly behind vertical surfaces, is very difficult to correct effectively.

Strengthening Many ingenious ways have been devised of strengthening structures to enable them to carry additional load. Additional structural members such as casings for columns, the attachment of steel plates by gluing and the redistribution of actions by

Extra costs were then calculated on a per surface area bonded overlay basis having various thicknesses. This demonstrated that the two low-slump concretes provided very economical protection when compared with the polymer-modified mixes. As an extension of the work, coatings were also considered as extra costs. They concluded that considerable reduction in chloride ingress could be achieved by using particular coatings and sealants together with the special concretes.

Replacement of Reinforcement

If reinforcement has been severely damaged, or has lost a substantial proportion of its cross-sectional area through corrosion, it may need to be replaced. In some situations, it may in fact be possible to remove the deficient steel and not replace it if its original function has passed and will not recur. For example, steel provided solely for handling purposes in precast components is not required after the component is in place. Some repairers have treated as dispensable the steel provided for the control of shrinkage and temperature. This approach does not seem advisable as, although shrinkage may be largely completed and early thermal effects are certainly long past, temperature movements and movements from wetting and drying usually continue and require steel for the control of further cracking. If any steel is to be removed, it is important to check that all aspects of the design have been considered before this step is taken.

Pullar-Strecker[7.15] suggests that replacement of steel is necessary if it has lost more than 20 per cent of its area but many specifiers require replacement if more than 10 per cent of the area is lost. As the loss of area is sometimes very hard to assess, without removing a sample of the corroded steel, the 10 per cent figure is to be preferred as it allows a greater tolerance on the precision of assessment. The damaged length of a bar can either be removed and replaced or it can be thoroughly cleaned and have bars added to it to bring it up to the required strength. In either case, it will be necessary to open up sufficient concrete to provide adequate lap lengths with the undamaged portion of the bar. A great deal of extra concrete has to be removed to provide for these laps. The replacement bar should be of the same quality as the original bar and should certainly not be so different that corrosion cells can be set up between the two adjoining bars. For example, stainless steel bars should not be used to replace or reinforce conventional steel.

In the side of a beam or column, where the damage is confined to links or ligatures, it may be possible to attach new bars by hooking them round existing main corner reinforcement (see Fig. 7.4). In other situations, where the steel may be required as extra reinforcement near joints, or where an extra thickness of cover has to be anchored on, it

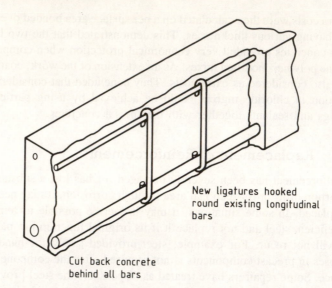

New ligatures hooked
round existing longitudinal
bars

Cut back concrete
behind all bars

may be possible to drill into sound concrete and anchor the bars in epoxy. If this is done, the cause of reinforcement-corrosion must be completely eliminated as a part of the repair procedure, since otherwise the new bar will be at risk where it emerges from the epoxy anchorage. If the two ends of a new link have both to be anchored, as seen in Fig. 7.5, it is usually better to install two separate pieces and lap or weld them together.

If it is convenient to saw into sound concrete, then new bars may be conveniently anchored in concrete placed in dovetailed slots (see Fig. 7.6). A generous length of anchorage, preferably rather greater than that used in conventional design, should be provided in all cases of replacement or addition of reinforcing material.

The replacement of prestressing tendons, even if they are unbonded, is almost impossible to achieve. Additional prestress may be provided by using external cables. In this case careful design is needed for the provision and attachment to the existing structure of anchorages and deflecting saddles. The final protection of external tendons is also difficult.

When new unstressed steel is inserted it will of course carry no load until stress is transferred to it through the surrounding concrete. Some of this transfer can be deliberately induced by jacking and propping before carrying out the repair but even then it will be difficult to decide how effective the transfer is going to be. Generous allowances should be made for this uncertainty when assessing the final capacity of a repaired structure.

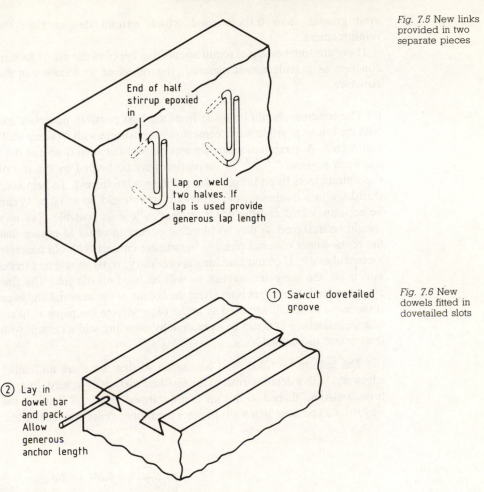

Fig. 7.5 New links provided in two separate pieces

End of half stirrup epoxied in

Lap or weld two halves. If lap is used provide generous lap length

① Sawcut dovetailed groove

Fig. 7.6 New dowels fitted in dovetailed slots

② Lay in dowel bar and pack. Allow generous anchor length

Methods of Placement

Formed Concrete

If a sufficiently large portion of concrete is removed, it can best be replaced with concrete placed in forms. This concrete can be placed without a bonding agent and without grout on the prepared surface of the old concrete. US Bureau of Reclamation[7.16] suggests that this method should be used (*i*) when the depth of the repair exceeds 150 mm, (*ii*) for holes extending right through the concrete section, (*iii*) for holes in unreinforced concrete with area greater than 0.1 m^2 and over 100 mm deep, and (*iv*) for holes in reinforced concrete which have an area greater than 0.05 m² and which extend deeper than the reinforcement.

There are some essential requirements that apply to the use of formed concrete as a replacement material, regardless of its location in the structure.

(*i*) The concrete should be made from the best possible materials and with the lowest possible water:cement ratio consistent with its being well-compacted. A maximum water:cement ratio of between 0.45 and 0.47 has been suggested.[7.15,7.16] Compaction may be helped by the use of superplasticizers (high-range water-reducing admixtures). To help keep shrinkage to a minimum, the aggregate size should be as large as can be accommodated and the water content as low as possible. The mix should be designed so that no bleeding occurs in order to ensure that the replacement material remains in intimate contact with old concrete located above it. If colour matching is necessary, trials should be carried out, using the same form finish as will be used on the job. The fine aggregate and the cement both affect the colour of the material and even if the same materials are used as in the old concrete the patch is likely to appear darker. This can be overcome by blending white cement with the cement to be used.

(*ii*) The hole to be filled must be shaped so that there are no feather edges and with a depth normal to the finished surface of at least 40 mm. It must also be shaped so that air is not trapped. Figure 7.7 shows the way this can be done in a wall. In other locations, vents may be needed.

Fig. 7.7 Preparation for repair in a wall (after *Concrete Manual*[7.16])

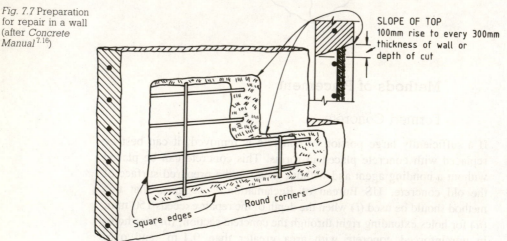

SLOPE OF TOP
100mm rise to every 300mm thickness of wall or depth of cut

Round corners

Square edges

(*iii*) Forms must be robust and firmly fixed so that they withstand any applied pressure and do not allow grout leakage.

(*iv*) Old concrete, against which new concrete is to be placed, must be sound, completely clean, and saturated and the surface must be free of moisture. Saturation and subsequent surface drying may take up to 24 hours and the program should be planned on this basis. In some examples only three hours saturation has been found to be adequate, but this should be checked by tests on site (see Chapter 8, p. 311). Sometimes concrete in an old structure, which appears to be sound when first exposed, will soften after a few days' exposure. The exposed concrete should therefore be left for a few days and re-examined to make sure that it is really sound, before the replacement proceeds.

Each repair situation needs special consideration to ensure that access is provided for the concrete and that compaction is possible. Some examples are given but they cannot cover all likely situations.

(*a*) If the replacement is at the top of a wall, or in any other situation where an open-top form can be used, the task is comparatively simple. If it is possible to arrive at this situation with the removal of some additional concrete, which is not damaged and therefore does not otherwise need to be removed, the likely economic and technical advantages should be studied closely.

(*b*) For walls and columns, where the concrete must be placed from the side of the structure, a satisfactory job can still be achieved but more preparation and careful planning is needed. To allow access for the concrete, the forms must be provided with chimneys which extend above the top of the opening to be filled. Provision should be made for the application of pressure to the concrete in the chimney. A possible arrangement is shown in Fig. 7.8 where the chimney cap can be tightened down with bolts. The forms must be robust enough to stand this additional pressure. For irregularly shaped holes, chimneys have to be provided at each top level. Forms should be stripped as soon as possible after casting so that the projections left by the chimneys can be easily removed. The time of stripping will be determined only by the need to avoid damage to the surface of the green concrete. US Bureau of Reclamation[7.16] recommend that the projections in the chimneys should not be removed until the second day after casting, as if done earlier the concrete tends to break-out in the repair. The projections should always be removed by working up from the bottom as working down tends to break concrete out of the repair. The rough surface left should be filled and stoned to match the surrounding concrete. The surface should not be plastered or rendered. Wet-curing should be provided as soon as possible after stripping the form.

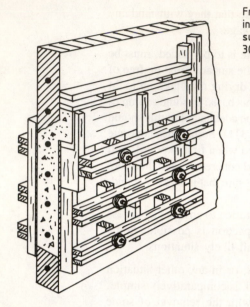

Front form is made
in sections for
successive
300mm lifts

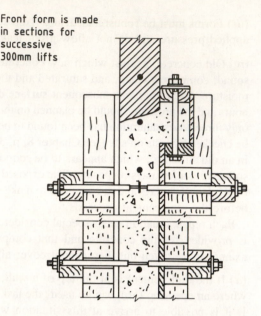

Back form may be
built in one piece

Fig. 7.8 Provision
for applying
pressure to wall
repair (after
*Concrete
Manual*[7.16])

(*c*) Repairs to the sides of beams can often be treated in the way described above. Most commonly, however, the damaged concrete includes the bottom corner of the beam. A bottom form will then have to be provided and must be sealed against the remaining sound concrete in the soffit of the beam. If the whole width of the soffit has to be replaced, the forms should be arranged so that the new soffit concrete is all placed from one side to ensure that no air is trapped. There should be a chimney, fitted with a chimney cap so that pressure can be applied, on each side of the beam.

(*d*) When the whole soffit of a slab has delaminated on account of corrosion of the bottom steel, the problem is acute. Proper cleaning of the corroded steel is difficult as it has to be done overhead. The same difficulty applies to cleaning and saturating the remaining old concrete and the process is further complicated by the presence of a complete mesh of bottom steel. The processes are highly labour-intensive and therefore expensive. The access platforms then generally have to be removed before setting the formwork as that will require more rigid support than can be provided from the platforms. The provision of vents and access chutes is always difficult. The simplest arrangement is often to drill through the slab from the top at a number of places. It is worth bearing this need in mind if test cores are taken from the slab as part of the investigation, as their location can often be adjusted to the subsequent needs. Another possibility is to place by pump from the furthest corner and to withdraw the delivery nozzle as the placement proceeds. The alternative of prepacking and grouting (see p. 267) should certainly be considered. If chimneys at the sides are used for placing, the side forms will have to be stripped early so that the excess concrete in the chimneys can be tidied up. The soffit forms will in all cases be left as long as is necessary for curing, that is between three and seven days.

Mortar and Dry-pack

Dry-pack is suitable for filling holes whose depth is at least equal to the smallest surface dimension of the repair area. For example, it may be used for filling she-bolt holes, holes left by removal of form ties, and narrow slots cut for repair of cracks. The holes should be at least 25 mm deep. Dry-pack is not suitable for shallow depressions where lateral restraint of the filling material is not possible, or for holes that go right through the concrete section where the filling can not be properly rammed.

Dry-pack mortar is usually a mix of one part of Portland cement to 2.5 parts of a fine sand, which passes the 1.18 mm screen. Sufficient water is added to produce a mortar that will stick together while it is being moulded into a ball in the hands. It should not exude water but should leave the hands damp. The proper amount of water will produce a mortar which is at the point of becoming rubbery when it is solidly packed. Any less water will not make a sound pack, as it cannot be properly rammed, and any more water will lead to shrinkage and a loose repair.

The holes should be prepared so that they are sharp and square at the surface edges but corners within the hole should be rounded. The internal surface should be roughened and, if possible, undercut slightly. As with all repairs, any defective concrete must be removed and the surface of the hole left clean. For most dry-pack repairs the surface should be dry. Immediately before starting to place the dry-pack, the surface should be brushed with a 1:1 cement-fine sand bond coat, which has been mixed to a fluid paste. Before this coat dries the filling material should be packed into place. Dry-pack should be packed in layers which have a compacted thickness of about 10 mm. Any thicker layers will not be well-compacted at the bottom. The surface of each layer should be scratched to provide a bond for the next layer. One layer may be placed immediately after another, unless an appreciable rubbery quality develops; if this occurs, work on the repair should be delayed for 30 to 40 minutes. Under no circumstances should alternate layers of wet and dry material be used. Each layer should be solidly compacted over the entire layer by a hardwood stick and hammer. A stick 200 to 300 mm long and not more than 25 mm diameter is appropriate. The stick is preferable to a metal bar, as the metal tends to polish the surface and reduce the bond between the layers. The compacting effort should be directed at a slight angle towards the sides of the hole. The hole should not be overfilled and can be finished by hammering on a piece of hardwood laid on the surface. Steel finishing tools should not be used and water must not be used to help the finish.

As most repairs have only a small volume of filling material and moisture is likely to be drawn into the old concrete, water curing, at least for the first 24 hours, is essential. Wet burlap or a plastic 'soakit' hose is suitable for this purpose. Further advice on dry-pack repairs can be found in US Bureau of Reclamation *Concrete Manual*.[7.16]

Mortar, which is not placed as dry-pack, may be used for repairing surface defects where the defects are too wide for dry-pack to be used, where they are too shallow for concrete filling and where they penetrate no deeper than the far side of the first layer of reinforcement. Mortar replacement is best done by a shotcrete method (see pp. 261–6), but under some conditions hand placement can be used. Hand placement is most successful if the new material is located so that it rests on the old concrete under gravity. Holes in vertical or overhead surfaces are not likely to be repaired successfully by this method and shotcrete or epoxy mortar is probably needed.

The area to be repaired should be cleaned and roughened and kept wet for several hours. The repair mortar should be mixed to a plastic consistency and left for as long as possible (often one to two hours) before it is placed, so that shrinkage is reduced. Before placing the pre-shrunk mortar, a small quantity of cement mortar should be scrubbed into the damp surface with a wire brush. The repair mortar should be compacted thoroughly and particular care should be taken to secure tight filling around the edges of the hole. Water curing should be applied as soon as possible and kept in place for at least seven days. If it is critical that there should be good bond between the old concrete and the repair, wet curing should continue for 28 days.

The use of expansive cements or expansive admixtures has been advocated for replacement mortar repairs. If the shape of the repair is such that compressive stress can be built up in the repair material during the expansive phase then this process may be successful in combating subsequent shrinkage of the repair. Any admixture used should not be of the type which relies on the corrosion of iron filings, as the chemical producing the corrosion is often one containing large quantities of chloride. Such expansive admixtures, used to grout in fixing dowels for precast facade panels, were the main cause of general corrosion throughout the reinforcement in the panels within a period of 10 years after fixing.[7.17]

Spray Placing

The process of applying mortar and concrete by pneumatic pressure produces a material to which many names have been given, including gunite, sprayed concrete, spraycrete and shotcrete. In spite of an inclination in the UK to use the term 'gunite' we shall here follow the American Concrete Institute term 'shotcrete', which is described as 'mortar or concrete conveyed through a hose and pneumatically projected at high velocity onto a surface'. The ACI Guide to Shotcrete is a valuable reference.[7.18] The force of the jet impacting on the surface compacts the material. If a relatively dry mix is used, as is commonly the case, shotcrete can support itself without sagging even when applied overhead. Shotcrete is useful for many repair jobs but its properties and performance are very much dependent on the skill of the crew, the type of equipment used and the conditions under which placing is carried out. In the commonly used dry process, the solid ingredients are dry mixed and delivered along the hose to the nozzle where water is added under the control of the nozzleman. The dry process includes both high- and low-velocity shotcrete. High-velocity shotcrete uses a small diameter nozzle and high air-pressure which produces a particle velocity at the nozzle of about 90 to 120 m s^{-1}, which is sufficient to produce high compaction. Low velocities through a larger nozzle produce less compaction. In the wet process, all the ingredients, including the water are mixed and introduced into the chamber of the delivery equipment. The mix is metered into the delivery hose and pneumatically conveyed to the nozzle, where additional air is added to increase the shooting velocity. With both processes, concrete with an aggregate size up to 20 mm can be used. The decision on which method to use may depend on the cost and availability of equipment and crews and on operational features, such as those listed in Table 7.2.

Plain shotcrete, consisting of Portland cement, sand, coarse aggregate up to 20 mm in size, water and possibly admixtures, is an effective material for replacing defective concrete. The water:cement ratio needs

Table 7.2 Operational features of dry and wet mix shotcrete[7.18]

Dry-mix process	Wet-mix process
Mixing water and consistency are controlled at the nozzle	Mixing water is controlled before delivery and can be accurately measured
Better suited for mixes containing lightweight porous aggregates	Better assurance of thorough mixing, possibly leading to less rebound
Longer hose lengths may be used	Less dust during gunning

to be kept low and test results have shown that water:cement ratios of 0.30 to 0.40 can be achieved. The exact determination is difficult since, with drier mixes in particular, up to 40 per cent of the material may be lost in rebound and this rebound contains very little cement. There is some difficulty in using a very dry mixture as, even if a test panel can be successfully gunned, sand pockets and large voids may occur in actual construction. Compressive strengths for good quality plain shotcrete have been reported from 28 to 54 MPa at 28 days.[7.19] For many repairs, early strength is desirable and at two days a strength of 27 MPa has been reported.[7.20]

Many repair jobs require very early strength which can only be achieved by adding accelerators. Common accelerators, which operate by the continuing production of ettringite in the hydration process, consist of alkali salts of which the best known are chlorides, silicates, carbonates and aluminates. Because of the danger of reinforcement-corrosion, chlorides should not be used. Alkali silicates produce the most pronounced acceleration, giving sets in a few minutes, but only at dosages as high as 20 per cent by mass of cement. The set acceleration is followed by a later loss of strength, increased shrinkage and efflorescence. Alkali aluminates may also produce a later loss of strength. With these accelerators the greatest acceleration is not always produced by the highest dose. There is quite a small dosage range to produce the greatest acceleration and this range depends on the nature of the cement. All the alkaline admixtures are liable to have a detrimental effect on the health of operators, producing for example skin irritation. To avoid this problem, an alkali-free accelerator has been developed and patented, as reported by Burge.[7.21] Compressive strengths of shotcretes containing different accelerators, all used at the same dose rate, are shown in Fig. 7.9. The alkali-free (inorganic neutral) accelerator not only gives high early strength but provides a continuing strength gain.

The durability of plain shotcrete is somewhat variable but some very durable repairs have been reported. The durability is highly dependent on the preparation of the bonding surface and the skill of the nozzleman. The conventional approach of providing resistance to freezing and thawing by entraining air is not certain of success in shotcrete. Air-entraining agents can be added in the wet process but the final air content, and more importantly the final bubble size and spacing, will not necessarily provide the protection sought after. The dry-mix process is not theoretically likely to provide air-entrainment as the mixing action is not present. However, a few users do specify an air-entraining admixture with dry-mix shotcrete in the hope that it will provide additional insurance for durability. There is some sketchy evidence that an air-entraining cement or admixture may appreciably improve the durability of dry-mix shotcrete,[7.22] but Schrader and Kaden put this information

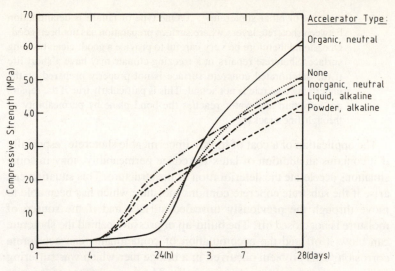

Fig. 7.9 Compressive strength of shotcrete with 5 per cent of accelerator (after Burge[7.21])

into perspective, when they say: 'Occasionally (and by chance) dry-mix shotcrete is produced which contains entrained air bubbles of a small enough size and with close enough spacing to be effective in providing freeze–thaw durability'.[7.23] In situations where the shotcrete can become saturated and it is not dense and impermeable enough to prevent this happening, then it will be susceptible to deterioration from freezing and thawing.

Reading[7.22] reports an example of a panel shot using the wet-mix process and air entrainment. The air content before gunning was 7.5 per cent. The shotcrete in the panel contained 6.4 per cent of air, but only about 2.5 per cent was in the desired size for freeze–thaw resistance. The air-void spacing factor was 0.23 mm and most of the voids were of irregular shape. In rapid laboratory freeze–thaw tests, these panels and others not air-entrained did not perform well. They were all considerably better in exposure tests at Treat Island on the coast of Maine, USA, where they were in the tidal range and exposed to two freeze–thaw cycles daily during the cold weather. Under these conditions, the air-entrained panels started to deteriorate significantly after six years. Other panels not containing air-entraining admixtures lasted up to 14 years, with the best panels having the lowest water:cement ratio of the series at 0.35. Reporting on the nature of durability failures in shotcrete, Reading says:

> In the writer's experience, most shotcrete durability failures do not involve failure of the material itself. Generally there is a peeling off of sound shotcrete because of bond failure. This occurs in spite of the fact that one of shotcrete's greatest attributes is its excellent bond to

concrete or other shotcrete. . . . One type of failure is delamination between shotcrete layers where surface preparation has not been good. One should therefore be very careful to provide a good, clean bonding surface. Shotcrete repairs in a freezing climate may have a short life if the deteriorated concrete surface is not properly prepared or the underlying concrete is not sound. This is particularly true if the repairs are very thin and water reaches the bond plane by permeability or through fine cracks.

The application of a coat of highly impermeable shotcrete, especially if it contains an addition of latex to reduce permeability, may in some situations accelerate the deterioration of the structure. This situation can arise if the substrate concrete contains moisture which has been able to move through the previously unsealed surface and if the source of moisture is not closed off. The build-up of pressure behind the shotcrete can blow it off and the accumulation of contaminants can accelerate corrosion. This problem occurred in a bridge pier which was suffering from corrosion caused by de-icing salts. The vertical faces of the pier were repaired by latex-modified shotcrete but the top of the pier was not sealed and salt-laden water continued to find its way into the pier, where it was trapped and could not get out.[7.23]

Since 1971, two major modifications to shotcrete have been introduced in the addition of fibres and more recently the addition of silica fume. Hooked steel fibres, in quantities up to 1 per cent by volume of the mix, have led to a shotcrete which has a substantial capacity to carry load after cracks have formed and which has excellent energy-absorbing characteristics. In the finished shotcrete, the fibres are largely oriented parallel to the sprayed surface as the end-on fibres fall off in the rebound. The finished material therefore has highly directional properties, which must be taken into account both in design and testing. Typical load—deflection curves for static loading of panels are shown in Fig. 7.10 where the capacity of hooked fibre shotcrete compares dramatically with that of shotcrete containing straight fibres and no fibres. To enable the fibres to be successfully mixed into the dry ingredients, the fibres were introduced in collated bundles glued together to give an artificially low aspect ratio of 20 to 25. When the glue dissolved in the mixer, the bundled fibres separated into individual fibres with aspect ratios of 60 to 75.[7.20] If individual fibres, not collated or bundled, are used the batching and mixing must be done carefully to avoid the formation of fibre balls. Two of the disadvantages of steel fibre, namely balling during mixing and surface rusting after placing, have been overcome by the use of stainless steel melt extract fibres. In a series of tests, the melt extract fibres were shown to give superior handling and mixing characteristics when compared with drawn wire fibres.[7.24] The problem of balling and of maintaining a uniform distribution of fibre in the various parts of the

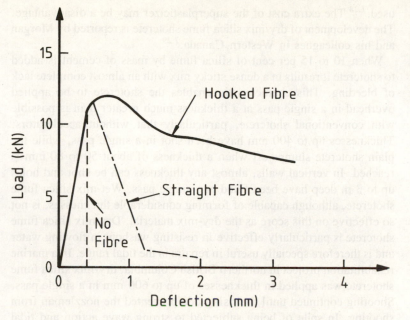

Fig. 7.10 Load–
deflection curves
for static loading
of shotcrete panels
(adapted from
Ramakrishnan
et al.[7.20])

repair has been overcome by a Swedish gun, in which the fibres are cut
from a reel of 0.5 mm wire which is fed continuously into the gun. A
high-speed cutter which can cut fibres up to 200 mm long operates in
the gun, which has a nozzle with a diameter less than the length of even
the shortest fibres. The fibres are therefore oriented in the direction of
the blast, a process which it is claimed reduces rebound. The gun allows
variations in fibre density to be achieved as required in different parts
of the work and permits the use of very long fibres.[7.25] Henager[7.26]
gives useful advice on the equipment and processes for the successful
production of fibre shotcrete. The ACI State-of-the-Art Report on Fiber
Reinforced Shotcrete is also a valuable source of information.[7.27]

The addition of silica fume to shotcrete has produced extraordinary
benefits in the properties of the plastic and the hardened material. Silica
fume (also called 'condensed silica fume' and 'microsilica') is a by-
product in the production of silicon metal and contains a very high
proportion of extremely fine amorphous silicon dioxide particles. The
specific surface of silica fume is about 20 000 m^2 kg^{-1}, which can be
compared with 400 to 700 for fly ash and 300 to 400 for Portland cement.
Because of its fineness and high glass content it is a very efficient
pozzolan. The earlier uses of silica fume in shotcrete were in Scandinavia,
where it was usually applied by the wet-mix process. Wet-mix silica fume
shotcrete can be made to give very high strengths and thereby a high
resistance to freezing and thawing if high doses of superplasticizer are

used.[7.28] The extra cost of the superplasticizer may be a disadvantage. The development of dry-mix silica fume shotcrete is reported by Morgan and his colleagues in Western Canada.[7.29]

When 10 to 15 per cent of silica fume by mass of cement is added to shotcrete it results in a dense sticky mix with an almost complete lack of bleeding. This characteristic enables the shotcrete to be applied overhead in a single pass at a thickness much greater than is possible with conventional shotcrete, particularly that without accelerators. Thicknesses up to 400 mm have been shot in a single pass, while the plain shotcrete slumps off when a thickness of about 50 to 80 mm is reached. In vertical walls, almost any thickness can be shot and holes up to 3 m deep have been filled at a single pass. Wet-mix silica fume shotcrete, although capable of forming considerable thicknesses, is not so effective on this score as the dry-mix material. Dry-mix silica fume shotcrete is particularly effective in resisting washout in flowing water and is therefore specially useful in repairs in the tidal range. In a marine rehabilitation project in northern British Columbia, dry-mix silica fume shotcrete was applied in thicknesses of up to 600 mm in a single pass. Shooting continued until the rising tide prevented the nozzleman from shooting. In spite of being subjected to strong wave action and tidal currents of as much as 10 knots, the freshly applied shotcrete remained well bonded and displayed virtually no washout. The stickiness of the material markedly reduces rebound, which has economic advantages in large applications.

Dry-mix silica fume shotcrete commonly has a 28-day compressive strength of up to 60 MPa and even higher for wet-mix material. Substantial strengths of about 15 MPa can be obtained at 24 hours, but if strength at only a few hours is needed then accelerators must be added. It has been suggested that for routine work the 28-day strengths are excessive and that economies might be achieved by reducing the cementitious content. This action is not recommended as leaner mixes tend to high rebound and less adhesion. The freeze—thaw resistance of silica fume shotcrete appears to be satisfactory, at least in the short term, as it is sufficiently impermeable not to become saturated.

The addition of silica fume to fibre shotcrete produces a material which has the advantages of both. If short, unhooked fibres are used the flexural strength is much improved when compared to the same fibres in cement shotcrete. In tests by Burge,[7.21] the bond in the silica fume shotcrete was found to be so good that flexural failure of test specimens did not occur until all the fibres had fractured. The flexural strength was almost double that found with short fibres in shotcrete without silica fume (see Fig. 7.11). Typical comparative properties of plain, silica fume and silica fume fibre shotcrete are shown in Table 7.3.

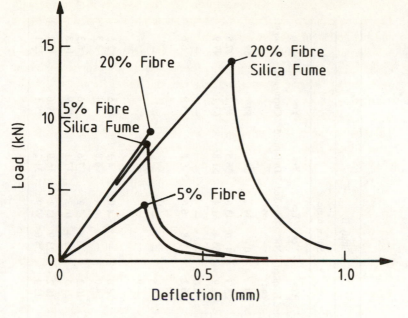

Fig. 7.11 Load–
deflection curves
for static loading
of fibre shotcrete
panels — with and
without silica fume
(adapted from
Burge[7.21])

Prepack

Preplaced aggregate concrete, sometimes referred to as prepacked
concrete or prepack, is made by forcing cement grout into the voids of
a compacted mass of coarse aggregate. Prepack is especially adaptable
to underwater construction and repair. It may also be used in many other
repair situations where it is difficult to place conventional concrete,
especially where air would easily be trapped with conventional concrete
procedures. Prepack differs from conventional concrete in that it contains
a higher proportion of coarse aggregate and this aggregate is in point-
to-point contact as placed. Drying shrinkage is therefore about one-half
that of conventional concrete made from the same materials.[7.30] Other
properties of the hardened prepack depend largely on the properties of
the grout, and the skill with which it is placed, but there is no reason
why strength, durability and thermal properties of prepack should not
be comparable to those of conventional concrete. Appropriate check tests
should be carried out to confirm the actual properties achieved. Bond
with existing concrete is often an important requirement in a repair. It
has been suggested that prepack has a better bond with old concrete than
is found with conventional concrete repair. This has been found to be
the case when the grout contains aluminium powder (an expansion agent),
grout fluidifier and pozzolan[7.31] but it should not be assumed to be the
case with all grouts. If bond is important, the old concrete must be

Table 7.3 Comparative evaluation of different dry mix shotcretes[7.29]

	Study A						Study B				
Matrix type	Plain	Silica fume					Silica fume				
Accelerator	No	No	No	No	No	Yes	Yes	Yes	Yes	Yes	Yes
Fibre type	Nil	Nil	Hooked end	Corrugated	Corrugated	Hooked end	Nil	Hooked end	Deformed	Corrugated	Straight
Fibre content (kg/m³)	0	0	60	60	75	60	0	59	59	59	59
Compressive strength (MPa) 7 days	44.5	42.5	42.4	43.0	43.0	25.3	41.0	43.6	36.5	46.7	27.5
Compressive strength (MPa) 28 days	49.6	51.8	55.9	60.0	58.0	41.9	63.4	61.4	51.8	63.1	47.9
Flexural strength (MPa) 7 days	—	—	5.4	4.9	5.1	4.0	5.8	—	—	—	—
Flexural strength (MPa) 28 days	—	—	7.9	8.0	7.4	6.2	—	6.6	6.8	5.9	6.6
Toughness index, ASTM C1018 7 days I_5	—	—	4.4	2.7	4.1	3.5	—	—	—	—	—
Toughness index, ASTM C1018 7 days I_{10}	—	—	6.2	3.6	6.7	5.6	—	—	—	—	—
Toughness index, ASTM C1018 28 days I_5	—	—	2.2	2.1	2.8	3.0	—	4.0	4.1	2.5	2.5
Toughness index, ASTM C1018 28 days I_{10}	—	—	3.1	2.3	4.0	4.0	—	6.4	6.8	3.4	3.3
Boiled absorption %	6.7	—	7.0	6.6	6.9	10.2	6.8	7.2	7.6	5.6	6.3
Permeable voids %	14.8	—	15.6	14.7	15.4	21.5	15.4	16.3	17.1	12.9	14.3
Rebound (%) Vertical	35.0	21.6	22.1	24.5	26.7	18.2	23.9	14.1	16.9	19.6	16.2
Rebound (%) Overhead	54.6	26.6	28.5	30.5	32.8	33.3	25.6	39.9	31.2	29.1	22.1

carefully prepared by sand-blasting or other appropriate clean-up which removes any unsound concrete and leaves clean aggregate and mortar exposed. Resistance to freezing and thawing can be achieved by air entrainment in the grout.[7.32] The dose of air-entraining agent must be checked, as the methods of mixing and transporting the grout have a different air entraining action from that in conventional concrete containing coarse aggregates.

The minimum size of coarse aggregate is dependent on the type of grout used. When grout contains conventional concreting sand with a maximum size of 5 mm, the minimum coarse aggregate size should be 40 mm. If a finer sand, such as a plastering sand, with maximum size of 1.2 mm is used the minimum size of coarse aggregate may be as low as 14 mm, and this small size is often needed in other than the very largest repair jobs. The maximum size of coarse aggregate is dependent on the availability of material and the usual limitations established for thickness of section and spacing of reinforcing bars. The aggregate should be sound, reasonably free of flat or elongated particles, and should not be liable to excessive wear and breakage while it is being compacted in the forms. The grout should contain sand, of the sizes discussed above, Portland cement, fly ash or other pozzolan, water-reducing set-retarding admixture and an expanding admixture. The admixtures may be combined in a special grouting admixture, but if a combined admixture is used it should be checked that it does not contain chlorides. The grout should be of a thick cream consistency which can be readily injected into the aggregate to completely fill the vacant space, but will not bleed, and so produce water voids under the aggregate particles, before setting takes place.

The proper design and installation of the grout pipe system is an essential part of successful prepack. The system not only provides a means of delivering grout but also serves to monitor grout elevation and provides vents for the escape of water and air. Grout pipes are commonly of 19 to 25 mm diameter and are spaced at about 1.5 m. For repairs which are essentially vertical, such as column and pier repairs, the grout pipe can conveniently be made up of jointed lengths which can be withdrawn as grouting proceeds; the lower end should be kept about 300 mm below the grout surface. Horizontal repairs, as for example the soffit of a slab, are more difficult to arrange. Generally it will be necessary to drill through the slab at a sufficient number of points so that grouting can proceed from one edge of the slab with an advancing sloped front. The slope will range from 1:4 to 1:8 depending on aggregate grading and other factors. Before grout is injected through any row of holes, there should have been an appearance of grout from the next hole in front. In some locations, as for example under metal bases, additional vents may also be needed. Grouting should continue until structural quality grout is forced from all the holes.

The preplaced aggregate, which must be clean and well-graded, is washed and screened to remove fines immediately before it is placed and compacted. The aggregate surface must remain moist while grout is being injected as otherwise water will be absorbed from the grout and incomplete filling may occur. If the aggregate does become dry, it should be flooded through the grout pipes, not just watered from the top, so that fines formed during the placing are not carried to the bottom. The coarse aggregate should, if possible, be placed mechanically into its final position. The free fall should be limited, to reduce breakage. A rubber elephant trunk may help. Where direct placement is not possible, the aggregate may be moved by a high pressure air jet or be transported and placed by a 'rock blower'.[7.33]

Any forms used must be rigid enough to resist the placing of the aggregate and to withstand the hydrostatic pressure of the retarded grout. They must also be well sealed to prevent loss of grout under the grouting pressure.

Details of equipment and methods can be found in References 7.16, 7.30 and 7.33.

Crack Repair

What is an appropriate method of crack repair depends on whether the crack is still actively moving or not. Cracks may appear as the result of defects in joint construction and if this is the cause the crack is almost certainly active. Other active cracks may arise as a result of a failure to provide adequate movement joints at all. The repair process in these cases involves converting the crack into a movement joint which will generally require to be sealed. Advice on the making and sealing of joints is given in *ACI 504R-77*.[7.34] If only small movement is occurring at the crack, it may be possible to restrain this movement by bonding with epoxy, provided the resultant stress does not exceed the strength of the concrete which is being bonded. Inactive cracks can be repaired by epoxy injection or by sealing with mortar. Whatever method of repair is used, the location of the repaired crack is almost certain to remain visible, unless the whole surface is subsequently coated.

Epoxy Injection

The injection of a low-viscosity epoxy is a possible repair method for cracks between about 0.02 mm and 6 mm in width.[7.35] There are wide variations in the properties of different low-viscosity epoxy systems and it is therefore necessary to choose carefully to match the individual job requirements in relation to such matters as temperature of application, capability of bonding to moist concrete, shrinkage, thermal and elastic properties of the hardened resin and other special needs such as fire-resistance and high temperature stability.

For epoxy injection to be effective, the crack must be free of dirt, grease or other contaminants. In relatively new work, satisfactory cleaning can often be achieved by vacuum cleaning ahead of the sealing operation. In older cracks, flushing with water or other solvents may be needed but any solvents used must be compatible with the concrete and the epoxy. Acids have been used but are not recommended. Blowing with compressed air or blasting with water or air/water mix have been suggested but the process tends to drive dust and contaminants into the bottom of the crack. If compressed air is used, then it is important that the equipment is fitted with efficient oil and water traps, to prevent surface contamination from the air supply.[7.36]

Before injection can begin, the crack, where it appears on the surface of the member, must be sealed, so that the liquid resin will not leak and flow out of the crack before it begins to gel and cure. Repair by injection may therefore be difficult or impossible if any of the faces at which the crack appears cannot be reached, even though in some cases the backfill or subgrade material may provide an adequate seal. Sealing is commonly achieved by providing a surface dam which completely bridges the crack. A non-sagging epoxy adhesive which sets rapidly is effective for this purpose. Entry ports must be provided through the seal at a spacing slightly greater than the thickness of the member being repaired. For very fine cracks, less than 0.15 mm in width, the entry ports should be spaced no more than 150 mm apart.[7.37] If the crack goes right through the member, then inspection or vent ports should be provided on the far side.

Three methods of providing entry ports are in general use:

(*i*)   drilled holes with fittings inserted and bonded in with the adhesive used for sealing;

(*ii*)   bonded flush fittings, attached by means of the sealing adhesive;

(*iii*) interrupted seal, using a gasket that covers the unsealed portion (the interruptions to the seal can easily be made by placing 6 mm wide strips of masking tape over the crack before the seal is placed).

The third of these methods is likely to have the least cost, but the particular method used depends on the personal preference of the operator involved.

If rapid application is needed, then equipment which provides for a continuous supply of resin mixed in the application head may be used. In-head mixing may, however, not give fully effective injection, particularly for sealing very fine cracks.[7.35] The preferred method is to use batch mixing, followed by injection from a pressurized pot. There is a risk that premature stiffening may occur in the pot as a result of the early exothermic temperature rise but this difficulty can be overcome by the use of refrigerated pots.

Injection should start at the lowest port and be continued until resin appears at the next higher port. The injection nozzle is then removed, the port sealed and the nozzle moved up to the next port. Complete filling is assured when resin appears at all the injection ports and vents. If pumping pressure can not be maintained in a crack that appears full, it is likely that the resin is leaking through a broken seal or port, or is draining into connected cracks, or is passing through the member into voids.[7.38] The action to be taken if this happens can only be decided on site.

Finally when the injected resin has cured, the sealing adhesive must be removed by grinding or cutting and the holes at ports must be made good with epoxy.

Mortar Repair

Wide cracks which are not active may be repaired by routing out to a width of about 20 mm and filling with drypack (see pp. 259−60) or epoxy mortar. Wide active cracks must be effectively converted into a movement joint. In water-carrying conduits transverse cracks which move with temperature and moisture conditions may be sealed by cutting out the crack to about 12 mm width and caulking it thoroughly with wick and mastic.[7.16]

Surface Impregnation and Coatings

Conditions of Use

One reason for the use of coatings on concrete is that of aesthetics. Painted concrete surfaces are more easily cleaned of graffiti and grime than untreated concrete finishes. Furthermore they may be such as to change entirely the visual appeal and aesthetic characteristics of concrete, particularly if gloss surfaces are achieved. A uniform appearance is often demanded after repair of concrete members, and this may not be achievable without resorting to complete over-coating of both the original and repaired surfaces.

A second reason for surface impregnation or coating is to afford some increased measure of protection to concrete and embedded steel under adverse conditions. Degradation processes which influence concrete and embedded steel almost invariably occur as through-solution mechanisms. Carbonation, sulphate attack, corrosion and alkali−aggregate expansions are all associated with ion transfer into or from the electrolytic pore-solution existent within the concrete. Aggressive ions such as chlorides, which depassivate embedded steel, carbonates or bicarbonates, which form calcite from the essential hydroxides, and sulphates, which react to form highly expansive products from calcium aluminate hydrate minerals, can be introduced *via* the exposed surfaces of the concrete. If the concrete is dry and impermeable, diffusion of such ions is generally slow, whereas if it is wet and permeable, diffusion rates may be significantly increased. Introduction of undesirable ions may reach maximum rates under conditions of cyclic wetting and drying, where pore suction rather than diffusion is the governing mechanism. A truly impermeable barrier placed on the concrete surfaces should therefore overcome most durability problems associated with external attack. Sometimes the presence of a barrier has undesired consequences as, for example, when water is trapped within concrete suffering alkali—aggregate reaction, or when incomplete sealing leads to differential oxygen concentration cells within a reinforced section.

Any concrete:coating interface presents a surface pitted with pores of various shapes and cross sections, often filled with highly alkaline pore-solution. The substrate is subject to continual dimensional change due to thermal and hygral variations, and these changes on the micro-scale are many times the overall measured changes on concrete specimens where aggregate restraints are active. The products of hydration and the residual unhydrated cement are never in chemical equilibrium with one another nor with the pore fluids which may be present. This leads to continual changes in the micro-structure of the interface to which the coating must adhere. The longevity and strong adherence of paint films on bright steel may never be duplicated in the case of concrete acting as a substrate, but certain coatings do have acceptable service lives. Once a concrete structural member is coated, regular inspection and recoating should be instituted as a matter of course in the same manner as for painted steel.

Some polymer coatings are susceptible to alkali attack, which may take several forms. An illustrative example is that of the softening of linseed oil based paints. Calcium hydroxide solutions lead to saponification of the linseed oil base to form an insoluble calcium soap layer, which may for a time protect the film. Alkali salts in the presence of calcium hydroxide form sodium and potassium hydroxides which in turn attack the saponified material, to form soaps which are now water soluble. This process, which is controlled by the concentration of alkalis and the presence of water, may cause the paint film to become sticky, soften and eventually to fail completely. Alkyd paints undergo similar changes. The effectiveness of a linseed oil based coat in protecting concrete depends on concentrations of active materials in pore-solutions as well as amounts of water available for soap removal. It may be satisfactory for use with a low-alkali cement, and yet fail to provide any protection at all when applied to a cement product made with a cement containing more sodium or potassium. If it is applied to a completely dried concrete, which subsequently remains essentially dry, it will continue to perform satisfactorily. Once water penetration occurs however, breakdown of the oil may result. Priming solutions of various compositions have been recommended to neutralize the calcium hydroxide before coatings are applied. Solutions of zinc sulphate and magnesium silico-fluoride have been used for this purpose, but they do not offer an effective alternative to complete drying and surface carbonation of the concrete. An alternate approach has been first to apply unpigmented binder solutions, silanes or other organic substances before final coating in order to create a hydrophobic layer extending into the concrete. This hydrophobic layer presents a barrier to the hydroxide bearing solutions so that reactions with the organic coatings are reduced or eliminated.

The pore structures of most *in-situ* concretes are such as to allow moisture diffusion under conditions of thermal gradients and differential pressures. Water vapour may accumulate and condense at the interface between the concrete and impervious coatings under differential temperature gradients or by semi-permeable membrane actions. Expansion of the water and water vapour can cause bubbling of the coating material unless the concrete pore system allows a very rapid rate of inward transport of the fluid. Once such bubbles occur, debonding continues at an ever-increasing rate and complete failure of the coating may result. A similar outcome may result from the presence of holidays or pinholes in the coat. The physical properties of the concrete or cement product play a dominant role as to whether or not debonding will occur; a rather more permeable concrete may allow diffusion to occur in such a manner as to preserve the coating intact, whereas a better concrete or cement product may lead to debonding. Under certain conditions even the dimensions of sections become important in coating longevity. Sheet products such as glass-fibre reinforced cement panels may require sealing on unexposed surfaces if water diffusion takes place from the rear to the front of the panel.

As much detailed consideration needs to be given to the particular properties of the concrete substrate when selecting coating types and procedures as is given to the organic materials themselves. Furthermore much laboratory testing of organic coats does not closely follow the conditions or variations in substrate physics and chemistry presented by *in-situ* concrete in service. Although field success of a product is the best guide to re-use, the service life may turn out to be quite different on a different substrate.

General Classification

Concrete sealers, whose task it is to reduce the permeability of the surface to water, sea salts, oxgyen, carbon dioxide, other aggressive atmospheric pollutants and sulphates in ground-water, are classified as either coatings or penetrants. Coatings may be defined as having 2 mm or less penetration, and penetrants as intruding the concrete to more than 2 mm in depth. Coatings form a film on the treated surface to which they are applied, and their greatest benefits are derived when, in service, the film retains both complete continuity and at the same time strong adherence to the substrate. Penetrant sealers are designed either to fill the pores thereby blocking them to ingress by other materials, or to line them with water repellent substance, which reduces the tendency of the concrete in the impregnated zone to absorb liquid water. Certain penetrant sealers produce little or no change in the appearance of the surface to which they are applied, and may be compatible with coatings used as top-coats. They may also bind any fine friable material present on the substrate surface.

Coatings are formed from an extremely wide range of organic substances, and new polymers are being added to the list which includes acrylics, polyvinyl acetates, styrene—butadiene, chlorinated rubber, polyurethane, epoxy resin and bituminous materials. Penetrant sealers include various derivatives from halide silanes, all of which polymerize to form silicone resin or polysiloxane. This material is then present in the concrete as a hydrophobic surfactant. Other penetrants include alkali silicates and wax emulsions. A special type of penetrant is represented by methylmethacrylate, which after penetration may be polymerized *in-situ* by thermal or nuclear radiation.

Waterproofing systems may be classified as thick or thin. Thin systems, which are usually fully *in-situ*-applied, may, for example, consist of a primer, a main coat of material such as urethane or neoprene and a top-coat of ultra-violet resistant polymer. Thick systems may be exemplified by rubberized asphaltic materials, or they may include a preformed membrane which adds to their bulk.

Actions of Coatings and their Assessment

The manner in which the coatings and sealants improve durability of concrete is by reduction of permeability to water, salts, oxygen and carbon dioxide. In the USA most work has been directed towards reducing the ingress rates of chloride ions derived from de-icing salts applied to bridge-decks, and hence fairly high-strength concretes are invariably involved as are air-entrained bubble systems. In Germany the problems relate to the decrease in alkalinity through carbonation and to expansions caused by reactive aggregates. Thermal and ultra-violet degradation processes acting on coatings are of considerably more importance in certain environments than those of Europe. Most routine tests for measurement of permeability-reducing properties are conducted on materials not subject to the extremes of heating and cooling nor the ultra-violet radiation dosages, which influence exposed coatings in such environments.

A somewhat artificial and perhaps misleading division of evaluation of concrete coatings has emerged from literature based on laboratory assessment. This division has been brought about in part by unclear concepts of molecular transport through concrete and of the associated driving forces; by the ease and convenience of measurement; and by disregard of the methods of coating employed in practice. Evaluations are generally made on the basis of individual tests of water ingress or repellency, and the efficiency in preventing oxygen, carbon dioxide and chlorides from passing through the barrier. Following on this, specific materials are considered as being carbonation barriers, chloride barriers and so on. Here it should be pointed out that, in practice, due to economic or other circumstances, the external or weather-side of a concrete member is often the only one barrier-coated. The hope of arresting active corrosion by decreasing oxygen diffusion rates from one surface only is forlorn. Only by complete encapsulation of concrete can the strong driving force behind oxygen diffusion, *viz* energetic corrosion reactions, be stifled. Furthermore the danger of actually creating differential oxygen cells becomes a possibility. It appears then, that in practice, the effectiveness which coatings display in controlling corrosion rates is due much more to limiting water ingress than to the cutting off of oxygen supply except under complete encapsulation. The prediction of the performance of some surface sealants, which have water-repellent properties, may be complicated by other actions. For example, such sealants will not reduce the permeability to carbon dioxide and oxygen. Under certain conditions the effect of their presence is to lower the water content of the surface concrete to the point where carbonation rates are at a maximum, leading to a possible decrease of the overall durability. Finally, if the pore structure of the substrate under test does not approximate closely that of the coated concrete, transfer of results may be meaningless. This is particularly so when filler particle size and pigment volume concentrations control the diffusion rates through the coating.

As yet standards for the assessment of coatings on concrete substrates have not been formulated in most countries. Perhaps the most well developed are those from Japan, where JIS A 6909 (Thin Textured Finishes), 6910 (Glossy Textured Finishes) and 6915 (Thick Textured

Finishes), are specifications for the quality of coating materials, and JASS 23, issued by the Architectural Institute of Japan, specifies application methods. The JIS Standards outline a series of tests and limits, some of which relate to the storage stability and change-in-flow properties of the coating materials *per se*, while others attempt to assess characteristics of the coat after application. The coat is tested for cracking under conditions of initial drying, adhesion both in the dry and after wetting, deterioration under cyclic treatment, water ingress, and its resistance to cleaning, alkali exposure and weathering. Minimum elongation, an additional requirement for waterproofing coatings, provides an assessment of crack-bridging capability.

The JASS 23 Standard Specification on the other hand outlines actual procedures to be adopted during spray finishing of a group of substrates, including concrete, cement mortar, autoclaved aerated concrete, plasters and cement sheeting by a whole series of coating materials, which, apart from the organics, include rock wool, fibrous and plain cement mortar materials. It deals with substrate preparation, using polymer modified cement paste, or emulsion putty to achieve surface flatness, and sealer application to attain required drying and adhesion properties. Coating according to the requirements of the standard, must generally be carried out at a temperature above 5 °C on the dry substrate. The materials for under-coating and final coating, or primer, main layer and top-coat are defined. The mix proportions and methods thereof are defined as those proposed by the manufacturers. Coating thickness is quantified as a mass of coating material per unit area of surface to be covered. The number of applications of each coat is defined, as are the intra- and inter-process time intervals and final curing period. These Japanese Standards do provide a basis for the development of methods of test for use in other countries, but they need extensive review prior to application, as many of the given limits will require modification to fit local exposure conditions.

Most laboratory research on coating materials relates to non-standard comparative tests of water penetration and gas diffusion rates, together with some degradation tests under heating, freezing and ultra-violet exposure. Often, however, the tests of this type are not done on comparable concrete substrates, or indeed even on concretes at all. Furthermore, either trade-names are used as identifiers, or, if compositions are given, they are presented in such general terms that no reader could be sure that he was purchasing precisely the same material. Much the same problem arises when papers on service condition assessment of coupon specimens are consulted, or when information is provided on the service lives of coatings on actual structures. The result is that much depends on the accuracy of information supplied by product vendors, who are generally backed by well-established laboratories.

The result of this situation is that any engineer, faced with specifying the use of coatings for concrete, needs to formulate his requirements before approaching the supplier, and to be prepared to press for further information relating to these requirements if they are not covered in the product literature. Advances in this area are so rapid, the field of products so vast and variations within a single manufacturer's product so subtle, that any attempt to give strong guidance on choice without first-hand knowledge of the materials and problems is impossible. Satisfactory use over a period of years is persuasive, but not infallible, evidence that a particular system will prove satisfactory. If all concerned subscribe to such a philosophy, and accept only those materials with documented histories of successful use, any future real advances will only slowly be accepted by practitioners.

Coating Systems and Application Methods

Bituminous Materials

These materials include asphalts and coal- tars, and have been used over a long period as concrete-coating materials. Asphalts are resistant to many mildly corrosive materials, but are readily dissolved by solvents, oils and petrol, and unless specially formulated may be water permeable. Asphaltic coatings are used to protect foundations and basements from water ingress. Coal-tar coatings are better waterproofers, and are suitable for use under conditions of continuous immersion. They crack and craze when exposed to ultra-violet light and high temperatures. These materials are used extensively for protection of underground pipelines. Mixed systems of epoxy-coal-tar formulations are extremely durable. When reinforced with fabric or wrapping tapes their thickness and impermeability are both increased. The major disadvantage in using bituminous systems is their appearance. Top-coating can often be used to deal with this shortcoming.

Oleo-Resinous Paints

Paint formulations based on drying oils can only be used on concrete surfaces which are well matured and dried prior to application, and where rewetting will not take place. Accumulation of moisture behind the paint film leads to blistering and rapid failure. The problem associated with alkali hydroxide and saponification of the oils has already been mentioned. The possibility of increasing resistance by using a tung oil priming coat, followed by linseed oil or alkyd top-coat, is recommended by some manufacturers on the basis of its reduced tendency to saponify.

Latex or Emulsion Paints

These general-purpose paints for concretes use emulsified binders, wherein resin or latex in the form of small discrete particles is dispersed in an aqueous phase. Certain of these paint formulations have good alkali resistance and weathering characteristics. When heavily filled or pigmented they form permeable coatings, and the development of water vapour pressures between the film and substrate is thereby prevented. A wide variety of resins is available to be used in this form. Examples are polyvinyl and acrylic resins. All of these resins form films by evaporation of the aqueous phase, during which process the formerly dispersed resin particles coalesce, to form an integrated film. Film formation rate, which is temperature dependent, is controlled by the addition of coalescing solvents. It is often found that very little surface preparation is required. It is generally considered to be sufficient merely to ensure that the surface is dry, and broomed free of loose matrix materials and dirt. Whether such preparations are sufficient to ensure the anticipated coating life depends on the conditions of exposure, the nature of the substrate and the paint formulation.

Solvent Solution Resin Coatings

Other than those in the emulsion group, synthetic resin coatings which are successfully used for concrete are generally applied as solvent solutions. These include, amongst other systems, chlorinated rubber and epoxy resins. The most commonly used are the epoxy-resin based coatings, being applied as solvent solutions, which may be water-based. Under exterior conditions the epoxy-resin systems tend to chalk and this makes them unsuitable as architectural coatings, unless top-coated. More recently developed acrylic-amine epoxy-resin systems are less affected by ultra-violet radiation, and are used externally, but are less chemically resistant than conventional epoxy-resin systems. Where protection against specific aggressive solutions is necessary, epoxy and chlorinated rubber paints are generally chosen, despite their relatively high cost. Epoxy resins are also the most widely used organic systems for bonding, patching and overlaying concrete. Since epoxy resins cure by cross-linking, shrinkage is lower than for systems such as the polyesters which cure by polymerization. The compressive, tensile and bond strengths of cured epoxy systems are all higher than those of concrete. Under severe exposure conditions, the solvent soluble systems such as chlorinated rubber or soluble urethane resins are often chosen. Chlorinated rubber systems have excellent chemical, water-exclusion and weathering characteristics, and may also be used as curing compounds. Such chlorinated rubber paints are bound by a non-saponifiable resin and are suitable for use on alkaline surfaces provided that no large amounts of water remain in the concrete. They can be formulated to be either glossy and impervious or matt and permeable. The latter should be used if moisture is to be allowed to escape. They form very tough coatings, but may require to be plasticized during formulation. Multi-coating may be necessary to reduce the incidence of pinholes. Certain of the soluble urethane resin formulations are used to achieve high-gloss finishes on cement products. Unless applied to surfaces which have been primed, such high-gloss finishes are very prone to bubbling. Their use may prove to be limited to interior situations, unless recoating is to be performed on a regular basis.

Silanes

Rather than using an external coating to ensure that water is prevented from ingress to a concrete, a more recent approach has been to make the outermost layer of the concrete water-repellent by treatment with silane. It is argued that this process does not in any way change the appearance of the concrete and, by allowing moisture transfer in the vapour but not liquid phase, problems of bond failure are avoided. Although the active material, polysiloxane, is formed within the concrete pores in all cases, its precursor may vary considerably in composition, catalyst and extender. With many variables to consider it is obvious that only by systematically following the supplier's recommendations with respect to application can an active hydrophobic layer of concrete be formed. It should be noted that carbon dioxide and oxygen are still free to enter such concretes without interference. On the other hand siloxanated surfaces have been used as improved substrates for the application of barrier coats and in particular should be considered if gloss finishes are to be achieved for architectural purposes. Most work on silanes has been conducted in the colder European climates and the post-application evaporation rates in warmer areas are not yet established, but will be relatively high unless a barrier coating is added.

Soluble Silicates

The impregnation of concrete using soluble alkali−silicate solutions, which react with the calcium hydroxides to form insoluble hydrated calcium silicates, is well established. Only by complete pore blocking can reduced permeability be attained, and this may require several applications. A net gain in available alkali after reaction is an added advantage.

Cement- and Lime-washes

An old solution, brought up to date when modern admixtures are used, these washes have the advantage of being fully compatible with the substrate materials. Regular recoating is necessary as carbonation causes embrittlement of the coat.

Sprayed and Hand-applied Cement Mortar

Again this is a somewhat dated material, but it is presently undergoing a revival in use due to the availability of chemical admixtures and the incorporation of fibres in the coat. In particular the use of fibres substantially to reduce rebound during spraying, and increase bond within the coat and hence its adhesion to the substrate, is permitting its use overhead and in other difficult places. The use of polymer-modified mortars in hand application has significantly increased work rates by plasterers and improved virtually all the required properties of the coat. The incorporation of pigments into the plaster is again not a modern step, but the range of these has been extended, and the appreciation of changes anticipated during the structure's life has improved the choice of hues.

Practical Aspects

Choice of type of coating should be dictated by the reasons for coating; the durability and serviceability conditions of the particular structure; the service environment to which it will be subject; and the anticipated time between periodic recoatings. Unless coating is planned as part of the original design, the choice of system and the method of application are often governed by immediate economic factors. For this reason choices are often based on relative product cost and ease of system application, and especially the ease of recommended pre-coating surface preparation. Probably the greatest problem facing the practitioner in selecting a coating system is the lack of consistent information on all aspects of coatings. In the case of waterproof membranes, for example, it can be said that they have been used extensively to minimize ingress of solvated chloride ions into concrete. Requirements for the application of such systems are that they should be easily applied, that they should have good bond to the substrate, and should be compatible with all components. Under service conditions, impermeability must be maintained under temperature extremes, concrete cracking, ageing, and imposed loads. Cost comparisons need to be made between preformed sheet types and materials applied as liquids. Sheet types may be difficult to install, usually require adhesives, and are highly vulnerable to quality of workmanship at critical locations. Liquid systems are more easily applied and are generally less expensive. Sheet systems generally show up better in laboratory tests as in such tests the overriding factor of workmanship is eliminated.

With such differences of opinion on virtually every product and system available on the market the practitioner is faced with the necessity of relying either on his personal experience of a product and applicator, or accepting the advice, backed hopefully by an enforceable guarantee, of an experienced operator in the field of concrete coatings. For this reason both operator and client should detail all jobs undertaken in such a manner as to allow easy comparisons to be made in future. Also it would be most valuable if a full quality assurance scheme were to be followed as a matter of course by the applicator. The extra cost to the client of such a procedure should be seen as a real investment, to be set against possible premature recoating. On the other hand impossible claims as to coating efficacy should be avoided. For example, no scheme of coating yet developed should be considered in itself a sufficient measure to overcome major problems besetting some concretes, such as reinforcement corrosion or alkali—aggregate expansion, although it may slow the rate of degradation.

In seeking the most up-to-date information on available materials and recommended procedures, the reader is referred to the abundant trade literature on the subject. As a guide to waterproofing, dampproofing, protective and decorative barrier systems for concrete, the Report of the ACI Committee 515[7.39] is recommended. A section on the use of coatings to enhance concrete durability is presented by ACI Committee 201.[7.40] The use of coatings in the repair of concrete damaged by reinforcement-corrosion is addressed in a Concrete Society Report.[7.41] The *Proceedings of a Workshop on the Durability of Reinforced Concrete Structures* provides information on the state of the art in Australia, but more importantly has a detailed English discussion of the Japanese Standards for Coatings and is therefore referenced.[7.42]

Special Reinforcement

Fibre-Reinforced Concrete

Fibre materials which have been tested or used to increase both the tensile strength and toughness of cement composites include glass, polypropylene, carbon steel and stainless steel, asbestos, cellulose, carbon and kevlar, apart from natural organic fibres such as bamboo and sisal. From the viewpoint of repair materials, only the first four materials are of importance. Two reports by ACI Committee 544 provide useful background to the properties and use of fibre-reinforced concrete.[7.43, 7.44] A publication by Johnston[7.45] gives a historical account of fibre use and also presents selected case histories of fibre use. Probably the greatest use of steel-fibre concrete is in pavement overlays (see Chapter 4, pp. 131—7). The use of fibre reinforcement in hydraulic structures is referred to in Chapter 8, pp. 323—7.

The use of stainless steel fibres has been mooted, but it would seem that their use is only really advantageous when appearance is of importance. The corrosion of black steel fibres progresses rather slowly in the rich mixes used with fibre reinforcement, and in any case the process does not lead to severe disruption of the concrete, as much of the corrosion product can be accommodated within the surrounding concrete without leading to spalling, because of the crack-arrest properties of the remaining uncorroded fibre. The special application of refractory concrete containing stainless steel fibres is mentioned in Chapter 4, p. 137.

Glass fibres are used in the manufacture of sheet products and thin-walled concrete pipe. Johnston[7.45] notes that in a pavement trial conducted in St Paul, Minnesota in 1974 at Snelling Avenue, glass-fibre concrete overlays failed due to heavy cracking after 24 months of service and were replaced. Steel fibre reinforced concrete at the same location was successfully used. A successful use of glass fibre reinforced mortar for the crack repair of precast concrete panels has been described by Pashina.[7.46]

Polypropylene fibre has been considered for use in structures where impact is of importance. It has also been used from time to time in overlays to control early plastic shrinkage cracking. As a rule control of this type of damage to concrete can be overcome by cheaper means.

Fibres are used in conjunction with organics for repair purposes. Scanlon[7.47] gives, as examples, the use of polymer-impregnated fibre-reinforced concretes in repairs to the cavitated and eroded stilling basins of the Dworshak and Libby Dams. He notes that the fibres do not increase the resistance of the concrete to normal erosion, but may help in resisting cavitation forces by the action of increasing the tensile strength of the PIC (see Chapter 8, pp.323–7).

Ferro-cement

The properties and use of ferro-cement are given in a State-of-the-Art Report of the ACI Committee 549.[7.48] However, neither this report, nor a detailed paper on ferro-cement by Shah,[7.49] notes the useful application of this type of material in repair. The probable reason for this is that the material is generally considered simply to form a thin wall reinforced concrete member, and work has been directed to study its properties and actions in that form. Nevertheless the composite material, formed of multiple layers of steel meshes embedded in a Portland cement mortar, is often ideal for surface repair. The properties of ferro-cement which are important from the repair viewpoint include a very high tensile strength:weight ratio and superior cracking behaviour in comparison with reinforced concrete. Furthermore, the fact that plastering may be done by hand or by pneumatic application, allows application of the material on quite different scales, ranging from minor patch repairs to reinstatement of water-tanks and pools. On vertical and overhead surfaces where deep, wide patches are to be placed, the use of thin wire mesh, anchored to reinforcement, allows plastering to proceed stepwise without the use of special processes to re-establish bond between each successive layer. Special meshes are being developed for different repair procedures and exposure conditions.

Romualdi[7.50] discusses the use of ferro-cement in the rehabilitation of water containment structures such as tanks and swimming pools, and in relining tunnels, culverts and chimneys. In discussing these examples he notes that the use of multiple layers of conventional mesh to form the reinforcement is avoided by the use of a special three-dimensional mesh developed in New Zealand. This mesh is not welded, but held in place by friction from undulating keeper wires. The flexibility of the mesh sheet coupled with the friction property also allows the mesh to be pushed into a particular shape, and remain there. Paul[7.51] chose the use of epoxy-coated shotcreting mesh in the repair of an historic concrete lighthouse. This type of mesh is used to delay corrosion of the steel located close to the surface. No details are recorded of bond properties of epoxy-coated bar to shotcrete, but properties such as this may determine the longevity of the repairs.

Recycled Concrete

The difficulty and cost of removing deteriorated concrete from a site may be somewhat mitigated by crushing and reusing the old concrete as aggregate in the new concrete construction. The most usual application is in the recycling of concrete pavements but there have in the past been examples of buildings constructed of recycled materials. The properties of concrete made from recycled-aggregate (the product of crushing existing concrete) depend on the properties of the concrete to be recycled and on whether only the coarse fraction or both the coarse and fine products are used. Hansen and Boegh[7.52] made up a series of concretes in three different grades, using recycled-aggregates from three qualities of existing concrete. They found that the modulus of elasticity of recycled-aggregate concrete was between 15 and 30 per cent lower than the modulus of conventional concrete of corresponding strength, and that the drying shrinkage was between 40 and 60 percent higher. In extreme cases, where recycled-aggregates obtained from poor quality concrete were used to make concrete of higher strength, the modulus might be 50 per cent lower and the drying shrinkage several times higher than that of conventional concrete. If building rubble which contains large amounts of mortar is used as the source of recycled-aggregate, the properties of the new concrete may be very poor and unpredictable. In all cases where recycled-aggregates are proposed, extensive testing is necessary before use.

The operation of recycling concrete pavement requires a substantial organization of specialist equipment and the job is further complicated if reinforcing has to be removed. An example of a successful operation of this sort was the recycling of 15 km of 250 mm thick concrete freeway pavement in a period of eight months.[7.53] Developments in recycling and crushing plant, including the control of noise and the improvement of recycled-aggregate quality, are discussed in papers at the Second International Symposium on Demolition and Reuse of Concrete and Masonry.[7.54, 7.55]

References

7.1   Uchikawa H, Tsukiyama K 1973 The hydration of jet cement at 20 °C. *Cement and Concrete Research* **3**(3)

7.2   Popovics S, Rajendran N, Penko M 1987 Rapid hardening cements for repair of concrete. *ACI Materials Journal* **84**(1): 64−73

7.3   ACI Committee 223 Standard Practice for the Use of Shrinkage-compensating Concrete *ACI 223−83* American Concrete Institute 1983 36 pp

7.4   ACI Committee 223 1970 Expansive cement concretes — present state of knowledge. *ACI Journal* Proceedings **67**(8): 582−610

7.5   ACI 1973 Polymers in Concrete *ACI SP−40* 362 pp

7.6   ACI 1978 Polymers in Concrete *ACI SP−58* 420 pp

7.7   ACI 1981 Applications of Polymer Concrete *ACI SP−69* 222 pp

7.8   ACI 1985 Polymer Concrete — Uses, Materials and Properties *ACI SP−89* 346 pp

7.9   ACI 1987 Polymer Modified Concrete *ACI SP−99* 214 pp

7.10 ACI 1989 Polymers in Concrete: Advances and Applications *ACI SP−116* 110 pp

7.11 Dikeou J T 1980 Development and use of polymer concrete and polymer impregnated concrete. *Progress in Concrete Technology* CANMET pp 539−82

7.12 ACI Committee 548 Polymers in Concrete *ACI 548R−77* American Concrete Institute (reaffirmed 1981) 92 pp (7 page Abstract in *Manual of Concrete Practice*)

7.13 Malhotra V M 1980 Sulphur concrete and sulphur infiltrated concrete: properties, applications and limitations. *Progress in Concrete Technology* CANMET pp 583−683

7.14 Pfeifer D W, Perenchio W F 1982 Coatings, penetrants and speciality concrete overlays for concrete surfaces. *National Association of Corrosion Engineers Seminar, Solving Rebar Corrosion Problems in Concrete* Chicago 1982. Reprinted in *ACI Seminar Course Manual* ACI Infrastructure Seminar — Rehabilitation of Concrete Structures (undated) pp 311−49

7.15 Pullar-Strecker P 1987 *Corrosion Damaged Concrete: Assessment and Repair* CIRIA Butterworths, London 99 pp

7.16 *Concrete Manual* 1975 US Department of the Interior, Bureau of Reclamation 8th edn Denver 627 pp

7.17   Campbell-Allen D, Roper H 1981 Durability of precast facades. *Symposium on Concrete* Adelaide 2 June 1981 Institution of Engineers Australia, National Conference Publication No 81/3 pp 67–72

7.18   ACI Committee 506 Guide to Shotcrete *ACI 506R–85* American Concrete Institute 1985 41 pp

7.19   Morgan D R 1988 Dry-mix silica fume shotcrete in Western Canada. *Concrete International Design & Construction* **10**(1): 24–32

7.20   Ramakrishnan V, Coyle W V, Dahl L F, Schrader E K 1981 A comparative evaluation of fiber shotcretes. *Concrete International Design & Construction* **3**(1): 59–69

7.21   Burge T A 1986 Fiber reinforced high-strength shotcrete with condensed silica fume *ACI SP–91* pp 1153–70

7.22   Reading T J 1981 Durability of shotcrete. *Concrete International Design & Construction* **3**(1): 27–33

7.23   Schrader E K, Kaden R A 1989 Durability of shotcrete *ACI SP–100* pp 1071–1101

7.24   Robins P J, Austin S A 1985 Sprayed steel fibre concrete. *Concrete* (London) **19** Part I (3): 17–20; Part II (4): 18–20

7.25   Cederqvist H 1987 Prefabrication of load bearing structures in steel fibre reinforced shotcrete *ACI SP–105* pp 367–74

7.26   Henager C H 1981 Steel fibrous shotcrete: a summary of the state-of-the-art. *Concrete International Design & Construction* **3**(1): 50–8

7.27   ACI Committee 506 State-of-the-art report on fiber reinforced shotcrete *ACI 506.1R–84* American Concrete Institute 1984 13 pp

7.28   Gilbride P J, Morgan D R, Bremner T W 1988 Deterioration and rehabilitation of berth faces in tidal zones at the port of Saint John *ACI SP–109* pp 199–225

7.29   Morgan D R 1988 Use of supplementary cementing materials in shotcrete. In Ryan W G (ed) *Papers, Concrete 88 Workshop* Concrete Institute of Australia July 1988 pp 403–32

7.30   ACI Committee 304 1969 Preplaced aggregate concrete for structural and mass concrete *ACI 304.1R–69 ACI Journal* Proceedings **66**(10): 785–97

7.31   *Final Report of Tests on Prepacked Concrete — Barker Dam Materials Laboratories* Report No C–338 US Dept of the Interior Bureau of Reclamation, Denver March 1949

7.32   Tynes W O, McDonald J E 1968 *Investigation of Resistance of Preplaced Aggregate Concrete to Freezing and Thawing* Miscellaneous Paper No 68–6 US Army Engineer Waterways Experiment Station, Vicksburg Sept 1968

7.33   Davis R E, Johnson G D, Wendell G E 1955 Kemano penstock tunnel liner backfilled with prepacked concrete *ACI Journal* Proceedings **52**(3): 287–308

7.34   ACI Committee 504 Guide to joint sealants for concrete structures *ACI 504R–77* American Concrete Institute 1977 57 pp

7.35   Warner J 1977 Methods of repairing and retrofitting (strengthening) existing buildings. *Workshop on Earthquake-resistant Reinforced Concrete Building Construction (ERCBC)* University of California, Berkeley July 11–15 1977. Reprinted in *ACI Seminar Course Manual* ACI Infrastructure Seminar — Rehabilitation of concrete structures (undated) pp 61–92

7.36   ACI Committee 546 Guide for repair of concrete bridge superstructures *ACI 546.1R—80* American Concrete Institute 1980 20 pp

7.37   Warner J 1973 Restoration of earthquake damaged concrete and masonry. *Proceedings of the Fifth World Conference on Earthquake Engineering* Rome pp 882—5

7.38   ACI Committee 503 Use of epoxy compounds with concrete *ACI 503R—80* American Concrete Institute 1980 33 pp

7.39   ACI Committee 515 A Guide to the use of waterproofing, dampproofing, protective, and decorative barrier systems for concrete *ACI 515.1R—79* American Concrete Institute 1979 41 pp

7.40   ACI Committee 201 1982 Guide to durable concrete *ACI 201.2R—77* American Concrete Institute (reaffirmed 1982) 37 pp

7.41   Concrete Society 1984 *Repair of Concrete Damaged by Reinforcement Corrosion* Report of a working party Concrete Society Technical Report 26 Oct 31 pp

7.42   *Proceedings of the Second Australia/Japan Workshop on Durability of Reinforced Concrete Structures Session 7 — Coatings* CSIRO Division of Building, Construction and Engineering, Highett Victoria Australia Nov 1988

7.43   ACI Committee 544 State-of-the-art report on fiber reinforced concrete *ACI 544.1R—82* American Concrete Institute 1982 22 pp

7.44   ACI Committee 544 Guide for specifying, mixing, placing and finishing steel fiber reinforced concrete *ACI 544.3R—84* American Concrete Institute 1984 8 pp

7.45   Johnston C D 1980 Fibre reinforced concrete. *Progress in Concrete Technology* CANMET pp 451—504

7.46   Pashina B J 1986 Crack repair of precast concrete panels. *Concrete International Design & Construction* 8(8): 22—6

7.47   Scanlon J M 1981 Applications of concrete polymer materials in hydrotechnical construction *ACI SP—69* pp 45—62

7.48   ACI Committee 549 State-of-the-art report on ferrocement *ACI 549R—82* American Concrete Institute 1982 26 pp

7.49   Shah S P 1980 Ferrocement: a new construction material. *Progress in Concrete Technology* CANMET pp 505—38

7.50   Romualdi J P 1985 Pool relining with ferrocement. *Concrete International Design & Construction* 7(10): 19—22

7.51   Paul M J 1987 Brandywine Shoal Lighthouse. *Concrete International Design & Construction* 9(6): 46—53

7.52   Hansen T C, Boegh E 1986 Elasticity and drying shrinkage of recycled aggregate concrete. *ACI Journal* Proceedings 83(6): 983—7

7.53   Pearson R I 1988 Recycling Detroit's Lodge Freeway. *Concrete International Design & Construction* 10(8): 17—19

7.54   Kasai Y, Fujii T 1989 Demolition and reuse of concrete and masonry. *Concrete International Design & Construction* 11(3): 24—8

7.55   *Demolition and Reuse of Concrete and Masonry. Proceedings of the 2nd International RILEM Symposium* Tokyo Nov 1988 2 Volumes Chapman and Hall, London 1989

Further suggested reading, p. 352

8   Examples of Repairs to Structures

As outlined in Chapter 7, there is available a great variety of materials and techniques which may be used for the repair of damaged concrete structures. The final choice of a method of repair can only be made after a full study of the causes and extent of damage and after an assessment of the degree to which repair is feasible and economical. As each repair job is individual, it is not often that a previous solution can be copied precisely. Nevertheless we believe that there is much to be gained from examining some of the ways in which repair jobs have been planned and carried out so that the strengths and weaknesses of individual operations in particular circumstances can be assessed.

With this object in mind, we have selected from the published literature some examples of interesting repairs and have summarized them under headings denoting the main sources of damage which have led to the particular repair. We do not claim that our examples set out the only way, or the best way, of carrying out a repair in the conditions described. But we do believe that there is much to be learned from a critical study of the examples chosen and we have added our own comments as a part of this process. Robert Stephenson is quoted as saying: 'Nothing was so instructive to the younger Members of the Profession as records of accidents in large works and of the means employed in repairing the damage. Engineers derived their most useful store of experience from the observations of those casualties which had occurred to their own and to other works'.[8.1] In spite of this early advice, there is not a large literature reporting repair jobs, especially repairs that are not wholly successful, and we acknowledge the valuable contribution that has been made by those few who have prepared descriptions of their repair jobs for the technical literature.

Low Member Strength

Flexural Members

There have been many situations in which flexural members, and especially bridge girders, have been found to have less than their desired strength. A widely used method of providing additional strength has been the introduction of prestress by means of external tendons. A reinforced concrete viaduct in Cambridgeshire, UK, was strengthened by this method. Crozier[8.2] provides simple clear details of the external stressing which used Macalloy bars covered with fabric and shotcreted. An old reinforced concrete bridge was strengthened and changed from six simply-supported spans to a continuous structure by external post-tensioning. The design and construction procedures are described by Vernigora *et al.*[8.3] The method has also been applied where prestress is already present. Special attention needs to be given to three points, namely (*i*) the provision of anchors, (*ii*) the installation of deviators and (*iii*) the protection of the stressed tendons. If the lack of strength has been brought about, to any extent, by deterioration of existing tendons and by loss of concrete then it becomes necessary to provide in the repairs for the diversion or removal of the causes that have led to this situation.

The Pancevo bridge structures over the river Danube in Belgrade were built in the early 1960s and were some of the first prestressed concrete bridges in Yugoslavia.[8.4] The form of a typical structure is shown in Fig. 8.1. Even before the bridges were brought into use, some were found to have cracked under test loads and a number of new tendons were added inside the boxes. In subsequent inspections after some years of use under heavy traffic, large numbers of cracks and other defects were found, including damaged concrete, corroded tendons, corroded bearings, damaged and inoperative joints, inadequate drainage, and rough and pot-holed carriageways. Over the life of the bridges, the weight and speed of vehicles has increased and dynamic loading resulting from the rough carriageway and the damaged joints has added to the applied loads.

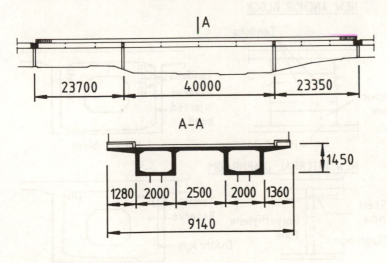

Fig. 8.1 Pancevo Bridge, Belgrade, Yugoslavia. Typical structure (after Pakvor and Djurdjevic[8.4])

A typical bridge

The repairs consisted of three separate phases:

(*i*)   injection of any cracks over 0.1 mm in width and complete replacement of badly deteriorated concrete;

(*ii*)   replacement of gullies and expansion joints, repair or replacement of bearings and other damaged elements, exclusion of water and birds from the inside of the boxes (10 cm of bird excrement had collected in the bottom of the boxes);

(*iii*) addition of new tendons.

In each structure the necessary amount of additional prestress was assessed and it was found in the majority of cases that the complete live-load and 10 per cent of the dead-load should be covered. In some cases, where corrosion of the existing tendons was very severe, even more than this additional stress was needed. The layout of the additional tendons is shown in Fig. 8.2. No provision had been made in the original design for extra tendons to be added and the original anchor blocks could not be reached from the outside. New anchor blocks were therefore built as additional thick diaphragms within the boxes, providing for access for jacks and for transmission of the anchorage forces (see Fig. 8.3).

Fig. 8.2 Pancevo Bridge. Additional tendons (after Pakvor and Djurdjevic[8.4])

Fig. 8.3 Pancevo Bridge. Additional anchor blocks and diaphragms (after Pakvor and Djurdjevic[8.4])

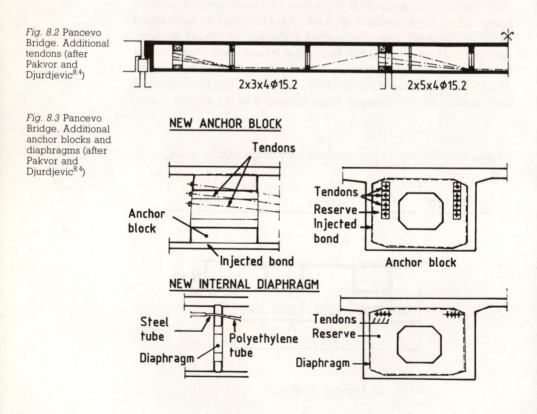

Special attention was paid to the bond between the old concrete and the new anchor blocks. The existing concrete was cut back to the depth of the cover and roughened. After the new block had been cast *in-situ*, the contact surface was injected with low viscosity epoxy resin under pressure, the injection being monitored ultrasonically. Some of the new tendons were deflected at existing diaphragms, reinforced if required. Where no diaphragms existed new diaphragms were cast *in-situ* and bonded to the box. The tendons consisted of four No. 15.2 mm strands in a polyethylene tube. At the deviators the tube was replaced by a steel tube preformed to the necessary radius. After tensioning cement grout was injected into the tubes to provide corrosion protection. The new anchor blocks and deviators included spaces for a further 25 per cent of tendons should they be needed in the future.

In view of the importance of the new anchor blocks to the success of the repair, we might have expected that dowel bars would be provided to connect the block to the existing concrete but no mention is made of this possibility and apparently what was done has been found to be successful. Presumably the basis of this success is the roughness imparted to the old concrete. Epoxy jointing between smooth concrete surfaces would be expected to deform over a period of time and to relax the stressed tendons.

The addition of external tendons was used in the repair of the Pier 39 parking garage in San Francisco, a concrete framed structure which was suffering from leakage and excessive deflection.[8.5] The interesting feature of this job was that it was completed without taking the garage out of service.

A second method of providing additional flexural strength is the bonding of mild steel plates to the concrete member.[8.6] The method was introduced on a major bridge on the Autoroute du Sud, France, in the late 1960s and has since been used in other parts of Europe, South Africa and Japan. Only limited use is reported from North America. Advantages claimed for the method are the ability to strengthen a structure while it is still in use, and low cost. The effect on headroom is minimal when compared with the introduction of supporting beams but on this score there is no advantage over additional prestress.

The behaviour of the resulting composite system depends largely on achieving a successful bond between the concrete and the plate. Proper surface preparation of both concrete and steel is essential. If the concrete member shows any sign of reinforcement corrosion or of spalling, or if there are high concentrations of chloride present, the technique should not be used. The choice of thickness of plate is important as if relatively thick plates are used to strengthen simply supported members, failure can occur by horizontal cracking and plate separation, which starts at the ends of the plates.[8.7–8.9]

Two bridges on the M5 Motorway at Quinton were the first in the UK to be externally strengthened. Cracks caused by poor reinforcement design were present. Loading tests before and after the strengthening demonstrated the effectiveness of the technique in reducing both deflection and cracking under load.[8.6] Other examples include the temporary strengthening of a prestressed hollow box beam skew bridge in Yorkshire, UK, so that it could carry an abnormal load; and the increase of load capacity of solid floor slabs in two telephone exchanges in Zurich, Switzerland, from 2 MPa to 7 MPa.

Swamy points out that although plate bonding has a significant effect on both deflection and crack widths, in relative terms the glued plates appear to contribute more to control of cracking than to control of deflection. He also warns against the use of glued plates for shear strengthening until further investigation has been carried out.[8.7]

Compression Members

An interesting example of an under-strength column in a building in Australia is reported by Grill.[8.10] During the course of construction of a 25-storey reinforced concrete building, one of the basement columns was mistakenly poured with 30 MPa concrete instead of the 50 MPa concrete specified. The error was discovered when only four levels of the structure had been erected and so no temporary propping was needed in spite of the low strength. To upgrade the column to its design capacity, the cross section was increased from 850 mm square to 1100 mm square, as shown in Fig. 8.4(a). Fortunately the space in the basement allowed this to be done. The new column, when constructed entirely of 30 MPa concrete, was calculated to have a strength 6 per cent higher than the column initially designed with 50 MPa concrete. To achieve this performance in the real column, it was necessary to ensure that (i) all the vertical bars were properly restrained against buckling; and (ii) the two parts of the final column, cast at different times, remained acting in a fully composite way.

The additional 24 mm main bars were restrained in one direction by 10 mm ties at 300 mm centres. Restraint in a second direction was provided by 6 mm links which connected the main bars to the ties in the existing column at suitable intervals. The ties were exposed as part of the preparation for the repair. As the column is located at the bottom of a tall building, it is a member with little flexural action and so the reinforcement is primarily to ensure that there are no large pieces of unreinforced concrete and to control creep and shrinkage in the added part. To reduce the possibility of differential elastic shortening and creep in the two parts of the repaired column, concrete of the same strength as that in the original column (i.e. 30 MPa) was specified. The faces

Fig. 8.4 Upgrading
the strength of a
column: (a) adding
to the cross section;
(b) casting concrete
(after Grill[8.10])

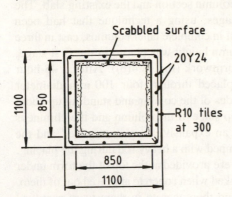

(a) Original and new cross-section

Scabbled surface

20Y24

R10 tiles
at 300

1100
850
850
1100

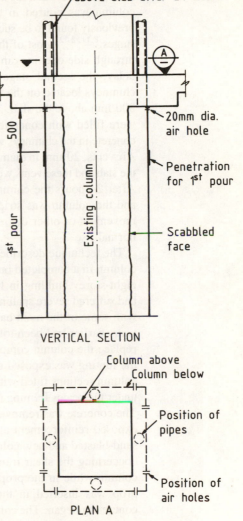

Concrete to be placed to top
of pipe in 150mm layers &
strongly tamped with a 32mm
bar each layer

100mm dia pipe. Remove conc.
above slab after 24 hours

20mm dia.
air hole

Penetration
for 1st pour

Scabbled
face

Existing column

1st pour

500

VERTICAL SECTION

Column above
Column below

Position of
pipes

Position of
air holes

PLAN A

(b) Method of placing concrete

of the original column and the soffit of the slab out to the perimeter of
the new column were roughened by mechanical tools and then cleaned
by air-blasting with oil-free air. The column faces were scabbled to the
depth of the ties. The new reinforcement was put in place with the

longitudinal bars extending between the basement slab and the soffit of the slab above. The oil-free blast was then repeated.

The method of placing concrete was devised to ensure that there was complete filling and that as near perfect bearing as possible was maintained between the added column section and the existing slab. The column was poured in two stages, using a technique that had been previously found to be successful in complete taller columns, cast in three stages.[8.11, 8.12] Most of the column height was placed in the first stage, through side openings in the formwork (Fig. 8.4(b)). After a 48-hour delay, the second stage was placed through four 100 mm diameter chimneys located on the four sides of the column and standing up about 800 mm above the slab. The top stage of the column and the chimneys were filled with concrete with an expansive admixture added and the concrete in the chimneys was tamped with a heavy rod during the placing. Air vents, 20 mm in diameter were provided at the top of the form under the slab and these vents were corked when concrete appeared out of them. After 24 hours the chimneys and the concrete in them were removed and the column was stripped after a further 48 hours. No differential movement or other distress has been noted in the finished building in normal use.

The technique described has also been adapted to replace a damaged column in a completed building. The corner ground floor column in an eight-storey building in Newcastle, NSW, Australia, close to the sea, had suffered severe spalling and advanced corrosion of the reinforcement, with some of the main bars completely lost. Local patching done some years earlier had been totally unsuccessful and a decision was made to replace the column completely for the full storey height. A section of the footing was exposed to support temporary props consisting of steel column sections fitted with base and top plates. The props were erected under the beams framing into the corner and bedded on expanding grout. The concrete was removed completely from the damaged column. The exposed reinforcement at the top and bottom was left in position and sand-blasted and new column bars were welded in. As there was doubt concerning the shear transfer into the edge beams from the first floor column while in the propped condition, an additional permanent steel prop was inserted in the centre of the replacement column before concreting began. The concrete was placed in the same way as described above, but there was only room for one chimney which was located as shown in Fig. 8.5. The temporary propping was removed after 28 days and no settlement of the upper floors has been noted in subsequent inspections, even after a recent earthquake.

More conventional column repairs are described by Janney.[8.13]

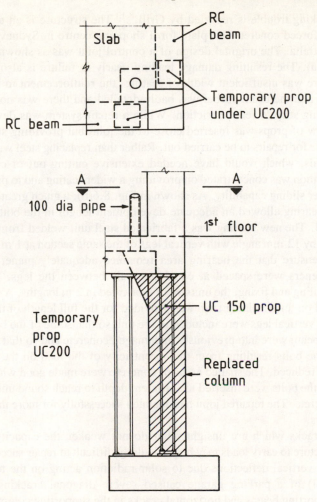

Fig. 8.5 Replacing
a damaged column
(after Grill[8.10])

Deflection and Cracking

Some of the repair methods described on pp. 288−92 have dealt with
deficiencies manifested as excessive deflection and cracking as well as
those relating only to low strength. In cases of cracking, the necessary
investigation may be more extensive as the causes of cracking may not
be as obvious as the cause of low strength or other deterioration. It
becomes necessary to resort to experience and informed engineering
judgement to find out the real cause of the failure and to devise structural
modifications which can correct built-in errors of concept.

An example of a successful repair to what is a common source of cracking trouble is reported by Grill.[8.10] The structure is an extended reinforced concrete flat plate for a shopping centre in Sydney, NSW, Australia. The original design of a control joint was as shown in Fig. 8.6(a). The resulting damage leading nearly to failure is also shown. There was insufficient width of seating, the reinforcement in both the slab and the beam corbel was badly detailed and there was no kind of sliding strip to reduce friction. While a repair system was developed, a row of props was inserted close to the joint but providing sufficient space for repairs to be carried out. Rather than replacing steel with better details, which would have needed extensive cutting out of concrete, attention was concentrated on providing a wider seating and to providing better sliding capability. As shown in Fig. 8.6(b), a much greater width of bearing allowed an adequate development length in the bottom slab steel. The new bearing was a fabricated steel unit welded from 152 by 152 by 12 mm angle with vertical legs of the same section at 1 m centres. To ensure that the bearing area remained adequately plane, 10 mm stiffeners were spaced at equal intervals between the legs. To help handling and fixing, the units were fabricated in 2 m lengths. A stainless steel on teflon sliding strip was provided for the full length of the joint. The vertical legs were included in the unit so that some of the bolts into the beam were into previously undamaged concrete and so that pull-out of the bolts resulting from the eccentricity of the load on the support was reduced. The spalled areas of concrete were made good with epoxy and the bolts were inserted to sufficient depth to reach sound unrepaired concrete. The repaired joint has operated successfully for more than three years.

Cracks which are unsightly but do not weaken the capacity of the structure to carry loads may be particularly difficult to repair successfully. The vertical deflections due to solar radiation acting on the top floor (roof) of a parking garage caused severe diagonal cracking in the supporting beams and horizontal cracks in the supporting columns. The garage which is located in Birmingham, Alabama, USA consists of post-tensioned slabs spanning 18.3 m supported on beams and columns. The investigation of the causes of cracking and the proposals for repair are reported by Fintel and Ghosh.[8.14] The garage was completed in 1976 and by late 1977 cracks had appeared in the interior beams and the exterior columns which supported the roof and the roof ramps. The general location of the cracks is shown in Fig. 8.7. Over the next seven years, the cracks were repeatedly repaired with various patching materials. In most cases the patches either cracked or spalled off and the cracks reappeared.

The measured upward deflections during a six-month period from May to November 1984 showed close correlation with the air and roof temperature fluctuations and calculated cambers due to solar radiation

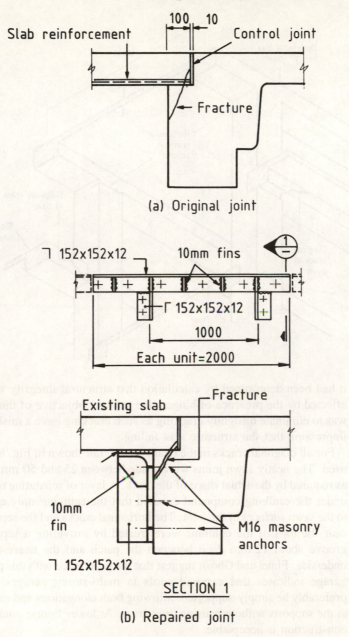

Fig. 8.6 Repair of a
control joint: (a)
damage to the
original joint; (b)
replacement joint
(after Grill[8.10])

were found to be close to those observed. The structure had throughout
its life responded to periodic movements by creating hinges. Any repairs
that did not consider these continuing movements could not be successful.
The final repair accepted the articulations created by the structure. Since

Fig. 8.7 Cracks in
parking garage
caused by sun-
camber (based on
Fintel and
Ghosh[8.14])

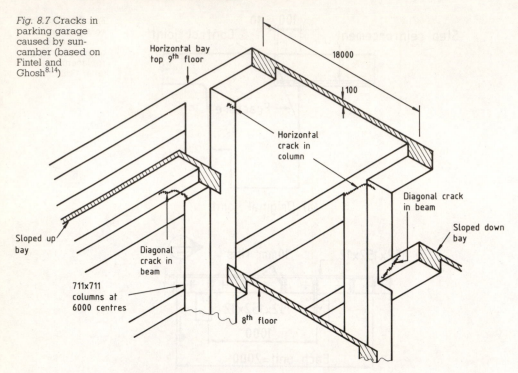

Fig. 8.7 Cracks in parking garage caused by sun-camber (based on Fintel and Ghosh[8.14])

it had been determined by calculation that structural integrity was not affected by the presence of hinges, the primary objective of the repair was to eliminate unsightly cracking as such cracking gave a misleading impression that the structure was failing.

For all diagonal cracks in the beams, the detail shown in Fig. 8.8 was used. The neatly sawn joints were made between 25 and 50 mm wide, as required by the actual shape of the crack. A layer of separating material under the caulking compound ensured that the caulking only adhered to the sawn surfaces of the joint. The horizontal cracks and the separation near the tops of the columns were treated by providing a separating groove about 12 mm deep between the patch and the nearest beam underside. Fintel and Ghosh suggest that the experience with this parking garage indicates that exposed roofs in multi-storey garages should preferably be simply-supported, allowing both elongations and rotations at the supports without causing overstress. At lower floors, continuous construction is acceptable.

These examples illustrate a general point in relation to repair of cracking. To provide adequate strength at a joint is often difficult in a repair as the additional stiffness that accompanies the additional strength frequently attracts further load to the connection. The alternative approach of relieving the stiffness, if this can be done without endangering the

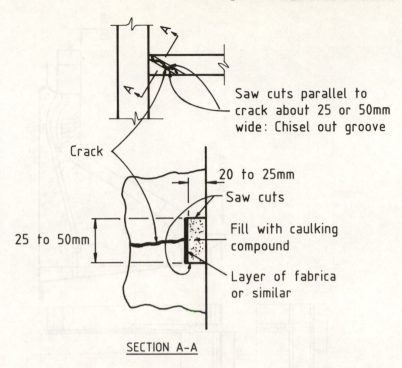

Fig. 8.8 Details of
repair to diagonal
cracks (after Fintel
and Ghosh[8.14])

Saw cuts parallel to
crack about 25 or 50mm
wide: Chisel out groove

Crack

20 to 25mm

Saw cuts

Fill with caulking
compound

25 to 50mm

Layer of fabrica
or similar

SECTION A-A

stability of the whole structural system, is more often likely to be
successful. These repairs have as their essence the modification of the
structural behaviour from the original design concept to the one chosen
by the structure in the process of cracking.

Chemical Disruption

Alkali–Aggregate Reaction

Three examples have been picked to demonstrate the types of problems
that can arise in dams affected by alkali-aggregate reaction and to show
ways in which remedial works can be approached.

Center Hill dam in Tennessee, USA, has a long history of operational
problems appearing as leaking horizontal joints, as binding of spillway
gates and as damage to the mechanisms for operating the gates.[8.15] The
dam consists of a 421 m long concrete gravity section on the right side
of the valley and a 237 m long earthfill embankment on the left side.
The layout of the concrete section is shown diagrammatically in Fig.
8.9. The spillway is controlled by eight tainter gates, each 50 ft (15.2 m)
wide, and is traversed by a highway bridge which consists of simply-
supported spans of steel girders and concrete deck.

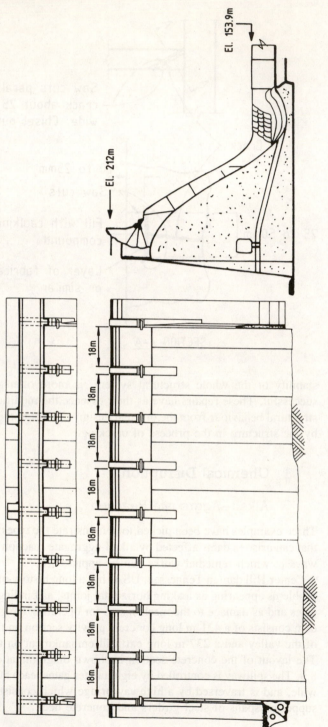

Fig. 8.9 Center Hill
Dam, Tennessee,
USA. Layout of
concrete section
(after
Hugenberg[8.15])

From the first filling in 1948 up till 1967, there were no serious deficiencies noted. In 1967, some horizontal lift joints near the centre of the spillway and near the crest were found to be leaking excessively. The leaks were put down to poor construction procedures and the joints were reinforced with anchors/bars. In 1974 several of the fixed supports of the bridge were found to be tilted and were reset. More leaking joints appeared between 1975 and 1980 and in 1980 a gate jammed in the raised position. It was not until 1983, after excessive joint movement in the bridge spans, buckling of torque shafts for the gates, buckling of electrical conduits, and severe binding of the end gates, that major investigations were undertaken. It was concluded from the observations, tests and instrumentation data that the concrete was experiencing an alkali—carbonate rock reaction. The expansion resulting from this reaction was causing the structure to grow and to move into the spillway opening. In 1984, the total spillway opening was 53.3 mm short of the design distance and the end gate bays (Nos 1 and 8) were both short of design distance by 27.2 mm at the level of the top of the dam. The right side of gate bay No 1 was leaning 47.6 mm into the opening and the left side of bay No 8 was leaning by 23.8 mm, also into the opening. The operational deficiencies in the bridge were overcome by shortening the bridge girders and deck and by resetting the supports and the expansion joints. The problems with the gates were fixed by shortening the two end gates and by building out to vertical the gate sealing strips in the adjacent monoliths.

No other remedial work was undertaken and it was recognised that the structure, even after 40 years, may continue to grow and that further corrections may be needed. No conclusion could be drawn as to the stage reached in the expansion process or as to the potential for continued expansion.

Val de la Mare dam in Jersey, Channel Islands, is a mass concrete dam completed in 1962.[8.16] The layout of the dam is shown in Fig. 8.10. In 1971, upstream movement of 6 to 13 mm was noticed in the crest walkway in some blocks, and darkening and damp patches, accompanied by surface cracking, were observed on the downstream faces of the same blocks. After an extensive investigation, lasting three years, it was confirmed that alkali—silica reaction was occurring and that some blocks were much more severely affected than others. It was concluded that, although the aggregate throughout the dam was somewhat reactive, the worst reaction was occurring in sections built in the course of a three-month period. During this period cement with an unusually high alkali content was used.

It was feared that expansive cracking could lead to an increase in uplift pressures which would endanger the stability of the dam. Piezometer measurements, taken in three selected blocks, are shown in Fig. 8.11 and revealed that in some of the blocks uplift pressures were indeed

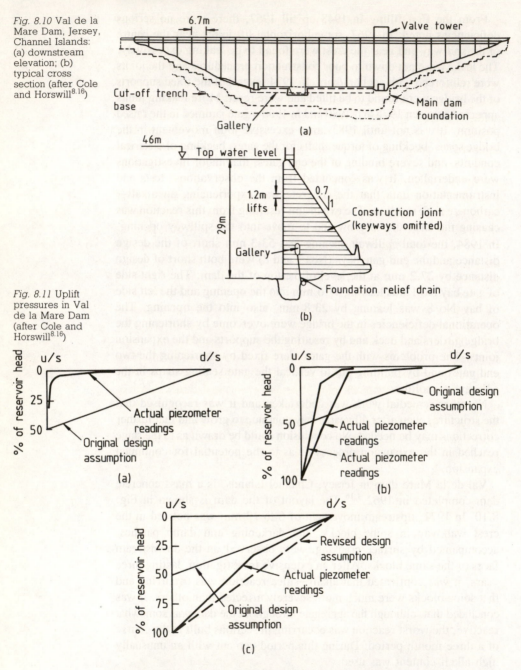

Fig. 8.10 Val de la Mare Dam, Jersey, Channel Islands: (a) downstream elevation; (b) typical cross section (after Cole and Horswill[8.16])

Fig. 8.11 Uplift pressures in Val de la Mare Dam (after Cole and Horswill[8.16])

greater than those assumed in the design. Remedial works were therefore undertaken to ensure the stability of the dam. The works were confined in the first instance to those areas showing problems. The dam is a major water-supply facility and it was necessary to maintain it in operation at all times, even while remedial work was going on. Two approaches were adopted to deal with the excessive uplift. In the worst block (No 6), three 40 mm Macalloy bars were inserted vertically, anchored well below the cut-off and stressed to 86 t each. In all the high blocks (Nos 10 to 16) relief drains, 75 mm in diameter, were drilled at 3.4 m centres so that leakage water could be collected in the drainage gallery. An attempt was made to seal the fissures in the upstream face by grouting with a low-viscosity oil-based chemical grout, but very little penetration was achieved. A monitoring procedure was put in place for action by the owner water company's staff and at intervals by the investigating engineers as shown in Table 8.1.

In 1977 it was found that one of the three Macalloy bars had broken with a brittle-type fracture. It did not appear that this failure was related to the alkali—silica reaction and was probably the result of a seismic shock. The two remaining bars were judged to be adequate for stability and no replacement of the broken bar was attempted. Over the years, expansion has continued and in 1982 a controlled destressing of the two bars was carried out, as the load in them had risen from 847 to 868 kN.

Generally no large flow of water has been found in the relief drains, though in one block a considerable increase in flow was noted during 1980. This appeared to be coming largely from a single lift-joint, which was therefore sealed under water. The sealing was achieved by inserting a 40 mm neoprene rubber gasket into a groove chased along the joint and up the vertical inter-block joints. The gasket was sealed in with epoxy putty. By this relatively simple technique the flow was reduced to only 7 per cent of its previous value. The drains above the gallery, which are not normally filled with water, have become blocked in some places by calcium carbonate encrustations. Regular clearing of these drains is

Table 8.1 Monitoring procedure at Val de la Mare Dam[8.16]

Monitoring activity	Frequency
Piezometer readings	2 weeks
Relief drain flows	Monthly
Anchor bar load cells	Monthly
Crest movement gauges	3 months
Visual inspection and survey of face cracking	3 months (later 6 months)
Sonic velocity survey	3 years (later 4 years)
All readings plotted on graphs and reviewed	Monthly by investigating engineers, unless significant changes occur which are notified immediately
Review reports with recommendations for immediate work	3 years
Inspection by independent panel	As engineer considers necessary

therefore necessary and they are at intervals inspected by closed-circuit TV.

The third example, Kamburu spillway in Kenya, contains features similar to those in the examples above but some aspects of the treatment are considerably different.[8.17] The spillway structure, shown diagrammatically in Fig. 8.12, was completed in 1974. It is set in a rock cut 50 m wide and 30 m deep and is controlled by three radial gates each 13 m wide. Eight years after completion, relative movement was noticed between the left bank stoplog store and the adjacent pier No 1. After three further years of observation it was found that pier No 1 had moved into the gate opening so much that the gate could not be opened. The

Fig. 8.12 Kamburu Spillway, Kenya: (a) transverse section; (b) cross section (after Sims and Evans[8.17])

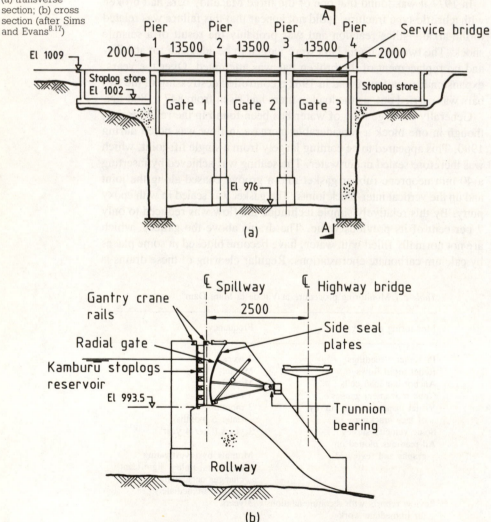

(a)

(b)

maximum deviation from the vertical was 40 mm. Pier No 4 was found to be similarly distorted. At the same time extensive cracking was observed in pier No 1, with cracks up to 20 to 30 mm wide. Water was seeping through the cracks and especially those that were horizontal. Cracking in the stoplog store and displacement of the crane rails confirmed that the left hand stoplog store was tilting.

The major cause of the distortion was determined to be alkali–silica reaction, which occurred at spots throughout the structure where there was a concentration of opal contaminant in the aggregate. The worst expansion occurred in the thickest sections of concrete in the piers, immediately below the level of the stoplog store floor, and in the sections most affected by water penetration. Movement of rock wedges behind the piers was considered a possible additional cause of trouble and the remedial works were therefore designed to include relief of any such forces, by installing rock anchors.

As at Center Hill, the gates and gate seals were modified to accommodate the movements that had occurred. 140 mm was trimmed off gate No 1 at the pier No 1 side. New side plates and seals were added and a new stainless steel guide was set vertically in the pier as shown in Fig. 8.13.

Diversion of water away from the affected concrete was judged to be important and a number of steps were taken to achieve this object. Additional curtain and rib grouting were carried out using cement grout and in some places also resorcinol-formaldehyde chemical grout. The upstream faces of the stoplog stores were coated with bitumen and flexible surface water-bars were inserted in the joints around the stoplog stores. The joints in the highway bridge were sealed with steel plates and water deflectors were put on the top of the piers. After the grouting was complete, relief drains were installed on the downstream side of the grout curtain. The larger cracks in the piers were injected with epoxide resin in the conventional way, but the smaller cracks were left untreated to allow some drainage from the piers.

Instruments were installed and a program of observations established so that careful management could continue to keep the spillway in operation (see Table 8.2).

In all three cases it has been recognized that there is no long-term cure for alkali–aggregate reaction and that continued observations are necessary even after remedial action has been undertaken. It is interesting to note that in only one of the three cases was a serious attempt made to reduce future expansion by keeping out water. The choice will in the end be determined by the physical possibilities and the economic advantages of doing this. Further advice on dam behaviour has been reported in relation to five US Bureau of Reclamation dams affected by alkali–aggregate reaction.[8.18]

Fig. 8.13 Modifications to spillway gates (after Sims and Evans[8.17])

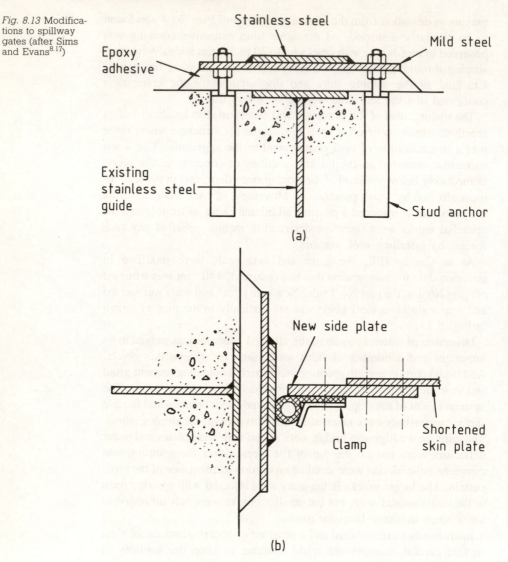

A host of structures other than dams have been seriously affected by alkali−aggregate reaction. Repair histories for a cross section of these are given in papers in the Proceedings of the 7th International Conference on Concrete Alkali−Aggregate Reactions.[8.19] Repaired structures include expressways in Japan, highway structures in South Africa, railway bridge piers in Canada, columns, beams and power-station members. Hoppe at the same conference discussed the rehabilitation of an arch bridge. After outlining methods adopted for the strengthening of the structure, the treatment of existing concrete affected by alkali−

Table 8.2 Instruments installed at Kamburu spillway[8.17]

Description	Type	Reading Frequency	Location							Total
			Pier				Abutment		Highway bridge	
			1	2	3	4	Left	Right		
Tape extensometer*	Mark II	1 month	Tape extensometer measurement points are distributed throughout structure							67
Rod extensometer	1.4 and 5 rod, dial gauge	1 month	7	3	3	7	3	3		26
Crack width gauge	Demec model MD	1 month	Crack widths monitored at critical points thoughout structure							30
Inclinometer	Mark IV	1 month	1			1	2	2		6
Piezometer	Acoustic and standpipe	1 week†					4	4		8
Seepage measurement	V-notch weir	1 week					3	3		6
Precise survey		6 months								

* Extensometers, inclinometers and piezometers were supplied by Soil Instruments Ltd
† Readings taken more frequently at times of rapid change of reservoir level

aggregate reaction was considered. The treatment included imparting hydrophobic properties to the concrete surface using silanes.

Sulphate Attack

Severe sulphate attack was experienced in the top slab of the Captain Cook Dry Dock, Garden Island, Sydney, Australia. A detailed description of the dock is given by Brown.[8.20] The dock has concrete gravity retaining walls, a concrete floor of varying thickness, and caisson gates; it has overall dimensions of 360 m length, 45 m width, and 12.3 m depth below low water at the sill. Drainage valves were incorporated in the floor to relieve ground-water pressure. Low-heat cement was used throughout the dock, when it was constructed, and by its nature (tricalcium aluminate content), such low-heat cement is also sulphate-resistant. Discharge of high-sulphate ground-water from drainage valves therefore did not cause deterioration to the concrete floor of the dock.

In 1964 the Australian Navy adopted an improved system of docking which required that the dock floor have a closer tolerance on surface flatness than it had when constructed. In order to meet this requirement, a concrete surfacing slab, 150 to 190 mm thick, was laid over the original dock floor. Incorporated in the floor was a series of 25 mm diameter PVC pipes to drain the flow from floor relief valves into the dockside drains. The concrete specification did not require a particular level of tricalcium aluminate in the cement, and hence an ordinary cement

(Australian Type A) was used, with adverse results. In the then current standard no limit was set for the permissible maximum tricalcium aluminate content. The complete failure of the slab was in part caused by sulphate attack; the accompanying expansions were responsible for crushing of the PVC drains, and docking became both difficult and costly.

In 1974 the surfacing slab was replaced. The most important features of this replacement are given in a report of the Commonwealth Department of Construction.[8.21]

(*i*)   Australian Type C (low-heat) Portland cement having a tricalcium aluminate content of less than 5 per cent was used throughout the reconstruction of the floor. The ready-mixed concrete used had the following characteristics:

Cement content 320 kg m^{-3}
Maximum water:cement ratio 0.45
Maximum slump 40 mm
Maximum aggregate size 37.5 mm

(*ii*)   Joints in the concrete slab were formed to match the jointing pattern in the original floor. All joints were fitted with 16 mm thicknesses of self-expanding cork.

(*iii*) Open channel drains fitted with removable precast-concrete covers were provided for draining ground-water from the relief valves.

(*iv*)   The floor surface was sand blasted clean, and prepared with a cement—sand grout immediately before the concrete was placed to ensure adequate bond between the original floor and the new slab.

(*v*)   A combination of sprayed membrane and trickle hose curing was used as most of the concrete was poured between October and March, during which summer period temperatures in the dock were considerably above ambient. These high temperatures could have resulted in excess loss of moisture, with resultant shrinkage cracking of the slabs, unless the concrete was adequately cured.

The dock has functioned successfully since that time. This example is of interest since initially the dock had been constructed with satisfactory materials considering the exposure conditions. For some reason, the aggressive nature of the ground-water, as opposed to the sea-water entering and pumped from the dock, was not considered in the design of the first topping slab. This led to the problem which required rectification. A further factor of interest is that the concrete which was used for the replacement slab would, by today's standards, be considered rather poor under the circumstances. Nevertheless, it has fulfilled its service requirements, chiefly because of the care and attention to detail when it was placed.

Aggressive Water

An example of concrete subjected to aggressive water comes from an investigation by the Engineering and Water Supply Department of South Australia.* As far back as the late 1960s corrosion problems with concrete in departmental water storage tanks in the metropolitan region were noted. Columns in these tanks were consistently in need of repairs. Some degree of softening of the internal surfaces of the concrete walls was also noted. The repair was generally confined to application of epoxy and chlorinated rubber to slow down the surface corrosion. Other coatings had not been found to be satisfactory.

The problem of concrete surface deterioration was given more recognition when it was realized that deterioration was occurring at Hope Valley Water Filtration Plant. In 1978 after a period of operation of only 12 months it was noted that some degradation of the concrete surface layer had occurred in the flash-mixer area and to varying degrees throughout the plant. At the Anstey Hill Water Filtration Plant the degree of deterioration in August 1980 was considered to be even more severe and had occurred after an operation period of only seven months. Subsequently it was found that all metropolitan tanks suffered deterioration of walls and columns to some extent. Comparison of the metropolitan tanks with tanks in country regions which were performing satisfactorily indicated that concrete deterioration was principally associated with the storage of River Murray water and some surface catchment waters. Supplies drawn from the River Murray required to be more highly chlorinated than waters from other sources. It was concluded that the level of treatment necessary to render River Murray water potable was a contributary factor to its aggressiveness, which was such as to leach lime to a depth of 4 mm below the surface of the concrete. From examination of the State Water Laboratory Testing Records it was noted that the Langelier Saturation Index was negative wherever significant corrosion was observed. As explained in Chapter 4, (see pp. 126−30), the Langelier Index is a measure of the capacity of the water to dissolve calcium carbonate.

The addition of aluminium sulphate to act as a coagulant to the raw water at the start of the filtration process in a water filtration plant increases the acidity and hence the aggressiveness of the water in the plant. Chlorination of the water further reduces the pH values and the relatively insoluble calcium carbonate is thereby converted to the more soluble calcium bicarbonate which results in a lowering of the Langelier Index. Since at times the pH of the water at the inlet of a filtration plant

* The permission of the Department to refer to this investigation and to quote from their report (author B van Zeeventer) is gratefully acknowledged.

had been as low as 6.0, the first step in protecting the concrete was to attempt to increase the pH and thus the Langelier Index. A maximum level of pH of the water was controlled by the need to prevent interference with the flocculation process. This level could be achieved by the addition of lime as part of the treatment of the water. Protective coatings were applied and are being assessed with respect to their cost and longevity.

It is important to note that unprotected concrete is attacked readily by pure distilled water at any temperature above ambient but can stand attack by brines at temperatures up to 120 °C. Under the latter conditions however limestone aggregates are not suitable.[8.22]

Weathering

De-icing Salts

Many bridges and parking structures in cold climates have been severely damaged by de-icing salt causing corrosion of reinforcement and requiring repair. The process is generally labour-intensive and costly as indicated in the following typical case. Camsley Lane Viaduct, in Cheshire, UK, is a six-span structure over a main road and a railway. It was built in 1963 at a cost of £263 000 and 20 years later it became necessary to spend almost £200 000 on repairs to the piers and cross-heads forming the supporting trestles.[8.23] Extensive delamination and cracking had developed in parts of the cross-heads and piers where water containing chlorides had leaked through from the deck. A survey of the worst affected areas showed that cover to reinforcement was less than that required by 1985 standards and that potentials indicated a high probability of active corrosion. Chlorides up to 5 per cent free chloride ion by weight of cement were found and even at depths up to 150 mm chloride contents of 1 to 2 per cent existed. Carbonation was found to be relatively low (3 to 5 mm).

As a first step in the repair process, the cause of the trouble was diverted by modifying the deck drainage system, re-waterproofing the verges and the central median and installing asphaltic joints over the piers. The trestle repairs were put out to tender with all quantities billed and repair materials and methods specified. The two concrete repair materials used were (i) a flowable concrete with 16 mm aggregate and containing a plasticizer and a shrinkage-compensating additive, to be cast against forms in heights up to 1.5 m, and (ii) a patching mortar to be applied by rendering, for areas less than 0.1 m². Laboratory and field trials were carried out. The octagonal piers were repaired first and to avoid overstress only three faces were tackled at one time. The specified sequence was:

(*i*)   Break-out in areas of delamination to 20 mm behind reinforcement. (This dimension allowed the 16 mm aggregate to penetrate). The area was extended as needed to expose 50 mm length of uncorroded steel, a requirement that significantly increased the amount of break-out.

(*ii*)   Square up edges with a 10 mm cut to avoid feather edges to patches.

(*iii*) Replace any reinforcement which had lost over 10 per cent of effective area.

(*iv*)   Grit-blast concrete and steel to expose coarse aggregate and to remove rust.

(*v*)   Coat all steel with an inhibiting primer if it would have less than 15 mm of cover after reinstatement.

(*vi*)   Erect formwork to provide pour depth no greater than 1.5 m.

(*vii*) Fully saturate repair areas. (Three hours were found to be adequate.)

(*viii*) Mix and place concrete.

As the reinforcement was very congested, small pneumatic hammers were used for breaking out the concrete. The only reinforcement that needed to be replaced was stirrups in the cross-head and this was done by hooking bars around the top and bottom corner bars. Two types of formwork were used on the piers. The simpler form consisted of plywood planks, strap-banded together, which was readily adapted to the variations in the existing sections. A rigid glass fibre form did not have this advantage and allowed grout loss.

Work on the cross-heads was limited by the fact that there were areas of repair which were directly under bearings. Load had to be transferred from the bearing by using jacking beams before break-out was started. The break out was shaped so that air would not be trapped when the repair concrete was poured.

To maintain control of the ingredients, when many small pours were necessary, the repair material was supplied pre-mixed in 50 kg bags and only water had to be added at the site. Each batch was tested for flow and cube strength. As the work continued from summer to winter, the strength was specified at two different temperatures: at 20 °C, 35 MPa at 24 hours and 50 MPa at 72 hours; at 5 °C, 15 MPa at 24 hours and 35 MPa at 72 hours. The requirement of early strength at low temperature was included to allow rapid repairs, but in the event was not required. Formwork was left in place for 38 hours and after stripping the repairs were sprayed with curing compound or wrapped in polythene sheeting until seven days old.

After repair, the trestles were coated with a water resisting compound, either a silane or a quartz sand cement slurry mix. Because of the limitations on the extent of the break-out, it was not certain that all the chloride contamination had been removed, but the repairs ensured that leaking from the deck had been almost completely eliminated; the corrosion process had been slowed down by reinstating the reinforcement in dense, highly alkaline concrete; the permeability of the existing concrete had been reduced by coating. Inspection of the repairs has continued.

The repairs outlined show all the essentials for this sort of work, which was carried out on this job with careful attention to the control of materials and procedures. There must, however, be some doubt about the effectiveness of coating reinforcement which is to finish with less than 15 m of cover. When a coat is applied, the part of the bar adjacent to the end of the coating may be in a more dangerous condition than it was before, since a local corrosion cell can be set up at this point. It would seem on this job that the necessary cover of more than 15 mm could have been provided at all repairs and this would, in our view, have been the preferred procedure. It is doubtful, also, whether the curing process used after the removal of the forms was effective or necessary. If any further curing was required it would have been better to have used water sprays.

In the case of parking decks and bridge-decks where corrosion is confined to the top steel, the repair may consist of an overlay. The process applied to a very severely deteriorated reinforced concrete parking garage is described by Pinjarkar *et al.*[8.24] The 15-year-old structure was 113 m by 101 m in plan and consisted of six staggered levels, two of which were on grade. The four supported levels consisted of 230 mm slabs with capitals, drop panels and columns. The minimum cover over slab reinforcement was specified as 19 mm. The actual cover varied from 3 mm to 50 mm. Severe deterioration had occurred on all supported levels, principally in the form of delaminations and severe corrosion, occurring primarily along the column strips which contained negative reinforcement. The floors had been patched extensively and most of the patches, including the epoxy mortar patches, had again spalled after only a few years of service. Core compressive strengths were found to be 28 to 41 MPa at 15 years but the chloride content was very high, particularly in the top 25 mm of slab, where it averaged 12.5 kg m^{-3}, well above a commonly assumed threshold value of 0.9 kg m^{-3}.

Specifications were prepared for a comprehensive rehabilitation of the entire garage, including epoxy injection of through-slab cracks, new expansion joints, appropriate shotcrete repairs, removal of all unsound concrete and steel, and installation of bonded fill and overlay concrete. Superplasticized concrete was selected for the overlay, and to ensure adequate impermeability a maximum water:cement ratio of 0.32 and a 28-day strength of 41 MPa was specified.

Delaminated areas were located by using sounding techniques such as dragging chains, and all unsound concrete was removed. The newly exposed sound concrete surfaces were cleaned with air-water blowing. The exposed steel was cleaned by sand-blasting. Additional bars were put in where the original were damaged or had lost significant area. Rails for screeding were set to ensure a minimum overlay thickness of 44 mm and to correct the drainage which had been badly affected by the original grading and by subsequent deflections of up to 100 mm. All concrete surfaces were wetted the night before casting. A neat cement grout was applied. Premixed concrete containing air-entrainment was delivered in 6 m^3 trucks with less than full capacity loads and the superplasticizer was added at the site. On average the specified 28-day strength was achieved after only three days, and at 28 days the strength averaged 52 MPa. Compaction and finishing was done with a vibrating roller screed, with some final hand finishing, using a bull float as required and dragging wet burlap. The concrete was cured under wet burlap for 48 hours. After three weeks of air drying, the surface was sprayed with two coats of a boiled linseed oil formulation penetrant.

In some situations, the additional dead-load imposed by a concrete overlay, or the loss of headroom arising from an overlay is unacceptable. Under these conditions, a thin polymer concrete overlay should be considered as described by Meinheit and Monson.[8.25] An eight-storey parking structure which contained reinforcing and unbonded prestressing tendons was repaired with an overlay of methyl methacrylate polymer concrete with a minimum thickness of 6 mm applied after a 6 mm skim coat of the same material had been applied to exposed reinforcement and prestressing tendons. In deeper patches the material had to be placed in lifts to reduce and control shrinkage and a limit of about 19 mm was used.

In both these examples, although some steps could be taken to divert future chloride attack by improving drainage, the essential process involved removing all corrosion products and damaged concrete and replacing the deck surface by as impermeable a material as possible. In the first example, the ineffectiveness of isolated patching is particularly brought out.

Carbonation

A reinforced concrete factory, built in Scotland in 1912, had continued in use since that date but had suffered such severe damage from steel corrosion and overloading that major repair had become necessary.[8.26] The building consists of five three-storey wings projecting from a main link building of three and four storeys which is about 170 m in length. The construction consists of exposed reinforced concrete frames based on a square grid of columns at 5.49 m (18 ft) centres with a floor-to-floor height of 4.27 m (14 ft). Internal primary beams and secondary beams at 1.83 m (6 ft) centres carry the 110 mm thick suspended floor slabs and roof. The exterior facade appears as concrete columns and perimeter beams supporting a double brick spandrel wall and windows at each floor.

Internally, suspended floors had suffered serious damage from heavily loaded trolleys and in some areas additional structural steel work had been installed to provide a supplementary framework. The external concrete frame showed serious and widespread deterioration. Spalling, and actual separation, of concrete had occurred in most perimeter beams and external columns. In the worst locations, spalling had occurred along the full length of the beam and some bars were so corroded as to be providing no strength. Previous repairs, using ordinary mortar in a piecemeal treatment of defects without proper preparation, had begun to spall again after only a few months.

A survey of the building was carried out and 17 cores were taken to assess the concrete quality. The compressive strengths ranged from 10 to 34.5 MPa, the depth of carbonation ranged from 26 to 34 mm and the chloride content was found to be very low. The original specified cover was 37 mm for the external beams and columns, 25 mm for the internal beams and 12 mm for the suspended floor slabs. Specified cover had not often been provided in the building as constructed, so that much reinforcement was in carbonated concrete and therefore unprotected. With this information, the structural capacity of the building was assessed and it was found that the safe distributed superimposed loads, on the basis of flexure and shear in the various members, ranged from 10.4 kPa for the secondary beams to 2.9 kPa for the main beams. It appeared likely that the original design had provided for a superimposed load of 1 cwt per square foot (5.38 kPa) but the capability of the building ever to carry this load was in doubt.

After a study of the possible repair programs, the owner decided that remedial works to the external members should be given priority and that these members should be reinstated so that they would be capable of sustaining a uniform liveload of 4.0 kPa over the floors. Any further repairs would be for this level of loading. Only visibly damaged external surfaces were to be repaired, although it was recognized that further work might later be necessary in other unrepaired parts.

A trial structural repair was carried out on one span of a perimeter beam, so as to aid in documentation and allow tenderers to see that the methods proposed were feasible. The repair proceeded in eight stages.

(i)   All unsound and carbonated concrete was removed and the cutting was extended to give a clear space of 15 mm round any exposed bars.

(*ii*)   The cutting out was extended along corroded bars to give a 40 bar diameter lap length of uncorroded bar.

(*iii*)   Badly corroded lengths of bar were cut out and the remainder blast-cleaned.

(*iv*)   New replacement bars were inserted with full lap lengths to existing sound reinforcement.

(*v*)   All exposed reinforcement was coated with a zinc rich epoxy primer. The formulation provided electrical contact between the bar and the zinc, which therefore acted as a protective anode. Pull-out tests confirmed that the bond was not affected by the primer.

(*vi*)   The edges of the cut concrete were trimmed square to a depth of 10 mm to eliminate feather edges.

(*vii*)   The exposed surfaces were treated with a polymer bonding agent, which was allowed to become tacky before the rebuilding began.

(*viii*)   The beam profile was rebuilt by trowelling in a polymer-modified cementitious repair compound in layers against forms. Each finished layer was treated with the bonding agent.

Over one half of the length of the beam, a more elaborate 'graft' repair was added to increase the strength of the beam, especially in shear. The basis of the graft was the inclusion of precut and shaped pieces of expanded metal mesh which were fixed to the sound concrete by drilled fasteners. The mesh was treated with the zinc rich primer and the exposed concrete with the bonding agent before trowelling in the cementitious repair compound. Doubts about trowelling the compound through the expanded metal were shown to be unfounded. The mesh was intended to provide additional flexural and shear reinforcement, to help in attaching the graft through the mechanical connectors and to provide crack resistance in the new surface.

A transparent two-coat system was selected as a surface treatment to reduce future carbon dioxide and chloride penetration while still allowing the structure to breathe. A final aesthetic coat of resin emulsion masonry paint was added.

After the acceptance of the trial by the owner, a bill of quantities for the complete repair work was drawn up to cover the 572 panels of the building, each panel consisting of at least one beam and one column. Five levels of repair were established and marked on the panel sketches, as shown in Fig. 8.14. The levels chosen were: 1A — patch repair, concrete only; 1B — patch repair, concrete and reinforcement to be replaced; 1C — crack injection; 2A — graft repair, additional mesh and concrete replacement; 2B — graft repair, additional mesh, concrete and reinforcement to be replaced. A contract which was effectively in the form of a Schedule of Rates was employed with the tenderer quoting rates per cubic metre of cementitious repair compound, per lineal metre of reinforcement, and per square metre of formwork, mesh reinforcement

Fig. 8.14 Repair
survey panel over-
marked to show
expected repair
(after Dinardo and
Ballingall[8.26])

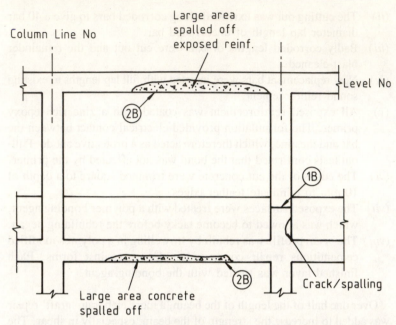

and surface treatment. As a preliminary to doing any repairs, steel props were inserted to pick up all the internal beams where they framed into the perimeter. These props were designed to carry the full factory dead- and live-loads. Presumably some method of transferring load into the props had to be provided but the engineers do not describe this process which is clearly critical in the repair procedure. The working space adjacent to the facade was limited by the need to continue the operation of the factory and it was necessary to completely prevent dust from entering the factory. Noise level restrictions also limited the type of equipment that could be used and the hours of working.

Additional reinforcement was provided in most members on account of the design deficiencies and badly deteriorated steel was removed and replaced using full lap lengths to sound bars. Where links were inadequate, the graft type of repair was called up. The shuttering needed for patch repairs and for graft repairs is shown schematically in Fig. 8.15. The column defects originated largely at the intersections of the beams and columns but in some columns the entire exposed face had to be cut back. One entire column had to be replaced over a storey height.

Two of the repaired beams were load tested, with loads up to twice working load, and were found to recover fully after unloading. There was no evidence of cracking or debonding of the repair material.

The total cost of refurbishing the building, including the internal work that still needed to be done on the floors, was estimated to be no more

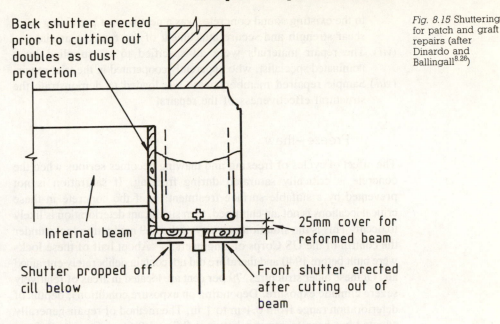

Back shutter erected prior to cutting out doubles as dust protection

Internal beam

Shutter propped off cill below

25mm cover for reformed beam

Front shutter erected after cutting out of beam

Fig. 8.15 Shuttering for patch and graft repairs (after Dinardo and Ballingall[8.26])

than 10 per cent of the cost of building an equivalent replacement. At the time of the report, the repairs had only been in place for up to two years and so any long- term evaluation had not taken place. At that stage the repairs seemed to be 'both effective and successful'. More details can be found in the referenced paper.[8.26]

We believe that this is an interesting repair job for a number of reasons.

(*i*)   As pointed out by the engineers responsible, the job is probably one of the largest structural repair jobs undertaken in the UK. Up to 200 m³ of concrete was reinstated.

(*ii*)   After a full structural assessment of the building, in the state it was in, an economical level of future live loading was agreed with the owner and repairs were planned only to reach that load capacity.

(*iii*)   Although the building was over 70 years old when repairs were carried out, it still proved very economical to repair rather than to replace and an aesthetically and functionally acceptable result was achieved.

(*iv*)   The contract documentation was developed in a way which made it relatively easy for tenderers to assess the work involved and made excessive contingency pricing unnecessary. The trial repair was available to tenderers and helped in accurate pricing.

(*v*)   Cost was saved by combining the functions of the propping, scaffolding and screening to a very considerable extent.

(*vi*)   The inclusion of expanded metal mesh, mechanically connected to the existing sound concrete, was a neat way of providing extra shear strength and secure attachment of the repair material.

(*vii*) The repair materials were all specified to be supplied by a nominated specialist, who apparently cooperated in the trial repair.

(*viii*) Sample repaired members were test loaded to demonstrate the structural effectiveness of the repairs.

Freeze—thaw

The effect of cycles of freezing and thawing becomes serious when the concrete is critically saturated during freezing. If saturation is not prevented by a suitable surface treatment, or if the concrete in these critical locations is not air-entrained, then significant deterioration is likely to occur. This is the situation in many of the 269 navigation locks under the control of the US Corps of Engineers.[8.27] About half of these locks were built before 1940 and therefore did not contain deliberately entrained air. Of these older structures, 79 per cent are located in areas of relatively severe climatic exposure. Depending on exposure conditions, depths of deterioration range from 0.1 m to 1 m. The method of repair generally adopted has been to remove between 0.3 and 0.6 m of concrete from the face of the wall and to replace it with new concrete designed to provide resistance to freeze—thaw. Unfortunately the repaired concrete has in all cases been found to crack.[8.28] The cracks, which in some repairs extend completely through the replacement concrete, are attributed to shrinkage, both plastic and drying, restrained by the bond with the existing substrate concrete. The cracks do not generally cause structural deficiencies but they are unsightly and may require additional maintenance as water can readily penetrate and freeze. Theoretical work has demonstrated that the only way totally to prevent cracking in the replacement surface layer is to use a bond breaker between the old and the new concrete. An alternative method of repair which uses precast concrete panels as permanent forms has been investigated on a prototype scale.[8.29]

Combined De-icing Salts, Freeze—thaw Cycles and Traffic

The combination of these three influences is common to bridges and parking structures throughout countries in the high latitudes, wherever de-icing of the concrete surfaces is accomplished by the spreading of chloride salts. As explained in Chapter 5, when de-icing salts in sufficient quantity reach the deck-reinforcing steel, the chloride ions at the steel surface depassivate the inherent passivity exhibited by steel embedded in concrete. The galvanic action commences and some areas of the steel corrode to produce products much greater in volume than can be accommodated within the concrete without bursting pressures being developed. When the tensile stresses so induced exceed the capacity of the concrete, under-surface fractures at the level of the top reinforcing mat develop. Traffic impact causes the concrete above the fracture to ravel, leading to a pothole in the deck surface and exposing the reinforcing steel. The potholes are hazardous to traffic, accelerate the process of deterioration and are very costly to repair. Their appearance is the first visual indication of serious deck deterioration.

Deterioration processes relating to corrosion and loadings, if unchecked, will continue until the bottom steel is affected, in which case a complete breakthrough of the slab may occur. Such an event is fortunately rare, but the failure of parking garage slabs, at least in part attributable to such corrosion, has been reported, an instance being a collapse of a small garage in Minneapolis in June 1984. The garage consisted of a slab on grade and a flat-plate slab elevated on columns spaced at 4 m intervals. The collapse of the 30 by 60 m slab left the building columns standing, and from these, long strands of bare reinforcing bar remained draped. It is suggested that the corrosion had proceeded to a degree where they had lost all bond to the concrete. No repair was possible.[8.30]

Extensive investigations of bridge-deck rehabilitation methods have been undertaken in the USA and Canada[8.31, 8.32] and more recently they are receiving attention in European countries.[8.33, 8.34] Findings from such investigations are not only applicable to the freeze—thaw problems, but may also be readily applied to certain marine structures such as wharfs and loading docks in more benign climates. Three alternative procedures for rehabilitation of bridge structures are often advocated. The first is the application of a concrete overlay of normal-slump, low-slump or latex-modified concrete. The second is to patch, waterproof and pave, generally with a bituminous concrete wearing course. The third is the installation of a cathodic protection system. Selection of the repair method is the crucial step in deck rehabilitation. It is a selection process which involves consideration of a large number of factors, which may be technical, economic and simply practical. Each of the repair methods has advantages and disadvantages as outlined in Table 8.3, which is taken from The Bridge Deck Rehabilitation Manual of the Ministry of Transportation and Communications, Ontario, Canada.[8.35]

The choice of examples illustrating the methods for cathodic protection of bridge structures is made complicated by the rapid changes in the technology. Perhaps the best manner of illustrating these is to trace a few examples which illustrate the changes to date. Probably the first and perhaps the simplest CP system for a bridge-deck was installed by Stratfull in 1973,[8.36] for the California Department of Transportation on one of two parallel spans at Sly Park Road in Northern California.

Table 8.3 Relative merits of rehabilitation methods[8.35]

Rehabilitation method	Advantages	Disadvantages
Concrete overlay	Structural component of deck slab Relatively impermeable Relatively long service life Well-suited to repair of badly spalled or scaled decks Increases cover to reinforcing steel Many qualified contractors	Less suited to decks with complex geometry Cannot bridge active cracks Difficult to provide adequate texture on low-slump concrete surface May not stop active corrosion
Waterproofing membrane with bituminous concrete wearing course	Bridges active cracks Relatively impermeable Provides good riding surface Applicable to any deck geometry Many qualified contractors	Performance highly variable Will not stop active corrosion Not suited to rough deck surfaces Service life limited by wearing course Non-structural component of deck slab Not recommended for grades in excess of 4% where heavy vehicles make turning or braking manoeuvres
Cathodic protection	Stops active corrosion Can be used on decks with active cracks Provides good riding surface Applicable to any deck geometry	Presence of wearing course without waterproofing may accelerate deterioration of the concrete Non-structural component of the deck slab Periodic monitoring required Service life limited by wearing course Specialized contractor and inspection required Electrical power source required

His system used carbon anodes embedded in an asphaltic layer containing aggregate of coked coal, termed coke breeze. The coke breeze is a highly conductive carbon and was used to spread the electrical current *via* the underlying concrete to the reinforcing bar. At the same time, a waterproofing membrane was applied to the deck of the other span, which was not given CP treatment, to prevent further intrusion of de-icing salts. Four years later, Caltran engineers compared the two methods. It was concluded that whereas the span which had been subject to CP protection had remained in the same condition as when overlayed, the span to which the waterproof membrane had been applied had continued to deteriorate. Based on the satisfactory results from Sly Park, Caltran installed CP systems on seven other bridges in 1974. Full details of these installations are given by Jurach in an unrestricted Report of the Federal Highway Administration dated 1980.[8.37]

By 1984 the Pennsylvania Department of Transportation had installed four generations of CP systems. The first installation consisted of iron-alloy anodes embedded in a 50 mm layer of coke—asphalt and topped by a 50 mm wearing course. Problems arose due to the lack of stability of the coke-based overlay. The second installation, undertaken in 1982, employed platinized niobium anode wires placed in longitudinal slots cut into the deck. These anode wires were surrounded by a conductive polymer developed by the Federal Highways Administration. With the slots spaced at approximately 600 mm centres the current demand was too high and hence in the third installation, in 1983, the distance between slots was reduced to 300 mm. At the same time secondary anodes composed of high-purity carbon strands were used to distribute the current. The fourth generation system used a grid of platinized wires and carbon strands with conductive polymer grout, and the decks were overlain with latex-modified concrete.

Concerns as to the stability of the coke-based materials used by Caltrans led to a refined system now used extensively by the Ontario Ministry of Transport and Communications.[8.38] Two significant refinements to that system were instituted. First, stone was added to the asphaltic layer, which allowed the material to withstand expressway traffic loading. Secondly, all of the CP hardware such as the anodes and test probes were moved as far as possible to the concrete kerb, and buried in the deck. This was done so that the wearing course could be replaced if required without replacing the CP system itself. The Ontario Transportation Manual[8.35] details the processes involved in the selection of the repair method, of which there are three acceptable to the Ministry. The manual then provides a contract preparation schedule, and contract documentation for the installation of a complete CP system. It also covers aspects such as typical placement of CP hardware (Fig. 8.16), conditions of use for tender items and details required to be shown on drawings.

Despite these and other examples which could be quoted, the system of CP for concrete structures is still in the process of evolution, and probably will continue to be improved. Nevertheless it is agreed that even at this stage of development, CP is the only procedure so far demonstrated which stops continued corrosion of reinforcement and hence

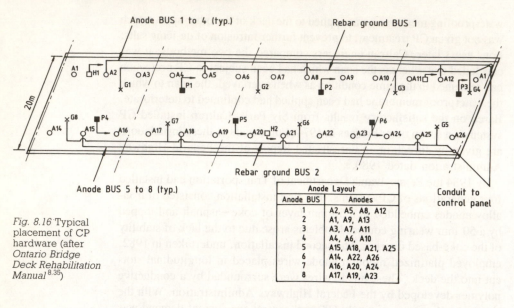

Fig. 8.16 Typical placement of CP hardware (after *Ontario Bridge Deck Rehabilitation Manual*[8.35])

Anode Layout	
Anode BUS	Anodes
1	A2, A5, A8, A12
2	A1, A9, A13
3	A3, A7, A11
4	A4, A6, A10
5	A15, A18, A21, A25
6	A14, A22, A26
7	A16, A20, A24
8	A17, A19, A23

general deterioration of structural members subject to de-icing salt use.[8.39]

CP methods have also been applied to substructures of bridges.[8.40] Such members are subject to chloride ingress from runoff water and also from the fine mist thrown up by traffic proceeding along highways adjacent to them. In the case of the substructure members, an anode system different from the conductive overlay used on decks is required. An example using an anode mesh consisting of a polymeric anode wire woven into a mat is selected. The site was a vertical face of a support of the east-bound off-rank at Spadina Street in Toronto, Canada, and was constructed on behalf of the Toronto Metropolitan Roads and Traffic Department. Drachnik[8.41] reports the details of the installation and results of this system from August 1983 to July 1984. The anode mesh consisted of 8 mm diameter conductive polymer wire woven into a mesh. The mesh was held together at intervals with plastic cleats that also served as anchor points to the structure. The anode wire had a 1.3 mm copper wire core which distributed current along the length of the anode and of the electrically conducted polymer outer jacket, which served as a site for the anode electrochemical reactions. The panels provided 9.8 m of anode per square metre of concrete surface area, thus providing uniform current distribution and an anode current density less than 100 mA m^{-2}. The anode mesh was fixed with plastic plugs and metal screws and covered with a 50 mm layer of shotcrete using a dry mix process. Potential surveys, polarization tests and tests of the decay of potential on turnoff of the power supply were conducted. In operation the anode provided adequate current to protect the reinforcing steel directly beneath it. The anode system polarized reinforcing steel to a level consistent with protection criteria determined by both polarization and depolarization tests.

The chief problems requiring research work in this area of concrete technology relate to the effects induced by the process itself, such as potential embrittlement of tendons, the potential for localized build-up of alkali salts and the severe corrosion at zones where chlorine gas is generated. This need is emphasized by the report that anode systems installed at the Burlington site in 1982 and 1983[8.40] experienced severe distress due to the debonding of the relatively thick shotcrete cover. Conductivity paint anodes also experienced problems of debonding during the operation of the system.[8.42]

Wear

Erosion in Hydraulic Structures

Where concrete has been damaged by erosion it is almost certain that any repaired section will again be damaged unless the cause of the erosion is removed. The best concrete made will not withstand the forces of cavitation or severe abrasion for a prolonged period. It may, however, be more economical to replace the concrete periodically rather than to reshape the structure to produce streamlined flow or to eliminate the solids which are causing abrasion. Some parts of the structure may be deliberately designed to dissipate energy and it is fruitless to talk of streamlining dentated sills and other obstructions to the flow of water in stilling basins.

A very severe case of cavitation erosion at Libby Dam is reported by Schrader and Munch.[8.43] Heavily reinforced high quality concrete (43 MPa at 90 days) in a low level discharge outlet was found after one year's operation to have suffered severe damage. One area of damage on the floor of the sluice extended from 41 m to 56 m downstream of the start of the sluice. The maximum width of damage was 2.1 m and maximum depth was 660 mm. The second area of damage was along a wall extending 12 m with a maximum height of 2.1 m and a maximum depth of 790 mm. Not only concrete but also a considerable amount of reinforcement had been removed from both damaged areas. Repairs were carried out by cutting back damaged concrete, replacing reinforcement where appropriate and filling with a concrete containing steel fibres. The batch quantities used led to a water:cement ratio of 0.376, air content of 4.8 per cent and fibre content 1.03 per cent by volume. Strengths above 67 MPa were obtained at 90 days. The authors state that after more than one year's further use initial inspections show that the fibrous

324 Concrete Structures

concrete, when considering overall characteristics of placeability, finishing, strengths and performance, is clearly superior to conventional concrete for this application. Further investigations of fibrous and polymerized concretes, intended for use at Tarbela Dam repairs, were carried out by Houghton and his colleagues at the Detroit Dam high head test facility. They concluded that the use of steel fibres increased significantly the resistance to cavitation erosion as compared to conventional concrete. Polymerization of conventional concrete also increased its resistance by an order approximately equivalent to that accomplished by the addition of fibres. Significantly superior test results were obtained by polymerization of steel fibrous reinforced concrete and by polymerization of monomer-filled sand patches in conventional concrete (see Fig. 8.17).[8.44]

Although fibre-reinforced concrete has proved to be a useful material for resisting cavitation erosion, it has not been at all successful in standing up to abrasion. Fibre-reinforced concrete has been used to repair abrasion−erosion damage on several US Corps of Engineers projects, but every such application 'has been regrettable' according to Liu and Holland[8.45] quoting from Liu and McDonald.[8.46] They challenge the view expressed by ACI Committee 210 that fibre-reinforced concrete has performed well in applications where it has been subjected to abrasion−erosion. Fibre-reinforced concrete has performed poorly in laboratory tests and these tests have been shown to predict accurately the performance in service in the case of the Kinzua Dam (see Fig.8.18).[8.47]

Fig. 8.17 Cavitation erosion resistance of concretes (after Houghton *et al.*[8.44])

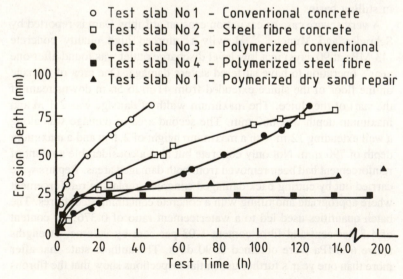

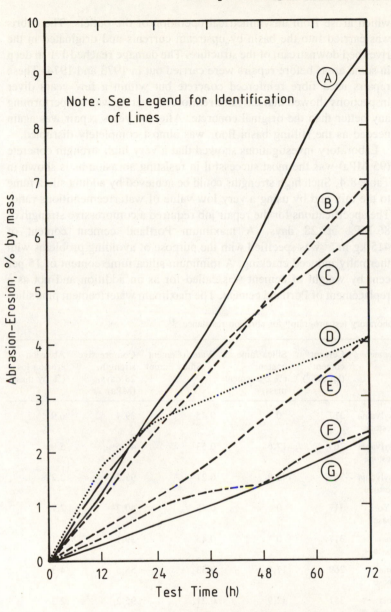

Note: See Legend for Identification of Lines

Abrasion-Erosion, % by mass

Test Time (h)

Fig. 8.18 Abrasion erosion resistance of concretes: (A) Fibre-reinforced concrete from Kinzua stilling basin; (B) Conventional concrete, Pennsylvania limestone; (C) Conventional concrete, Virginia diabase aggregate; (D) Conventional concrete, Mississippi chert; (E) Silica-fume concrete, average from actual construction; (F) Silica-fume concrete, Virginia diabase; (G) Silica-fume concrete, Pennsylvania limestone (after Holland et al.[8.47])

The Kinzua Dam has a concrete gravity spillway section which discharges into a stilling basin. The original slab in the base of the stilling basin was completed in 1967 and was 1.5 m thick and contained aggregate with a maximum size of 150 mm. By September 1969 divers reported that the stilling basin was damaged apparently as a result of severe abrasion–erosion associated with debris being moved by eddy currents, which arose from unsymmetrical operation of the outlets. The debris was carried into the basin by upstream currents and originated in the river bed downstream of the structure. The damage reached 1.1 m deep in some areas before repairs were carried out in 1973 and 1974. These repairs used fibre reinforced concrete but within a few years diver inspections showed that the fibre reinforced overlay was not performing any better than the original concrete. After 10 years repair was again needed as the stilling basin floor was almost completely disrupted.

Laboratory investigations showed that a very high strength concrete (95 MPa) was the most successful in resisting abrasion as is shown in Table 8.4. Such high strengths could be achieved by adding silica fume to the mix and by using a very low value of water:cementitious ratio. The specifications for the repair job required a compressive strength of 86 MPa at 28 days. A maximum Portland cement content of 415 kg m^{-3} was specified with the purpose of avoiding problems with thermally induced cracking. A minimum silica fume content of 15 per cent by weight of cement was called for as an addition and not as a replacement of Portland cement. The maximum water:cement plus silica

Table 8.4 Laboratory test program* for abrasion resistance[8.47]

Mix	Aggregate	Cement content (kg m^{-3})	Silica fume content (% by cement mass)	Water/(Cement + Silica Fume)	Compressive strength, 28 days, (MPa)	Abrasion– erosion loss (% by mass)
1	Pennsylvania limestone	317	0	0.45	39.4	6.9
2	Pennsylvania limestone	269	17.6	0.53	49.5	5.0
3	Pennsylvania limestone	351	42.9	0.21	95.5	2.2
4	New York diabase	317	0	0.45	40.7	7.7
5	Virginia diabase	317	0	0.45	39.1	6.1
6	Virginia diabase	269	17.6	0.53	58.5	4.3
7	Virginia diabase	351	42.9	0.21	95.2	2.3
8	Chert reference concrete	346	0	0.45	32.7	4.1
9	Kinzua FRC	NA	NA	NA	NA	9.4

* The fibre reinforced specimens were prepared from a large fragment of concrete taken from the Kinzua stilling basin

fume ratio was established as 0.30. To avoid favouring any proprietary product, the Corps of Engineers specified the silica fume and the water-reducing admixture as two separate components. In the event the contractor elected to use silica fume and admixture combined as a slurry. In trial placements at the dam, the contractor found that placing by crane and bucket was too slow and resulted in stiffening of the concrete before finishing could be completed. The result was a rough surface and plastic shrinkage cracking. In the stilling basin, the concrete was placed by pump, finished by early screeding using two vibrating screeds in tandem, 1.5 m apart, and coated with curing compound immediately behind the second screed. A total of 1540 m³ of silica-fume concrete was placed in the overlay, which was made up of 54 slabs. During construction, cracks developed in the slab two or three days after placement and divided each slab into five to 10 portions. The surface width of the cracks was initially 0.3 to 0.5 mm with the crack width decreasing with depth. The cracks usually went through aggregate particles. These cracks were ultimately attributed primarily to restraint of thermal volume changes. There had been no comparable cracks in the fibre-reinforced concrete when it had been put down. Various attempts were made to stop the cracks forming but overall no solution was found and no cracks were repaired. Part of the repair involved adding a debris trap and so subsequent performance could not be directly compared with the two previous conditions. The authors suggest that further investigation would be worth while into the use of fibre reinforced silica fume concrete for the control of thermal cracking in an overlay of this sort. Other examples of repairs of erosion damaged structures are given by McDonald.[8.48]

Traffic Wear and Abrasion

In many situations, the surface of a concrete pavement or slab may become so badly worn that it is no longer effective for the purpose for which it was intended. If such wear is not associated with corrosion of the top reinforcing steel, or with exposure of the steel to any great extent, an overlay of epoxy-modified concrete, or polymer concrete, may be a good repair method. A deck slab in a water-treatment plant in Lewiston, Idaho, USA, was severely deteriorated after many years of service.[8.49] At some point in the past, the concrete surface had received a cosmetic repair job, consisting of a thin (less than 6 mm) cementitious topping trowelled on in a manner to produce a clean looking face. In some places tar had been used as a fill material under the cementitious topping. Some of the surface had not been treated at all. With the continued deterioration of the slab and its cosmetic topping and with the development of plans for the expansion of the capacity of the plant, a more permanent repair job became necessary.

The method adopted was to remove all deteriorated material, which even included vegetation in some spots, to expose sound concrete and to cover the surface with an epoxy-modified Portland cement concrete. The surface preparation was done by chipping out unsound concrete, sand-blasting and removing all debris by blowing with compressed air. When the preparation was complete, screed bars were laid down to control levels and to ensure a minimum thickness of 25 mm. The screed bars were accurately set on patches of epoxy mortar or Portland cement grout at regular intervals. Existing joints were respected and joints in the overlay matched exactly the old joints.

A two-part epoxy was delivered to the site in 220-litre drums and 20-litre pails. Pre-mixed concrete, consisting of a 390 kg m^{-3} mix with pea-gravel aggregate was delivered in ready-mix trucks containing 3 m^3 of concrete. 260 l of epoxy was pumped into the truck and thoroughly mixed for five minutes. The two-part epoxy did not have to be pre-mixed before it was delivered into the truck. Water was added, if needed, to give a slump of 120 mm. At the same time, the epoxy bond coat was mixed and applied to the prepared surface by brush or roller. The concrete from the truck was placed by pumping, a method which was selected to allow prompt placement in temperatures that reached up to 38 °C. Rapid placing and finishing and a prompt application of curing compound was judged to be essential. The finish was achieved by a bull float followed by brooming to give a skid-resistant surface and to ensure that the topping was not overworked. The section was thin, the temperature was high and the humidity was low but almost no shrinkage cracks appeared, the concrete gained strength rapidly and could be used within hours of placement.

The use of epoxy-modified concrete in this situation provided an overlay which remained firmly bonded to the substrate, did not peel and provided a skid-resistant surface. The surface was also resistant to water penetration and to attack by chemicals in the water-treatment plant. The method is claimed to be cost-effective when judged against a cheaper conventional concrete overlay.

Some highway pavements lose enough of their initial tyre friction capacity to cause concern for traffic safety long before the end of their structural life is reached. To overcome this difficulty on Indiana State Road 37, south of Indianapolis, Indiana, USA, a thin mortar overlay for friction was applied in 1980.[8.50] The repair was in the form of an experimental section of overlay applied at the approach to a stop light on a slight downhill gradient in a major four-lane highway — the scene of frequent skidding accidents. Among various methods of surface preparation, steel-shot-blasting provided the easiest procedure and the best subsequent performance when compared with sand-blasting and with roto-milling. The last was especially poor as dust packed into the rolled surface and was not effectively removed. There must be no dust or free water on the surface when the mortar is laid. The thin overlay consisted of sharp hard fine aggregate, in this case blast furnace slag sand, Portland cement and liquid latex. The consistency of the mortar was adjusted to allow proper brooming and spreading of the mixture to achieve a depth of 5 mm $\pm$ 1.5 mm (3/16 in $\pm$ 1/16 in). Any thickness greater than 6.5 mm (1/4 in) was found likely to delaminate. The mortar was batched several hundred metres from the job site and was carried to the job in a tractor mounted paddle mixer. Slurry spreading equipment with moving brushes was used to broom the mortar vigorously into the prepared surface. It was screeded off by a squeegee at the rear of the spreader. Satisfactory friction characteristics of the repaired surface were reported up to five-and-a-half years after placing. A nine-year life was expected at that stage.

Fire

Most fire-damaged structures can be repaired. A survey of over 100 fire-damaged buildings in the UK, reported by Tovey and Crook,[8.51] showed that most of the structures were repaired and that others that were not repaired were demolished for reasons other than the structural damage sustained.

The forms of structural damage most commonly seen after a fire are

(i)   spalling;
(ii)   strength reduction in concrete and steel;
(iii)   loss of anchorage of reinforcement;
(iv)   excessive deflection of beams and slabs;
(v)   distortion of the whole structural framing.

The damage included in (iii), (iv) and (v) may be so severe that demolition and replacement is the only possible solution. Malhotra, reporting on five major fires in commercial concrete buildings,[8.52] includes three examples where the damage was so great that partial or complete replacement was selected as the course to be followed. One of these was a 14-storey office building in Rotterdam, Netherlands, of which a typical floor plan is shown in Fig. 8.19. The reinforced concrete floor slabs were 210 mm thick and were supported on all four sides by beams, 400 mm by 500 mm. The fire started on the 9th floor and spread upwards, involving all the upper floors except the very top one, which was set back. At the end of the fire the building was still standing, but there was spalling of the concrete on about 20 per cent of the exposed soffits of slabs and beams and exposed faces of columns. There was large residual deformation in the beams and slabs. The deformation at the centre

330   Concrete Structures

Fig. 8.19 Floor
plan of building in
Rotterdam fire
(numbers indicate
slab deflections
in mm) (after
Malhotra[8.52])

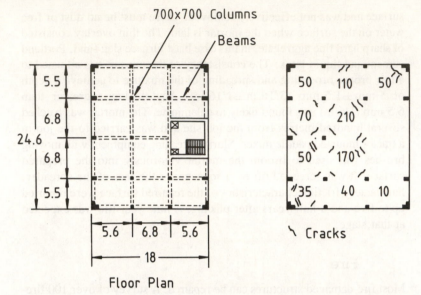

Floor Plan

of slabs varied from 15 mm to 210 mm, with the worst appearing in
the 10th floor in the 6.8 m span slab. The worst deformation in a beam
was 45 mm. No attempt was made to correct this amount of deformation
and above the 9th floor the building was demolished and rebuilt.

A fire in a 13-storey office block in London, England, started in two
places on the 9th floor and spread externally to the upper floors so that
almost the entire combustible contents of floors 9 to 11 were burnt. The
typical plan of the building and the floor slab details are shown in Fig.
8.20. The major damage consisted of spalling of concrete, particularly
on the external mullions. The internal columns were protected by the
plaster finish and were not damaged. There was no noticeable residual
deformation of the floors and they were considered to be structurally
sound. The construction was repaired without difficulty and the building
was back in full use in three months. If only spalled concrete has to be
replaced, then any of the methods discussed in Chapter 7, pp. 255–70,
may be appropriate; the question of future fire resistance may need to
be looked at if epoxy or other resin mortars or concretes are used. It
is probably better in general to avoid these materials.[8.53]

The repair to a fire-damaged structure in which cold-worked
reinforcing bars lost a lot of their strength has been described by Arioglu
et al.[8.54] The fire occurred in a single storey underground shopping
centre in Istanbul, Turkey. The structure, which is shown diagrammatically in Fig. 8.21, consisted of a flat slab, depth 550 mm, supported
on columns at 8 m centres with 850 m depth drop panels, 3.2 m in
diameter. The slab was designed with hot-rolled high strength bars but

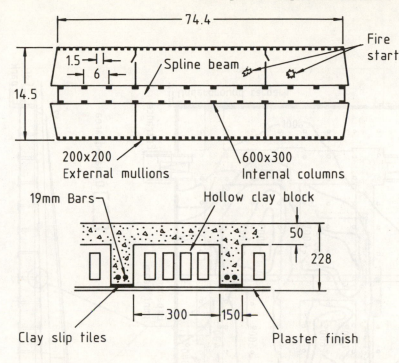

Fig. 8.20 Floor plan and detailed section of building in London fire (after Malhotra[8.52])

as these were unavailable cold-worked bars of the same strength were substituted. The drop panels, columns and footings contained mild grade hot rolled bars. All the concrete was made with limestone coarse aggregate and was generally designed with a strength of 22.5 MPa.

The fire started shortly after midnight on Christmas Day. In spite of the efforts of the brigades, it was judged by 2.30 am that the fire could not be extinguished and the shopping centre was sealed. After burning vigorously for 36 hours the fire continued until 63 hours had elapsed. Extensive surveys carried out after the fire enabled estimates of the compartment temperatures reached to be compiled (see Fig. 8.21) and the extent of the damage to be assessed. The criteria used for assessing the damage are listed in Fig. 8.22 and the extent of each classification of damage is shown in Fig. 8.23.

The columns were not generally severely damaged but 13 were found to have damage to the external concrete. This concrete was cut back to sound concrete, a mesh of reinforcement was welded to the main steel and the concrete replaced by shotcreting. Weak concrete in the drop panels was removed and replaced by shotcreting. The major damage was in the slab and the four damage classifications were treated. For classifications I and II, cleaning of the reinforcement by wire brushing, removing weak concrete and replacing by shotcreting was judged to be

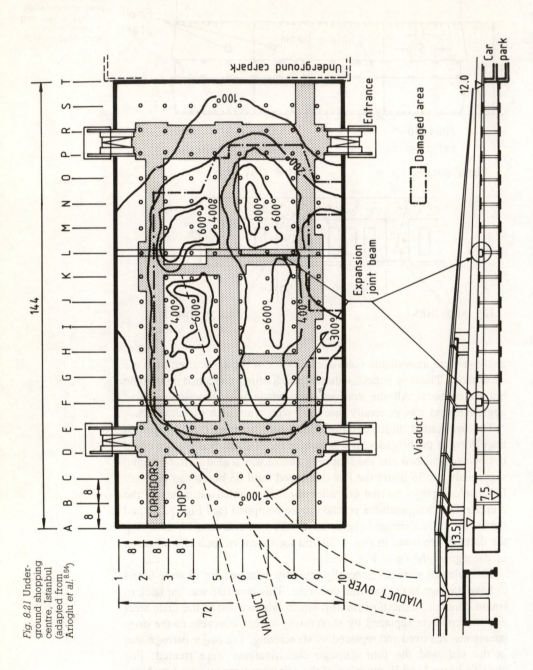

Fig. 8.21 Under-
ground shopping
centre, Istanbul
(adapted from
Arioglu et al.[8.54])

Damage classification	Spalled concrete area m²	Spalled concrete depth cm	Hammer sound	Percentage of deformed reinforcement %	Change of mechanical properties of reinforcement %	Fire space temperatures °C	Repair decision
No damage	—	—	—	—	—	<200	No repair
I. Degree damage	<10	<3	Sound	—	<5	<400	No repair
II. Degree damage	<60	<7	Sound	—	<10	<400	Cosmetic repair
III. Degree damage	>60	<12	Weak	10	<20	<600	Structural repair
IV. Degree damage	>60	<16	Weak	10	>20	<800	Structural repair and strengthening

Fig. 8.22 Damage criteria for Istanbul fire (after Arioglu et al.[8.54])

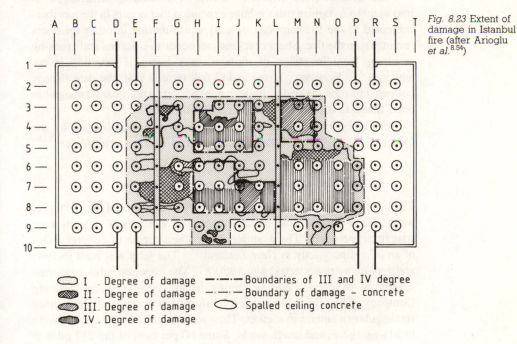

Fig. 8.23 Extent of damage in Istanbul fire (after Arioglu et al.[8.54])

I . Degree of damage   — · — Boundaries of III and IV degree
II . Degree of damage   — — Boundary of damage – concrete
III. Degree of damage   Spalled ceiling concrete
IV . Degree of damage

sufficient. For classification III, the cold-worked bars which had lost strength and had sagged were cut out and replaced by welding in hot-rolled high-strength bars. The concrete was cut back as required and made good by shotcreting. In the areas of classification IV, the repair method used for classification III was used but in addition the spans were reduced to one-half by inserting additional columns. The method used for this insertion of columns is shown in Fig. 8.24.

The loss of strength in the cold-worked bars was so severe in this case that in designing the repair it was assumed that the cold-worked bars left in the damaged areas had only the properties of the mild steel bars. The systematic classification of the extent of structural damage is well demonstrated in this fire damage case. The effective transfer of load to new supporting members is a notable feature of the repair. The repair process was apparently slow as the shopping centre was not reopened until almost five years after the fire. The reasons for this long delay have not been revealed but it can be supposed that much time was devoted to economic assessments, to insurance matters, and to meeting the requirements of building authorities, in addition to the time required for a full technical study of the condition of the building and the possible methods for repair.

When assessing the effects of a fire on a building structure, it is important to recognize that the huge expansion that occurs in the members subjected to the fire temperature may cause damge in other members remote from the fire. Shear cracking can occur in columns and cracking resulting from inversion of moment may occur if detailing is not adequate.[8.55] However, Tovey and Crook remark that 'the data already collected has not yielded a single instance where significant damage has been noted in a concrete frame remote from the fire'.[8.51]

Procedures for assessment and repair have been published by the Concrete Society in a report which contains much useful advice.[8.56]

Marine Exposure

The reinstatement of concrete piles is a typical example of marine repair work. As a general rule damage occurs in the splash zone of such structural members. The example chosen is a repair of concrete piles of an oil refinery jetty in New Zealand.[8.57] The jetty was built in 1963 and repairs were undertaken in 1980/81. The concrete piles supported a jetty at Marsden Point belonging to the New Zealand Refining Company. The piles were precast reinforced concrete and were either rectangular or square in section. They were manufactured on site using local aggregates and beach sands. Some 60 per cent of the 282 piles in the jetty showed some degree of cracking at the corners. The cracks occurred between the midtide level and the underside of the pile caps

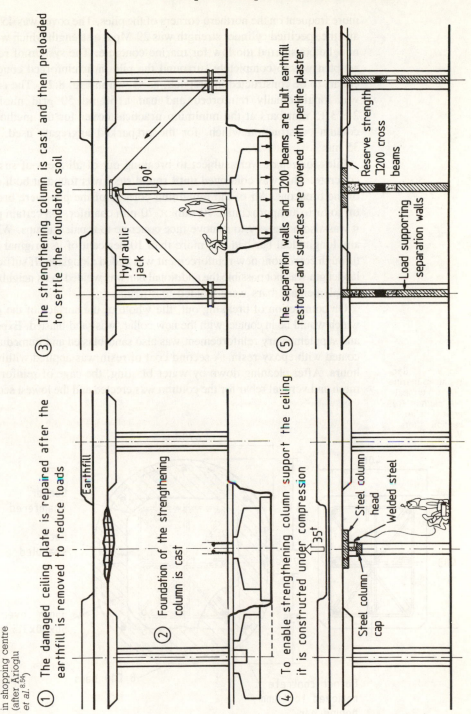

Fig. 8.24 Inserting additional columns in shopping centre (after Arioglu et al.[8.54])

① The damaged ceiling plate is repaired after the earthfill is removed to reduce loads

Earthfill

② Foundation of the strengthening column is cast

③ The strengthening column is cast and then preloaded to settle the foundation soil

90^t

Hydraulic jack

④ To enable strengthening column support the ceiling it is constructed under compression

35^t

Steel column head

Welded steel

Steel column cap

⑤ The separation walls and $\Box$200 beams are built earthfill restored and surfaces are covered with perlite plaster

Reserve strength $\Box$200 cross beams

Load supporting separation walls

more frequent on the northern corners of the piles. The cover was 45 mm and the specified cylinder strength was 29 Mpa, a strength which would now be considered too low for marine concrete. The system of repair adopted was to completely surround the pile in a reinforced concrete collar, to be constructed in prepack, as shown in Fig. 8.25. The collar was longitudinally reinforced and had a 50 × 50 grid mesh of 2.8/3.15 mm bars at the minimum practical cover for the method of column construction which, for the prepacked aggregate used, was 35 mm.

Piles for repair were subject to breaking out of all areas of suspect concrete. This was continued until sound steel was found at both ends of the breakout. All of the remaining corners of the pile were broken off to form an approximate 70 mm × 70 mm chamfer. On certain piles it was also necessary to remove face concrete to repair stirrups. Where any longitudual bar had lost more than 10 per cent of its original area through corrosion, new reinforcement was added alongside. If sufficient lap length was not possible the additional bar was welded to its neighbour. Supplementary bars were welded to stirrups.

On completion of breaking out, the whole of the surface of the pile, which would be in contact with the new collar, was sand-blasted. Exposed and supplementary reinforcement was also sand-blasted and immediately coated with epoxy resin. A second coat of resin was applied within 24 hours. After cleaning down by water blasting, the cage of reinforcing mesh and vertical rebar for the column was erected and the lowest section

Fig. 8.25 Ccllar repair of marine pile: (a) typical pile section; (b) typical collar section (after Wilson[8.57])

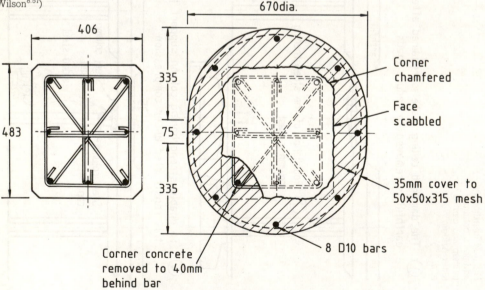

of mould placed. This first section was filled with coarse aggregate, the next mould was placed and the process repeated until the whole of the column was moulded to within 75 mm of the pile cap soffit. Grouting then took place, or if any delay occurred, the whole assembly was flushed with fresh water prior to grouting. The grout was a 1:1 cement to sand mix, with a water-reducing admixture, standard Portland cement and having a water:cement ratio of 0.43. The cement content of the prepack concrete thus produced was between 400 and 410 kg m^{-3}. During grouting, light vibration was applied to the moulds and an average pile collar took about 20 minutes to grout. The mean strength of the concrete was found to be 34.5 MPa. On stripping, the concrete was water-cured for seven days and then coated with a proprietary marine coat. The top 75 mm of collar between the top of the newly formed repair and the pile cap was formed by dry packing. It is reported that the pile repair work generally proceeded without significant problems. No shrinkage cracking was observed in any of the repair collars with the exception of four minor instances, some of which were found to be in locations where no coarse aggregate was present. These affected areas were repaired by patching with epoxy. In total, 174 piles were repaired with collars and coated.

An example of repair to railway bridge piers in a marine environment is given by Houde *et al.*[8.58] In this case location of the bridge precluded partial dewatering for repair purposes, and the upper parts of the piers were too badly cracked by a combination of alkali–aggregate reaction and freezing and thawing effects to be confined by steel collars or repaired by mere removal and replacement of concrete. The railway bridge was almost a kilometre long and had 44 simple spans of 22 m built with two steel-plate girders. Each girder was simply supported on 100 mm thick bearing plates resting on piers measuring 1.8 m by 3.6 m. The height of the piers from top to bedrock varied between 10.5 and 17.4 m, and an average length of 13 m was under water.

The damage to the concrete piers was visible as polygonal surface fracturing aggravated by subsequent freezing and thawing. The part of the concrete piers above water level was heavily cracked. In certain cases cracks opened to 35 mm in winter. Coring confirmed the low bearing strength capacity of the concrete and the bridge was closed. Solutions investigated included the building of a complete new bridge at a cost of 60 million Canadian dollars; demolition and repair of part of the concrete piers after removing all girders; and finally, the adopted method of lifting the spans through specially designed supports bearing on the pier whilst damaged concrete was removed and replaced.

The temporary support system which was adopted is shown in Fig. 8.26. Four holes were drilled into the pier down to a level where the concrete would be satisfactory to support the loads. Heavy steel inserts

Fig. 8.26 Railway
bridge repair,
Canada (after
Houde *et al.*[8.58])

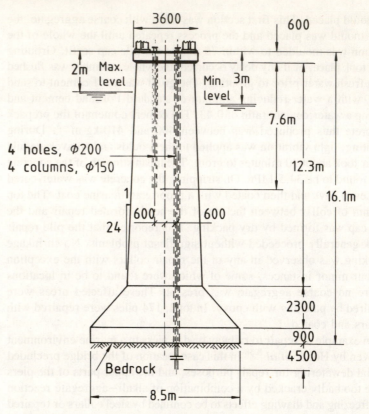

grouted in to these holes extended above the pier to serve as temporary
supports during repairs. Petrographic analysis confirmed that alkali–
aggregate reaction had occurred and drilling confirmed that anchorage
to the bedrock was not always acceptable. Core test results showed that
the concrete strength fell below 30 MPa for the upper 3 m of the pier,
whereas below water level all tests were above 35 MPa. Moduli of
elasticity varied between 13 000 MPa and 22 800 MPa, with an average
value of 17 700 MPa.

Hot-rolled steel axles 150 mm in diameter were found to be the most
convenient sections to serve as temporary supports to lift the spans during
concrete repairs, as they were easily obtainable and were threaded so
that various supporting collars could be moved up and down for easy
and precise levelling operations. Allowable bond stress values were
determined as between 3 and 4.5 per cent of the concrete compressive
strength. Bond tests between grout and steel and grout and concrete gave
acceptable results. Despite the alkali–aggregate reactivity of the concrete,
the material below the water level was such that it was possible to embed
and grout the steel axles into the lower levels of the concrete. This solution

had the advantage that the bridge could serve as a working platform from which drilling and grouting procedures could be carried on. The two upstream steel columns were extended 4.6 m into rock and grouted to increase the resistance for ice loading. The design allowed for 20 per cent of the load to be carried by bearing stresses and 80 per cent to be transferred to concrete by bond. The total load used in design was 2 200 kN per steel column. The bridge girders were jacked up and lowered on temporary support beams resting on sleeves threaded to the column ends. Concrete was removed to a depth of at least 1.5 to 1.8 m to allow sufficient space to install the lifting mechanism. A load transfer beam was then inserted and the lifting mechanism removed. Reinforcing bars were then set up. Formwork reached below the water level in order to replace a certain portion of removed concrete around the sides of the pier. The grouted steel columns were used to support the loads of ready mix trucks delivering concrete for the rehabilitation.

This example was chosen as it demonstrates that, despite certain durability processes such as alkali—aggregate reaction affecting concrete, they may not be sufficiently disadvantageous as to preclude the retention and use of certain portions of the structure. The repair technique also demonstrates the importance of establishing a working system which minimizes disruption and closure of the structure during rehabilitation. Finally this repair procedure incorporates aspects of materials selection, bond testing and the complete replacement of large segments of concrete members.

Two further examples of repairs to marine structures are given by Roper et al.[8.59] They are on wharves in Sydney Harbour and northern Western Australia. The Sydney Harbour structure was a beam and slab construction, 214 m long by 14 m wide, supported on circular concrete piles, and with a large rectangular service duct extending under the deck along the full length of the wharf (Fig. 8.27). Most of the corrosion damage was located on the underside of the wharf and in the splash zone. It had been repaired prior to examination. Widespread damage to the wharf was observed, with failure of cementitious patches applied to deteriorated concrete some years previously. Failure of these patches was generally considered attributable to the following.

(i)   The protection against renewed chloride penetration was inadequate due to insufficient cover thicknesses. (Maximum cover thickness of the patches ranged from 25 to 40 mm.)

(ii)   Properties of the patch mortar appeared to be incompatible with those of adjacent concrete.

(iii) 'Feathering' at the edges of the patches resulted in a reduction of the patch thickness, thus promoting more rapid drying and shrinkage drying.

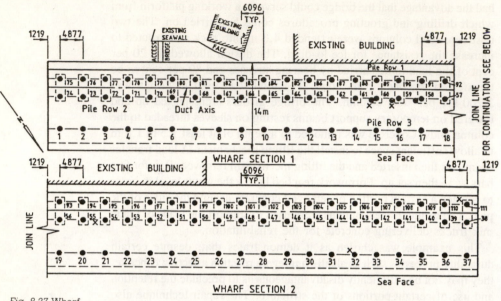

Fig. 8.27 Wharf structure, Sydney, NSW (after Roper et al.[8.59]) (•) core location; (x) drilling location

(*iv*) Insufficient care was taken in removal of chloride contaminated concrete behind the bars and the bars were not thoroughly cleaned prior to repair.

Surface coating materials were proposed for use. Tests showed that certain tar epoxy and urethane coatings provided effective barriers to chloride penetration when applied on new concrete, but did not arrest corrosion when placed on concrete that was already heavily contaminated with chlorides. Where coating was only done above the waterline corrosion continued to take place. Application of low-permeability coatings was found to result in build up of vapour pressures from entrapped moisture, loss of adhesion and blistering. In this case, because insufficient funds were available to carry out major restoration work, repairs were limited to those areas that were significantly damaged. Sections of corroded reinforcement were cleaned back to sound metal and patching was carried out with a cement-based mortar containing a styrene—butadiene admixture. Shotcrete was then applied over the repaired areas. It is unlikely that this type of repair work will provide a permanent solution. However, it was done on the basis of cost.

In the case of the wharf in northern Western Australia the deck, 184 × 34 m, was constructed in 1986 and consisted of three reinforced concrete slabs separated by two contraction joints. The slabs varied in thickness between 255 mm and 360 mm. The concrete slabs were supported on a system of longitudinal steel beams mounted on cross girders, which in turn rested on a series of cathodically protected steel tube piles. The wharf is used to bulk load sodium chloride and significant spillage of this material occurs on the deck.

Figure 8.28 presents details of cracking and other phenomena noted on the wharf slabs. Evidence was noted which suggested that spalls on the underside of the deck may have related to structural loadings, exacerbating corrosion deterioration. Spalls other than those relating to highly stressed zones of the slab are generally either related to contraction joints, through which strong saline solutions flow during washing down procedures, or to potholed patched areas on the top surfaces of the slabs, where again strong saline solutions accumulate. After considering all influences on the slabs, and based on knowledge of available repair procedures, it was decided to completely remove and replace the slabs. Recommendations were for use of epoxy-coated reinforcing bars in a

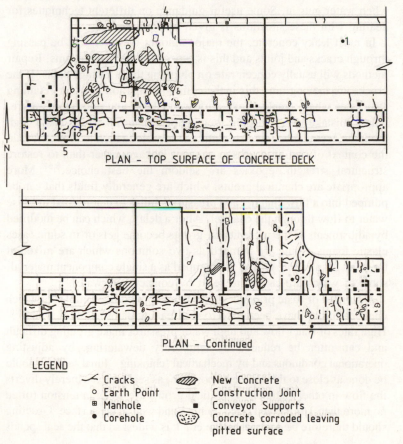

Fig. 8.28 Concrete deck and inspection summary, Western Australian wharf structure (after Roper et al.[8.59])

PLAN - TOP SURFACE OF CONCRETE DECK

PLAN - Continued

LEGEND

⌁ Cracks   ▨ New Concrete
o Earth Point   — Construction Joint
⊞ Manhole   ▪ Conveyor Supports
● Corehole   ⬡ Concrete corroded leaving
                    pitted surface

water-reduced, high-slag content cement concrete. When the repair was completed, cracking of the concrete was observed. This cracking, it is understood, appeared very early in the slab's life, and was probably due to plastic deformations later accentuated by drying shrinkage. It is understood that cracks are often wider over epoxy-coated reinforcement than when uncoated bar is used, and this will in future be a disadvantage to the durability of the slabs. The eventual overcladding of these slabs with a polymer concrete may become necessary. No attempt was made to use a cathodic protection system on this structure, as the replenishment of chlorides would have been too rapid, and the generation of chlorine gas too severe to accept.

Leakage

Water-retaining structures which leak are not adequately fulfilling their function and repair often becomes necessary. Equally, structures designed to exclude water, such as roofs and basement walls, may need repair when water gets in. Some useful guidance on different techniques for dealing with these situations is given by Perkins.[8.60]

In most leaky concrete, the major amount of water will be passing through cracks and joints and this is especially the case in dams. Repair methods will usually concentrate on plugging these leakage paths. If the cracks are moving, and this is the common situation in water-retaining structures where cracks move with water level and with temperature, it is a mistake to try to plug them with a material which has a high modulus, a strong bonding strength and low elongation to break. Within the context of sealing cracks to prevent leakage rather than to restore structural strength, epoxies are seldom the best choice.[8.61] More appropriate are chemical grouts, which are generally fluids that can be pumped into a flow channel at a pressure similar to that needed to cause water to flow through the channel. After a delay, which can be modified by adjustments to the ingredients, grouts become gels or in some cases elastic foams. Some are pumped as two solutions which are mixed at the point of injection. Others are pumped as a single component material, some of which are caused to react by contact with water.

A number of basic principles have been applied in examples in which chemical grouts have been used to seal leaking dams and lock walls.[8.61] High rates of water flow and high water pressures make grouting difficult and can often be reduced by temporary dewatering, by adjusting operational conditions and by mechanical 'chinking'. Final sealing should be done as close to the source as possible, as sealing exits merely diverts the flow to other channels. Chemical grouts are weak in tension (often no more than 3 MPa) and often do not bond well to the surface. Grouting should therefore be done when the crack is widest so that the seal spends most of its life in compression. The lack of bond is, as Waring points out, not a liability but a characteristic that makes chemical grouts capable of filling cracks without interfering with the natural movement of the structure.

In Hells Canyon Dam, Idaho, USA, a vertical crack in the non-overflow monolith, halfway between joints, needed repair. Grouting had to be done from internal galleries. Temporary flow control was achieved by drilling relief drains and by bolting gasket-backed steel plates to the walls. Each hole was fitted with a valved pipe. The crack was progressively grouted by drilling a series of holes. At each hole dye tests and flow rate tests were conducted to determine necessary gel times in the grout, and to assess pumping rates and travel paths. After each stage of grouting the tests were repeated as the flow dynamics were changed.

At Willow Creek Dam, Oregon, USA, which was placed as roller-compacted concrete, some leakage developed between lifts, which had purposely not been specially treated during construction. Although the seepage at any one joint was quite small, the total flow into the drainage gallery, which was 610 m long, amounted to over 4000 l min^{-1}. An attempt was made to draw chemical grout into the cracks by injecting the grout into the water in front of the face while a membrane of limited permeability was held a short distance away from the face (Fig. 8.29). Although moderate reductions in flow were measured, the method did not work well and was not cost-effective. More success was achieved by the more orthodox method of drilling diagonal holes from a barge and pressure grouting behind packers in the holes.

Lake Harriet Dam, USA, built in the 1920s, was in extremely poor condition and leaking badly. Chemical grouting was used to reduce the flow so that cracks which were very wide could be filled with cement grout. The method of treating the downstream end of cracks is shown in Fig. 8.30. The cracks at the upstream face were pasted over by divers to keep the cement grout from running out into the storage water. After

Fig. 8.29 Crack grouting at Willow Creek Dam (adapted from Waring[8.61])

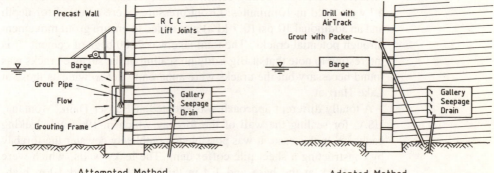

Attempted Method   Adopted Method

Fig. 8.30 Crack
repairs at Lake
Harriet Dam.
Cracks were first
drilled with 37 mm
rock drills to allow
drainage (3).
Chinking was
placed at (1). Short
holes were drilled
for chemical
grouting (2) (after
Waring[8.61])

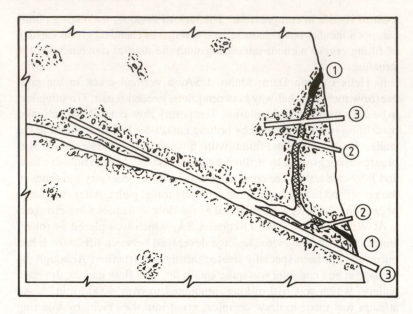

grouting the leakage flow had been reduced from 20 000 l min^{-1} to less than 40 l min^{-1}. The dam was then buttressed to restore structural stability.

Big Eddy Dam, Ontario, Canada, also built in the 1920s, was subjected in 1978 to an extensive study which revealed that additional stability was needed, that seepage was bad and that the downstream face, particularly where it had been previously repaired with shotcrete, was severely deteriorated.[8.62] The seepage was cured by a very extensive program of pressure grouting with cement grout. 325 grout holes, all 64 mm in diameter, were drilled and mechanical packers were installed at various levels. A thin water–cement grout (mix 4:1) was used but if a particular hole showed no reduction in grout take the mix was progressively thickened to 1:1. Grouting continued until refusal, i.e. until less than 4 l were used in 10 minutes. Grout pressures were adjusted for depth and an additional 10 psi (0.7 kPa) was added to ensure grout movement through potential cracks. The total take was 2600 bags of cement. It is interesting to note that at Big Eddy no sealing of the surface cracks was found necessary but the cracks were apparently not so wide as those at Lake Harriet.

A totally different approach was adopted at Hauser Dam, Montana, USA, for sealing the wall of the forebay which in 1967 was leaking badly.[8.63] In this case, it was possible to completely dewater the forebay by constructing a sheet pile coffer dam. The leaky walls, which were 3.5 m thick at the base and 1.4 m thick at the top and 10 m high, contained numerous cold joints, random cracks, rock pockets and miscellaneous area of concrete deterioration. About 60 lineal metres of cold joints and random cracks were injected with some 400 l of liquid epoxy adhesive to stop water intrusion and to restore structural strength. The upstream face was sand-blasted to remove loose debris and minor surface deterioration. Areas of serious deterioration were chopped back to sound concrete, the surfaces sand-blasted and an epoxy mortar trowelled on. One large rock pocket was treated by placing a temporary surface seal and pressure filling with 60 l of liquid epoxy. Finally the whole wall area of 700 m^2 was sprayed with epoxy in a single pass, finishing with a film which when dry was between 0.5 and 0.6 mm thick. From 1969 to 1975 the repair work was successful except for one leaky patch which needed further attention in 1972.

The common feature of all these examples is that leakage is almost entirely through cracks and that, to be effective, the seal must be provided at the upstream end of the crack.

References

8.1   Taylor R S 1985 The influence of research and development on design and construction, James Forrest Lecture. *Proceedings of the Institution of Civil Engineers* Part 1 **78**: 469–97

8.2   Crozier A C 1974 Strengthening a fifty-year-old viaduct. *Concrete (London)* **8**(7)

8.3   Vernigora E, Marcil J R M, Slater W M, Aiken R V 1969 Bridge rehabilitation and strengthening by continuous post-tensioning. *Journal of the Prestressed Concrete Institute* **14**(2): 88–104

8.4   Pakvor A, Djurdjevic M 1989 Causes of damages and rehabilitation of the Pancevo bridge. *IABSE Symposium, Lisbon, Durability of Structures* Reports Vol 57 pp 671–6

8.5   Aalami B O, Swanson D T 1988 Innovative rehabilitation of a parking structure. *Concrete International Design & Construction* **10**(2): 30–5

8.6   Mays G, Calder A 1988 External plates extend reinforcement's reach. *Concrete (London)* **22**(11): 25–8

8.7   Swamy R N 1989 Strengthening structures, Letters to the Editor. *Concrete (London)* **23**(3): 1

8.8   Swamy R N, Jones R, Bloxham J W 1987 Structural behaviour of reinforced concrete beams strengthened by epoxy-bonded steel plates. *The Structural Engineer* **65A**(2): 59–68

8.9   Jones R, Swamy R N, Charif A 1988 Plate separation and anchorage of reinforced concrete beams strengthened by epoxy-bonded steel plates. *The Structural Engineer* **66**(5): 85–94

8.10   Grill L A 1985 Building failures — the use of experience for diagnosis and repair. Concrete Institute of Australia, *12th Biennial Conference Proceedings*, Melbourne

8.11   Grill L A 1983 Strengthening and/or repairing of existing structures *IABSE Symposium, Venice, Strengthening of Building Structures — Diagnosis and Therapy* pp 329—36

8.12   Grill L A 1984 Strengthening and repairing of existing structures. Concrete Institute of Australia *CIA News* July 1984 pp 5—6

8.13   Janney J R 1983 Maintenance, repair and demolition of concrete structures. In Kong *et al* (eds) *Handbook of Structural Concrete* Pitman Chapter 25

8.14   Fintel M, Ghosh S K 1988 Distress due to sun camber in a long-span roof of a parking garage. *Concrete International Design & Construction* 10(7): 42—50

8.15   Hugenberg T L 1987 Alkali—carbonate rock reaction at Center Hill dam, Tennessee *ACI SP—100* pp 1883—1901

8.16   Cole R G, Horswill P 1988 Alkali—silica reaction: Val de la Mare dam, Jersey, case history. *Proceedings of the Institution of Civil Engineers* Part 1 **84**: 1237—59

8.17   Sims G P, Evans D E 1988 Alkali—silica reaction: Kambura spillway, Kenya, case history. *Proceedings of the Institution of Civil Engineers* Part 1 **84**: 1213—35

8.18   Stark D, DePuy G W 1987 Alkali—silica reaction in five dams in southwestern United States *ACI SP—100* pp 1759—86

8.19   Grattan-Bellow P E (ed) 1986 Concrete alkali—aggregate reactions. *Proceedings of the 7th International Conference on Alkali-Aggregate* 509 pp

8.20   Brown J G 1947 The Captain Cook graving dock. *Institution of Engineers Australia Journal* **28**: 19

8.21   Commonwealth Department of Construction Australia, *Deterioration of Concrete Floor in Marine Environment* Construction Technical Bulletin 15 Mar 1976

8.22   Graham J R, Backstrom J E 1975 Influence of hot saline and distilled waters on concrete *ACI SP—47* pp 325—41

8.23   Ambrose K 1985 Chloride contamination of Camsley Lane viaduct. *Construction Repairs & Maintenance* 1(2): 7—9

8.24   Pinjarkar S G, Osborn A E N, Koob M J, Pfeifer D W 1980 Rehabilitation of parking decks with super-plasticized concrete overlay — A case study. *Concrete International Design & Construction* 2(3): 62—7

8.25   Meinheit D F, Monson J F 1984 Parking garage repaired using thin polymer concrete overlay. *Concrete International Design & Construction* 6(7): 7—13

8.26   Dinardo C, Ballingall J R 1988 Major concrete repairs and restoration of factory structures — Uniroyal Ltd, Dumfries, Scotland. *The Structural Engineer* **66**(10): 151—60

8.27   McDonald J E 1987 Repair and rehabilitation of civil works concrete structures *ACI SP—100* pp 645—63

8.28   McDonald J E 1987 *Rehabilitation of Navigation Lock Walls: Case Histories* Technical Report No REMR—CS—13 US Army Engineer Waterways Experiment Station Vicksburg MS Dec 1987 316 pp

8.29   McDonald J E 1988 Pre-cast concrete stay-in-place forming system for lock-wall rehabilitation. *Concrete International Design & Construction* 10(6): 31—7

8.30   Anon 1982 Salt flattens old garage. *Engineering News Record* (24) June
       21 1982 pp 11—12

8.31   Transportation Research Board Washington 1982 *Resurfacing with
       Portland Cement Concrete* NCHRP Synthesis No 99 90 pp

8.32   Transportation Research Board Washington Concrete overlays and inlays.
       Transportation Research Record No 924 pp 1—19

8.33   Cavalier P G, Vassie P R 1981 Investigation and repair of reinforcement
       corrosion in a bridge deck. *Proceedings of the Institution of Civil Engineers*
       Part 1 **70**: 461—80

8.34   Vassie P R 1984 Reinforcement corrosion and the durability of concrete
       bridges. *Proceedings of the Institution of Civil Engineers* Part 1 **76**: 713—23

8.35   Manning D G, Bye D H 1984 *Bridge Deck Rehabilitation Manual, Parts
       1 and 2* Ministry of Transportation and Communications, Ontario 56 pp

8.36   Stratfull R F 1974 Experimental cathodic protection of rebar in concrete
       bridge decks. *Transportation Research Record No 500* pp 1—15

8.37   Jurach P J 1980 *An Evaluation of the Effectiveness of Cathodic Protection
       on Seven Bridge Decks* Report No FHWA/CA/SD—8Q/1 California
       Department of Transportation 57 pp

8.38   Fromm H J 1976 Cathodic protection of bridge decks in Ontario.
       Corrosion/76 National Association of Corrosion Engineers Paper 19 pp
       19/1—19/23

8.39   ACI Committee 222 Corrosion of metals in concrete *ACI 222R—85* 1985
       30 pp

8.40   Schell H C, Manning D G, Clear K C 1984 *Cathodic protection of
       substructures — Burlington Bay Skyway test site: Initial performance of
       systems 1 to 4* 63rd Meeting Transport Research Board, Washington 33 pp

8.41   Drachnik K J 1984 Application of a polymeric anodemesh for cathodic
       protection to a reinforced concrete structure. *ASTM STP 906 Corrosion
       Effect of Stray Currents and Techniques for Evaluating Corrosion of Rebars
       in Concrete* pp 31—42

8.42   Schell H C, Manning D G 1989 Research direction in cathodic protection
       for highway bridges. *Materials Protection* Oct pp 11—15

8.43   Schrader E K, Munch A V 1976 Fibrous concrete repair of cavitation
       damage. *ASCE Journal of the Construction Division* CO2: 385—99

8.44   Houghton D L, Borge O E, Paxton J A 1978 Cavitation resistance of some
       special concretes. *ACI Journal* Proceedings **75**(12): 664—7

8.45   Liu T C, Holland T C 1987 Discussion of ACI 210.R—87. *ACI Materials
       Journal* **84**(6): 576

8.46   Liu T C, McDonald J E 1981 Abrasion-erosion resistance of fiber-
       reinforced concrete. *Cement, Concrete & Aggregates* **3**(2): 93—100

8.47   Holland T C, Krysa A, Luther M D, Liu T C 1986 Use of silica-fume
       concrete to repair abrasion-erosion damage in the Kinzua Dam stilling
       basin *ACI SP—91* pp 841—63

8.48   McDonald J E 1980 *Maintenance and Preservation of Concrete Structures,
       Report 2, Repair of Erosion Damaged Structures* Technical Report No
       C—78—4 US Army Engineers Waterways Experiment Station Vicksburg
       MS 1980

8.49   Murray M A 1985 Epoxy modified Portland cement concrete overlays.
       ACI Seminar *Structural Repair — Corrosion Damage and Control,*

Seminar Course Manual SCM-8(83) pp 65—76

8.50 Scholer C F 1986 Thin mortar overlay for restoring friction on concrete pavements *ACI SP—93* pp 159—67

8.51 Tovey A K, Crook R N 1986 Experience of fires in concrete structures *ACI SP—92* pp 1—14. Also *Concrete* (London) **20**(8): 19—22

8.52 Malhotra H L 1982 Some recent experiences of fires in concrete buildings. *Proceedings of Ninth FIP Congress* Vol 3 Stockholm pp 165—73

8.53 Tovey A K 1986 Assessment and repair of fire-damaged concrete structures — an update *ACI SP—92* pp 47—62

8.54 Arioglu E, Anadol K, Candagan A 1983 An underground shopping center fire and after-fire repair project *ACI SP—80* pp 279—91

8.55 Krampf L 1982 How to obtain structural behaviour of concrete buildings under fire attack. *Proceedings of Ninth FIP Congress* Vol 3 Stockholm pp 144—51

8.56 Concrete Society 1978 *The Assessment of Fire Damaged Concrete Structures and Repair by Gunite* Technical Report No 15 The Concrete Society

8.57 Wilson G E B 1986 The repair of the concrete piles of an oil refinery jetty in New Zealand. *Marine Concrete '86* The Concrete Society, London pp 349—60

8.58 Houde J, Lacroix P, Moreau M 1986 Rehabilitation of railway bridge piers in a marine environment. *Marine Concrete '86* The Concrete Society, London pp 373—82

8.59 Roper H, Heiman J L, Baweja D 1988 Site and laboratory evaluation of repairs to marine concrete structures and maintenance methodologies — Two case studies *ACI SP—109* pp 563—86

8.60 Perkins P H 1986 *Repair, Protection and Waterproofing of Concrete Structures* Elsevier Applied Science Publishers, London Chapter 9 pp 214—60

8.61 Waring S T 1986 Chemical grouting of water-bearing cracks. *Concrete International Design & Construction* **8**(8): 16—21

8.62 Gore I W, Bickley J A 1987 Big Eddy dam. *Concrete International Design & Construction* **9**(6): 32—8

8.63 Anon 1976 Projects revisited: Hauser Dam, Montana. *ACI Journal Proceedings* **73**(8): 439

Further suggested reading, see p. 352.

Standards Cited in the Text

International

ISO 9000   Quality Systems

Australia

Standards Association of Australia, Sydney, NSW

AS 1012.14−1973   Methods of Testing Concrete — method of securing and testing cores from hardened concrete for compressive strength or indirect tensile strength
AS 1315−1982   Portland Cement
AS 1480−1982   SAA Concrete Structures Code
AS 1481−1978   SAA Prestressed Concrete Code
AS 3600−1988   Concrete Structures
AS 3900−1987   Quality Systems — Guide to selection and use

Germany

Deutsches Institut für Normung

DIN 1045   Structural Use of Concrete. Design and Construction
DIN 1048   Testing Methods for Concrete (several parts)
DIN 51951   (Draft standards no longer available)
DIN 55355   Bases for Quality Control Systems

KTA 1401   General Requirements for Quality Control, Kerntechnischer Ausschuss, c/ Gesellschaft fur Reaktorsicherheit

Japan

JASS 23   Japanese Architectural Standard Specification, Finishing Works by Use of Spraying Materials

JIS A 6909   Japanese Industrial Standard, Wall Coatings for Thin
             Textured Finishes
JIS A 6910   Japanese Industrial Standard, Multi-layer Coatings for
             Glossy Textured Finishes
JIS A 6915   Japanese Industrial Standard, Wall Coatings for Thick
             Textured Finishes

Norway

NS−ISO 9001   Quality Systems. Model for Quality Assurance in
              Design/Development, Production Installation and
              Servicing

Switzerland

SN 029 100   Requirements for Quality Assurance Systems (1982)

UK

British Standards Institution, London

BS 1881:   Part 120: 1983 Method for the determination of the
           compressive strength of concrete cores
BS 1881:   Part 5: 1970 Methods of testing hardened concrete for
           other than strength
BS 5337:   1982 The structural use of concrete for containing
           aqueous liquids
BS 5750:   Quality Systems (several parts)
BS 5882:   1987 Specification for a total quality assurance
           programme for nuclear installations
BS 8007:   1987 Code of practice for the design of concrete
           structures for containing aqueous liquids
BS 8110:   1985 Structural Use of Concrete

USA

American Concrete Institute, Detroit, Michigan

ACI 318−89   Building Code Requirements for Reinforced
               Concrete (and commentary)
ACI 503.4−79   Standard Specification for Repairing Concrete
               with Epoxy Mortars
ACI 506−66   (Revised 1983) Recommended Practice for
               Shotcreting

American Society for Testing and Materials, Philadelphia, Pennsylvania

ASTM A775−81   Standard Specification for Epoxy-Coated
                Reinforcing Steel Bars
ASTM C150−81   Standard Specification for Portland Cement
ASTM C494−81   Standard Specification for Chemical Admixtures
                for Concrete
ASTM C779−82   Test Method for Abrasion Resistance of
                Horizontal Concrete Surfaces
ASTM C944−85   Test Method for Abrasion Resistance of
                Concrete or Mortar Surfaces by the Rotating-
                Cutter Method
ASTM C1081−85   Test Method for Flexural Toughness and First
                Crack Strength of Fibre-Reinforced Concrete
                (using Beam with Third-Point Loading)

AASHTO T277   Rapid Determination of the Chloride
                Permeability of Concrete, American Association
                of State Highway and Transportation Officials,
                Washington, DC

ANSI No. 4.5.2 Quality Assurance Regulations for Nuclear
                Safety-related Structures, American National
                Standards Institute

Further Suggested Reading

[Entries are arranged chronologically by chapters]

Chapter 1

1978 Liu T C, O'Neil E F, McDonald J E *Maintenance and Preservation of Concrete Structures, Report 1* Annotated Bibliography 1927–1977, Technical Report C-78-4, US Army Engineers Waterways Experiment Station, Vicksburg, MS, 418 pp

1983 Kong F K, Evans R H, Cohen E, Roll F (eds) *Handbook of Structural Concrete* Pitman, London

1985 *Design Life of Buildings* Proceedings of a symposium organised by the Institution of Civil Engineers in association with the Concrete Society and the Royal Institute of British Architects, Thomas Telford, London 284 pp

1985 *Reducing Failures of Engineered Facilities* Proceedings of a workshop sponsored by the National Science Foundation and the American Society of Civil Engineers, ASCE 108 pp

1985 Masters L W (ed) *Problems in Service Life Prediction of Building and Construction Materials* Martinus Nijhoff Publishers (in cooperation with NATO Scientific Affairs Division) 289 pp

1986 *Improvement of Concrete Durability* Proceedings of the seminar 'How to make today's concrete durable for to-morrow' Thomas Telford, London 164 pp

1986 Carper K L (ed) *Forensic Engineering: Learning from Failures* Proceedings of a symposium sponsored by ASCE Technical Council on Forensic Engineering and the Performance of Structures Research Council of the Technical Council on Research, ASCE 98 pp

1986 Schoumacher B *Engineers and the Law — An Overview* Van Nostrand Reinhold, New York 337 pp

1989 Simm J D, Fookes P G Improving reinforced concrete durability in the Middle East during the period 1960–1985: an analytical review. *Proceedings of the Institution of Civil Engineers* Part 1 **86**: 333–58 (236 references)

1989 Bien J M J M *Maintenance and Repair of Concrete Structures* Heron, Vol 34, No 2 82 pp

1990 Nawy E G *Reinforced Concrete* 2nd edn Prentice-Hall

Chapter 2

1983 Comité Euro-International du Beton Quality assurance and quality control for concrete structures. *Bulletin d'information* No 157 98 pp

1990s Taylor M *Quality Assurance in Building Design* Longman Concrete Design and Construction Series, to be published

1990s Beale A M *Quality Assurance* Longman Concrete Design and Construction Series, to be published

Chapter 3

1986 Malhotra V M (ed) Fly ash, silica fume, slag, and natural pozzolans in concrete *ACI SP-91*, 2 vols 1609 pp

1988 Condensed silica fume in concrete. *FIP State-of-the-Art Report* Thomas Telford, London 37 pp (206 references)

1988 Hobbs D W *Alkali Silica Reaction in Concrete* Thomas Telford, London 183 pp

1989 *Analysis of Hardened Concrete — A Guide to Tests, Procedures and Interpretation of Results* Concrete Society Technical Report No 32 117 pp

1989 American Concrete Institute Committee 224 Control of cracking in concrete structures *ACI Report 224R-89*, American Concrete Institute 43 pp

Chapter 4

1982 Cryogenic behaviour of materials for prestressed concrete. *FIP State-of-the-Art Report* Ref 904/128 Thomas Telford, London 84 pp

1985 Design and construction of concrete sea structures. *FIP Commission on Concrete Sea Structures* 4th edn Thomas Telford, London 29 pp

1984 *Developments in Testing Concrete for Durability* Proceedings of a one-day symposium, London, 1984. The Concrete Society 88 pp

1987 Scanlon J M (ed) Concrete durability — Katherine and Bryant Mather international conference *ACI SP—100* 2 vols 2179 pp

1988 Malhotra V M (ed) Concrete in marine environments — Proceedings second international conference, St. Andrews by-the-Sea, Canada *ACI SP—109* 739 pp

1990 *Assessment and Repair of Fire-Damaged Concrete Structures* The Concrete Society, Ref TR 033 80 pp

Chapter 5

1986 Corrosion and corrosion protection of prestressed ground anchorage. *FIP State-of-the-Art Report* Ref 103 28 pp

1987 Gibson F W (ed) Corrosion, concrete, and chlorides — steel corrosion in concrete: causes and restraints *ACI SP—102* 169 pp

Chapter 6

1985 *Concrete Bridges: Investigation, Maintenance and Repair* Proceedings of a symposium, London, 1985. The Concrete Society 98 pp

1987 Nowak A S, Absi E (eds) *Bridge Evaluation, Repair and Rehabilitation* The University of Michigan, Ann Arbor 705 pp (Papers from a US—European workshop on bridges, June 1987)

1989 *Concrete Bridges: Management, Maintenance and Renovation* Papers for a one-day conference, 23 February 1989. The Concrete Society (8 papers)

1989 Comité Euro-International du Beton Diagnosis and assessment of concrete structures. *Bulletin d'information No 192* 120 pp

1989 Repairs of concrete structures — assessments, methods and risks. *American Concrete Institute Seminar Course Manual, SCM-21(89)* 514 pp

1990 Harding J E, Parke G A R, Ryall M J (eds) *Bridge Management — Inspection, Maintenance, Assessment and Repair* Elsevier Applied Science, London 870 pp (67 papers)

Chapter 7

1981 Shotcrete applications. *Concrete International Design and Construction* **3**(1): 23—108 (12 papers)

1986 Sasse H R (ed) *Adhesion Between Polymers and Concrete: Bonding, Protection, Repair* Chapman and Hall 759 pp

1987 Yoshihiko Ohama *Bibliography on Polymers in Concrete* University of Texas at Austin, EJC 5.200, Austin, Texas 78712-1077 344 pp (2722 references in chronological order)

Chapter 8

1987 Allen R T L, Edwards S C (eds) *Repair of Concrete Structures* Blackie 204 pp

1988 *Concrete Repair — Problems, Questions and Answers* Papers for the first annual conference and exhibition on construction repair, Palladium Publications 82 pp

1989 Repair, rehabilitation. *Concrete International Design and Construction* **11**(9): 37—72 (8 papers)

1990 Repair and rehabilitation. *Concrete International Design and Construction* **12**(3): 27—78 (9 papers)

Author Index

The entries in this Index are the page numbers in the text on which reference is made to each author's work or on which an author is mentioned by name (excluding Figure captions, Reference lists and Further Reading.

Subject Index